Soft Matters for Catalysts

Soft Matters for Catalysts

edited by
Qingmin Ji
Harald Fuchs

Published by

Jenny Stanford Publishing Pte. Ltd.
Level 34, Centennial Tower
3 Temasek Avenue
Singapore 039190

Email: editorial@jennystanford.com
Web: www.jennystanford.com

British Library Cataloguing-in-Publication Data
A catalogue record for this book is available from the British Library.

Soft Matters for Catalysts

Copyright © 2020 Jenny Stanford Publishing Pte. Ltd.

All rights reserved. This book, or parts thereof, may not be reproduced in any form or by any means, electronic or mechanical, including photocopying, recording or any information storage and retrieval system now known or to be invented, without written permission from the publisher.

For photocopying of material in this volume, please pay a copying fee through the Copyright Clearance Center, Inc., 222 Rosewood Drive, Danvers, MA 01923, USA. In this case permission to photocopy is not required from the publisher.

ISBN 978-981-4774-66-6 (Hardcover)
ISBN 978-1-351-27284-1 (eBook)

Contents

Preface   xi

1. Introduction to Catalysts   1
Qingmin Ji and Katsuhiko Ariga

    1.1 Importance of Catalysts   1
    1.2 Nanostructured Catalysts   6
    1.3 Soft Matters for Nanocatalysts   7

2. Structural Design for Molecular Catalysts   11
Qingmin Ji, Qin Tang, Jonathan P. Hill, and Katsuhiko Ariga

    2.1 Introduction   11
    2.2 Organometallic Molecules as Catalysts   12
        2.2.1 Transition Metal Complexes for Cross Coupling Reactions   14
        2.2.2 Non-Noble Metal-Based Compounds as Catalysts   20
        2.2.3 Multimetallic-Based Compounds as Catalysts   23
        2.2.4 High-Valent Organometallic Catalysts   25
    2.3 Metal-Free Organocatalysts   30
        2.3.1 Catalysis by Enamines and Imines   31
        2.3.2 Brønsted Acid Catalysts   36
        2.3.3 N-Heterocyclic Carbenes   41
    2.4 Ionic Liquids for Catalysts   47
        2.4.1 Ionic Liquids and Their Physical Properties   49
        2.4.2 Ionic Liquids for Catalytic Reactions   51
        2.4.3 Biphasic Ionic Liquids   53

		2.4.4	Task-Specific Ionic Liquids for Catalysis	55
		2.4.5	Supported Ionic Liquids for Catalysis	58
	2.5	Surface Molecular Ligands for Catalysts		61
		2.5.1	Surface Modification of Nanoparticles and Nanoclusters	62
		2.5.2	Functional Ligands for Anchoring of Nanoparticles	66
	2.6	Summary and Perspectives		69
3.	**Supermolecular Catalysts**			**93**

Shangbin Jin, Jiang He, Qingmin Ji, Jonathan P. Hill, and Katsuhiko Ariga

	3.1	Introduction			93
	3.2	Biocatalysts: Enzymes			94
		3.2.1	Engineering Artificial Enzymes for Selective Catalysis		96
		3.2.2	Genetic Engineering to Enzyme Technology		100
		3.2.3	Integrated Multifunctional Enzyme Systems		103
	3.3	Metal-Organic-Framework-Based Catalysts			107
		3.3.1	Metal Framework for MOF Catalysts		109
		3.3.2	Post-Functionalization of MOFs for Selective Catalysis		114
		3.3.3	Encapsulation of Catalytic Components in MOFs		119
		3.3.4	MOFs for the Construction of Inorganic Catalysts		126
	3.4	Covalent Organic Framework-Based Catalysts			129
		3.4.1	Classification of COFs as Heterogeneous Catalysts		130
			3.4.1.1	Post-modification COFs as catalysts	131
			3.4.1.2	Built-in catalytically active COFs as catalysts	134

		3.4.2	Catalytic Reaction Types by Using COFs as Catalysts	138
			3.4.2.1 Cross-coupling reactions	138
			3.4.2.2 Oxidation reaction	140
			3.4.2.3 Reduction reaction	141
			3.4.2.4 Photocatalytic reaction	141
	3.5	Supermolecular Nanoarchitectures for Catalysts		142
		3.5.1 Assembled Vesicles for Catalysts		143
		3.5.2 Dendritic Architectures for Catalysts		146
		3.5.3 Layered Architectures for Enzymes		151
	3.6	Summary and Perspectives		155

4. Organic Polymers, Oligomers, and Catalysis — **173**

Gaulthier Rydzek and Amir Pakdel

	4.1	Introduction		173
	4.2	Definition and Characteristics of Polymer-Supported Catalysts		174
	4.3	Soluble Polymer Supports for Homogeneous Catalysis		176
		4.3.1 PEG-Bound Catalysts		177
		4.3.2 Non-Cross-Linked Polystyrene-Bound Catalysts		181
		4.3.3 Other Soluble Polymer-Based Catalysts		183
		4.3.4 Dendrimers		184
			4.3.4.1 Peripheral grafting of organocatalysts	185
			4.3.4.2 Incorporation organocatalysts in the core or backbone	188
			4.3.4.3 Dendrimer-supported metal–ligand complexes	189
	4.4	Polymers and Heterogeneous Catalysts		192
		4.4.1 Polymer for Immobilizing Homogeneous Catalysts		193
			4.4.1.1 Polystyrene-supported catalysts	193

		4.4.1.2	Azlactone-containing polymers	194
		4.4.1.3	Ion exchange approach	195
		4.4.1.4	Porous organic polymers and rationalized polymer design	195
	4.4.2	Polymers for Designing Heterogeneous Catalysts		197
		4.4.2.1	Ex situ synthesis of polymer-supported catalysts	198
		4.4.2.2	In situ synthesis of polymer-supported catalysts	199
4.5	Emerging Materials			203
	4.5.1	Thermoregulated Strategies for Homogeneous Catalysis		203
	4.5.2	Poly(Ionic Liquid)-Supported Catalysts		205
	4.5.3	Polymer-Supported Nanoclusters		206
	4.5.4	Nanostructured Polymeric Materials		207

5. Carbons as Supports for Catalysts   229

Shenmin Zhu, Chengling Zhu, Yao Li, and Hui Pan

5.1	Introduction			229
5.2	1D Carbon Nanotubes for Catalysis			231
	5.2.1	Oxidized CNTs for Catalysis		232
	5.2.2	Doping CNTs for Catalysis		234
	5.2.3	Polymer Coating on CNTs		234
	5.2.4	Grafting Organometallic Catalysts on CNTs		238
5.3	2D Graphenes for Catalysts			239
	5.3.1	Graphene-Based Materials as Photocatalyst		241
		5.3.1.1	Photocatalysts for oxygen and hydrogen generation	242
		5.3.1.2	Photocatalysts for degradation of pollutants	246
		5.3.1.3	Graphene-based materials for CO_2 photocatalytic reduction	247

		5.3.2	Graphene-Based Materials for Electrocatalyst	249
			5.3.2.1 Graphene-based composite electrocatalysts	250
			5.3.2.2 Doped graphene as electrocatalysts	252
		5.3.3	Catalytic Oxidation Reaction	256
		5.3.4	Electrochemical Biosensing	258
	5.4	3D Architected Carbons for Catalyst		261
		5.4.1	Classification of 3D Architected Carbons	261
			5.4.1.1 Hard carbon	261
			5.4.1.2 Soft carbon	261
			5.4.1.3 Graphitized carbon	262
		5.4.2	Synthesis of 3D Architectured Carbons	262
			5.4.2.1 Hard template method	263
			5.4.2.2 Soft template method	263
			5.4.2.3 Nature template/precursor method	265
		5.4.3	Applications of 3D Architected Carbons for Catalysts	265
			5.4.3.1 Porous carbons for heavy metal removal and reduction	266
			5.4.3.2 Porous carbons for nitrogen oxide and sulfur oxide removal	268
			5.4.3.3 Porous carbons for photocatalysis	268
	5.5	Summary and Perspectives		270

6. Porous Inorganic Nanoarchitectures for Catalysts — 291

Qingmin Ji, Jiao Sun, and Shenmin Zhu

6.1	Introduction	291
6.2	Mesoporous Nanoarchitectures	292
6.3	Inorganic Porous Nanoarchitectures for Enzymes	298
6.4	Porous Nanoarchitectures for Gas-Phase Catalysis	302

	6.5	Porous Nanoarchitectures for Electrocatalysis	307
	6.6	Summary and Perspectives	310
7.	**Catalytic Reactions on Solid Surfaces**		**319**

Huihui Kong, Xinbang Liu, and Harald Fuchs

	7.1	Introduction	319
	7.2	Dehalogenative Coupling Reactions Assisted by Metal Surfaces	320
	7.3	Dehydrogenative Coupling Reactions Assisted by Metal Surfaces	325
	7.4	Summary and Perspectives	335
8.	**Soft Matters for Future Catalysts: A Perspective**		**341**

Qingmin Ji, Harald Fuchs, and Katsuhiko Ariga

	8.1	Multiforms of Catalysts	341
	8.2	Mimic from Nature	342
	8.3	New Challenges	343

Index   347

Preface

With the introduction of increasingly strict regulations to protect human health and the environment, the invention and development of green chemical processes has become a key subject of industrial manufacturing and requires production processes to be sustainable and safe with little or no waste materials. Adapting to this new situation for chemical reactions, the development of novel catalysts became an essential subject to achieve the safe, effective, and economic production. Catalyst can accelerate chemical reactions, while not getting consumed itself by the overall reactions. With the extensive developments in nanoscience and nanotechnology, the exciting prospects of nanoscience and nanotechnology have also infiltrated into the fields of catalysts. Nanocatalysts have shown their potentials for promoting the green synthesis and also emerged as one of the most attractive and exciting aspects of nanoscience and nanotechnology. The specific nanometric features of nanocatalysts open up a wide range of possibilities for reaction control through tunable chemical activity, specificity, and selectivity.

This book offers a comprehensive overview of the recent developments in nanocatalysts by various novel synthetic strategies, emphasis on the designs from the self-assembly of nanostructures, and hybridization with soft matters. The introduction of soft matters into nanocatalysts is illustrated from the precise control on the nanostructures of catalysts, the confined supports by supramolecular architectures, and great potentials of molecular assembly for creative catalysts. The novel catalysts with soft matters, including homogeneous and heterogeneous catalysts, are mainly investigated for various catalytic reactions.

The topics outlined are described in terms of catalyst preparation, characterization, reaction chemistry, and process technology.

Qingmin Ji
Harald Fuchs

Chapter 1

Introduction to Catalysts

Qingmin Ji[a] and Katsuhiko Ariga[b]

[a]*Herbert Gleiter Institute of Nanoscience,
Nanjing University of Science and Technology,
200 Xiaolingwei, Nanjing 210094, China*
[b]*WPI Center for Materials Nanoarchitectonics,
National Institute for Materials Science,
1-1 Namiki, Tsukuba, Ibaraki 305-0044, Japan*

jiqingmin@njust.edu.cn

1.1 Importance of Catalysts

It is widely acknowledged now that there is a growing need for more environmentally acceptable processes for the well-developed industries and sustainable society. Efforts to reduce the wastes from chemical manufacturing and abate pollution hinge on the invention of catalysts. Catalysis is an indispensable tool in the production of bulk, fine chemicals, and in the prevention of pollutions. It lies at the heart of countless chemical protocols, which are of direct relevance to many aspects of our daily life. Actually, catalysis has been the cornerstone of chemical manufactures for decades. Over two thirds of global economic outputs

---

Soft Matters for Catalysts
Edited by Qingmin Ji and Harald Fuchs
Copyright © 2020 Jenny Stanford Publishing Pte. Ltd.
ISBN 978-981-4774-66-6 (Hardcover), 978-1-351-27284-1 (eBook)
www.jennystanford.com

are created by the processes employing catalysts in the chemical industry, including food, pharmaceuticals, fuels, and polymers. We can say that the chemical industry cannot exist without catalysis. Living matters also rely on enzymes, which are the most specific catalysts in biological processes.

Catalysts improve the rate or selectivity of chemical reactions without being consumed during the processes. A catalytic reaction always begins from the binding of reactants at the surface of catalysts. It can significantly reduce the activation energy of the reaction compared with the cases without catalysts. Catalysts will not form chemical bonds with reactants, thus allowing the detachment of products after the reactions. They can maintain their chemical nature in the reaction and make themselves available for the next reaction. We can describe the catalytic reaction as a cyclic event in which the catalyst participates and is recovered in its original form at the end of the cycle.

In a spontaneous reaction, catalysts have multi-forms, varying from atoms, molecules and larger supermolecules. In addition, the catalysis may occur in various mediums, in liquids, in gases, or at the surface of solids. According to the states of phases during catalysis, catalysts can be classified as homogeneous and heterogeneous. In homogeneous catalysis, both the catalyst and the reactants are in the same phase, gas phase or liquid phase. An advantage of homogeneous catalysis is that it allows a very high degree of interaction between catalyst and reactant molecules. One of the simplest examples is ozone destruction in atmospheric chemistry, which is important for the prediction of the ozone hole. Via chlorine (Cl) atoms as catalysts, the decomposition of ozone can be accelerated tremendously to produce oxygen as shown in Scheme 1.1. In the gas phase, the reaction can occur spontaneously in cycles unaltered. Industry also uses a multitude of homogenous catalysts to produce chemicals.

Organometallic catalysis is a significant fraction for homogeneous catalysis. The success story of organometallic chemistry is also derived from homogeneous catalysis. Karl Ziegler and Giulio Natta won the Noble Prize in 1963 for discovering organometallic polymerization catalysts. Ernst Otto Fischer and Geoffrey Wilkinson later further deciphered how such compounds function as catalysts and won the 1973 Noble Prize.

$$Cl + O_3 \Rightarrow ClO + O_2$$
$$ClO + O \Rightarrow Cl + O_2$$

Scheme 1.1 The catalytic reaction of ozone destruction.

Bulk and fine chemicals and even natural products are being produced via homogeneous organometallic catalysis. Unlike metal catalysts, organometallic compounds are soluble in the catalytic reactions. Some important examples of organometallic catalysts include the following:

(i) Hydroformylation, the reaction of an olefin with CO and H_2 in the presence of a metal carbonyl to form aldehydes
(ii) Wacker process, in which an olefin is oxidized to an aldehyde or ketone in the presence of a soluble palladium salt
(iii) Ziegler Natta process, in which olefins are polymerized using an organo-aluminum–titanium catalyst to form stereoregular polymers
(iv) Fischer–Tropsch reactions, the reductive polymerization of CO to form straight-chain hydrocarbons, olefins and alcohols

Although homogeneous catalysts have the disadvantage of irrecoverability after the reaction completed, it is relatively easier to study the catalytic mechanism in a homogeneous system by chemical analysis methods such as NMR and UV-vis. Therefore, new mechanistic models may be developed from the design of homogeneous catalysts, which also offer the possibility of understanding more complex heterogeneous catalytic systems on a molecular level.

Heterogeneous catalysis refers to the phase of the catalyst differing from the reactants. In most of the cases, heterogeneous catalysts are solids and the reactants are gases or liquids. The solids are often expensive metals (e.g., Pt, Au, Ni, Pd). Therefore, to economically use the catalysts, nanometer-sized particles with well-dispersed state are preferred. The related studies about heterogeneous catalysis have won the Nobel Prizes several times. For example, in 1912 Paul Sabatier was awarded the Nobel Prize in chemistry for performing the first hydrogenation of ethylene using nickel (Ni) as a catalyst. In 1918, the award went to Fritz Haber, who developed the catalytic formation of ammonia from atmospheric nitrogen and hydrogen with the help of iron as a catalyst. Irving Langmuir (1932) and Sir Cyril Norman Hinshelwood (1956) were awarded for their research studies on the surface catalytic mechanism of chemical reactions. These achievements greatly promoted and expanded the industrial catalysts for a large range of chemicals.

Heterogeneous catalysis has been of paramount importance in many areas of the chemical and energy industries, such as production of transportation fuels, pollution control, and production of low-cost and high-quality raw materials. In an introductory example reaction for the catalytic oxidation of CO, which is applied for cleaning automotive exhaust, it begins with the adsorption of CO and O_2 on the surface of noble metals such as Pt, Pd, and Ru. The O_2 dissociates into two O atoms and reacts with the adsorbed CO molecule on the surface to form CO_2 (Scheme 1.2). CO_2 interacts weakly with the surface of catalysts and desorbs almost immediately after formation.

$$CO_{(gas)} + Solid_{(catalyst)} \leftrightarrow CO_{(adsorption)}$$
$$O_{2(gas)} + Solid_{(catalyst)} \rightarrow 2O_{(adsorption)}$$
$$CO_{(adsorption)} + O_{(adsorption)} \rightarrow CO_{2(gas)}$$

Scheme 1.2 The catalytic reaction of CO oxidation.

Enzymes act as biocatalysts for biochemical reactions. Enzymes are one of the keys to understand how cells survive and proliferate. In general, most of enzymes are large protein molecules, derived from living cells. Enzymes possess a very shape-specific active site, i.e., the molecule has just the right shape and functional groups to bind with one of the reactant molecules. Enzymes may be classified to be heterogeneous catalysts. Similar in principle to other types of chemical catalysis, they speed up reactions by providing an alternative reaction pathway with reduced activation energy. However, enzymes take a little different "route". When a reactant binds to an enzyme, it must collide in the right orientation with sufficient energy to form a reaction intermediate (referred to as substrate) (Fig. 1.1). When another reactant reacts with the intermediate, the catalytic reaction occurs and the enzyme is reformed.

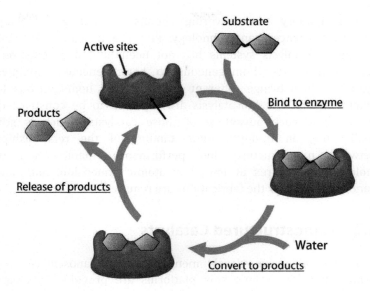

Figure 1.1 The catalytic cycle of an enzyme.

The study of enzymes has immense practical significance. The activity of certain enzymes in the blood plasma or tissue may be used to diagnose disease. Not only for the applications in medicine, enzymes are also practical tools in food processing, for cleaning, in agriculture, and in the chemical industry. Due to this specific binding pathway of enzymes, the most important advantage of biocatalysts is their high selectivity in contrast to other catalysts. The selectivity can be stereo-specific, regio-specific, and chemo-specific, which are very desirable in chemical synthesis processes. It may offer more green processes with mild reaction conditions, minimized side reactions, simplified synthesis steps, easy separation, and less environmental pollution. However, enzymes may also easily lose the catalytic activity by inhibitors, which can block or distort the active site of enzymes. Aqueous soluble enzymes may also cause the problem of hardly recovery. To overcome these drawbacks, enzymes are now always used by immobilized in solid substrates. By this way, the stability and life-time of enzymes for biocatalysis can be greatly improved.

The important roles of catalysts in the modern society greatly promote the development of this area. Catalysis has become a multidisciplinary field covering chemistry, materials, biology, computer science, nanotechnology, etc. The design of creative and novel catalysis systems has not been limited to only one type of catalysts (homogeneous to heterogeneous). Complex systems which bridge different catalysts from homogeneous to heterogeneous and biocatalysis are expected to be constructed. To combine each advantage of those catalysts, which is still challenging, an in-depth understanding of the relationships between nanostructure and performance, manipulation of molecular structures at nano- or atomic dimension, and new nanotechniques for the fabrications are required.

1.2 Nanostructured Catalysts

With the extensive developments on the nanoscience and nanotechnology, various new platforms are provided to create new novel systems in applications. Nanotechniques related to nanoscience greatly forward the steps toward the development to the perfect catalysts, which means to drive the reaction to produce only the desired form. Since most catalysis is carried out by solids and the reaction occurs only when the reactants are adsorbed onto the catalyst surface nanostructuring of catalysts will offer huge surface-to-volume ratios, favorable transport properties, altered physical properties, and confinement effects from the nanoscale dimensions. A large contact area between active sites of the catalyst and reactants will lead to higher catalytic activity or selectivity. The reduction of size down to a few hundreds or dozens of atoms will decrease the density of states in the valence and conduction bands and dramatically change the electronic properties. The quasi-continuous density of states is replaced with quantized levels with a size-dependent spacing (quantum size effect), which may also make it possible to modulate the catalytic activity.

However, nanosized catalysts are always weak on their thermal stability according to their critical sizes. That means the smaller the crystallite size, the lower thermal stability. Suitable support materials are thus needed to achieve an optimal dispersion of the catalytically active components and to stabilize them against sintering. The lifetime of nanocatalysts, thus, will also be increased. In several reactions, the support is even not inert and the overall process is actually a combination of two catalytic functions from both the catalysts and the support. As some catalysts are made of expensive metals, the supported catalysts may reduce the quantity of noble metal employed and increase cost-efficacy.

Research in nanotechnology and nanoscience is expected to have a great impact on the development of new catalytic systems. In particular, the innovative ideas on the design, synthesis, manipulation and characterization of catalysts. Designing catalysts that are more efficient, more selective, and more specific to certain reactions can lead to significant savings of manufacturing expenses. The higher activity can be reflected from the high productivity using relatively smaller reactors and less amount of catalysts or in mild operating conditions. Reactions with high selectivity can not only generate no waste products but also reduce energy and simplify the separation or purification procedures.

1.3 Soft Matters for Nanocatalysts

Soft matters can be described as the systems which have structural flexibility according to external conditions. They include liquids, colloids, polymers, gels, liquid crystals, biological materials, and also assembled molecular structures. The roles of soft matters in the design of novel catalysts may be reflected from the nanostructuring of catalysts, manipulation of organized nanostructures, and the controls on surface properties, etc. For solid catalysts of metals or metal oxides, the variety of solvent media may result into the formation of different crystalline structures of catalysts; various molecular ligands on the surface may form different organized morphologies of nanocatalysts; self-assembled porous structures may provide precise confined space for nanocatalysts. These effects can lead to greatly enhancement on the catalysis performance.

On the other hand, soft matters themselves can also directly be catalysts. For example, despite ionic liquids can be used as "green" solvents for homogeneous metal catalysts, acidic ionic liquids with chloroaluminate ions proved to be effective

Friedel–Crafts catalysts. For example, phosphonium halide melts were used successfully in nucleophilic aromatic substitution reactions. As a wide range of molecular components can be chosen from nature or by organic synthesis, the organic molecular systems may provide large design space for the nanocatalysts. Not only the precise molecular design on organic components but the further bottom-up assembly may bring more intelligence on the construction of catalytic systems. The keep growing catalysis systems by organometallic molecules, MOF and COF exhibit the potentials of high-level strategies for the fabrication of creative catalysts with better catalytic activity, longer lift-time and easier recyclability. Also for the biocatalysts, due to the deeper understanding on the activity of enzymes, protein structures can be rationally designed to tailor the properties of biocatalysts for particular chemical processes. Rational design involves rational alterations of selected residues in a protein to cause predicted changes in function, mimicking the natural evolution processes in the laboratory and generation of a library of different protein variants with the desired functions. Enzyme properties such as stability, activity, selectivity, and substrate specificity can now be routinely engineered in the laboratory.

For many years, the only way to develop new or improved catalysts was by empirical testing experiments. However, the fundamental understanding of the relationships among catalyst synthesis procedures, active site structure on the atomic and nano-scales, chemical reaction mechanisms, and catalyst activity, selectivity, and lifetime may ultimately lead to the first-principles design of novel catalysts for specific chemical reactions and rational catalytic schemes in predetermined two- and three-dimensional configurations. Nowadays, researchers start the ab initio simulation of nanoscale materials to predict the structural, morphological, compositional, electronic, and chemical aspects of a catalyst, to achieve specific guidelines for improved reactivity, selectivity and stability. In addition, the nanotechniques of surface science also play a key role in exploring the basic behaviors of catalysts at the molecular or atomic level.

In this book, we summarize the concepts and frontiers that have been developed for nanocatalysts in the past decade, emphasize the roles of soft matters for the creation of novel catalysis systems, and analyze the nanofeatures of catalysts to improve their catalytic performance. The content provides comprehensive information about the soft matters for catalysts and insights for further innovations to perfect catalysis systems.

Chapter 2

Structural Design for Molecular Catalysts

Qingmin Ji,[a] Qin Tang,[a] Jonathan P. Hill,[b] and Katsuhiko Ariga[b]

[a]*Herbert Gleiter Institute of Nanoscience,*
Nanjing University of Science and Technology,
200 Xiaolingwei, Nanjing 210094, China
[b]*WPI Center for Materials Nanoarchitectonics,*
National Institute for Materials Science,
1-1 Namiki, Tsukuba, Ibaraki 305-0044, Japan

jiqingmin@njust.edu.cn

2.1 Introduction

Of the extensive library of available catalysts, molecular species show promise because of their high tunability and activity, as well as their ability to be integrated into sophisticated molecular assemblies. The best homogeneous systems provide turnover numbers in thousands to tens of thousands.

Molecular catalysts are mostly based on the use of small organic molecules or organometallic compounds to catalyze organic transformations. These organic molecular-based catalysts are especially essential in a new and popular field for the

---

Soft Matters for Catalysts
Edited by Qingmin Ji and Harald Fuchs
Copyright © 2020 Jenny Stanford Publishing Pte. Ltd.
ISBN 978-981-4774-66-6 (Hardcover), 978-1-351-27284-1 (eBook)
www.jennystanford.com

enantioselective synthesis. Although chemical transformations that use small organic molecules as catalysts or organometallic compounds have been known for more than a century, it was not until the late 1990s that the field of organo-catalysis was paid attention with explosive research interests. During the past decades, the organic molecular-based catalysts have proved to be applicable for more than 130 discrete reaction types. It is now widely accepted that organic molecular-based catalyst is one of the main branches of enantioselective synthesis and those molecular catalysts involved in the synthesis of chiral molecules are also considered to form fundamentals to understand the catalysis processes [1–3].

In this chapter, we will present the success of different synthesis strategies for molecular catalysts or catalytic active ligands from organometallic, organic non-metallic, and biological groups. We intend to introduce the development on the constructions of molecular catalysis systems, and their advantages and significance on catalytic research. We hope to inspire the considerations of efforts for developing more effective and innovative molecular catalytic systems, and also employing molecular catalysts more effectively and innovatively.

2.2 Organometallic Molecules as Catalysts

Organometallic compounds are chemical compounds containing at least one metal–carbon bond. The metals include alkaline, alkaline earth, transition metal, etc. The field of organometallic chemistry combines the aspects of traditional inorganic and organic chemistry. In comparison with purely organic compounds, organometallic compounds show unique chemical properties, including structural diversity, variety of available interactions, and possibility of ligand exchange [4–6].

The productions by using organometallic catalysts include commodity chemicals, polymers, and also fine medicinal chemicals [7–9]. The first successful introduction of organometallic compounds for catalysis can be traced back from 1900s. Victor Grignard discovered organomagnesium reagents for the formation of carbon-carbon bonds and was awarded the 1912 Nobel Prize in chemistry. These catalysis reagents open an extremely powerful pathway in organic synthesis and provide a breakthrough in the utilization of organometallic catalysts in organic synthesis. Significant discoveries and developments of organometallic compounds were particularly reported in their applications in catalysis. New knowledge regarding structure and reactivity of organometallic compounds has created new catalytic processes in industry or has reestablished old catalytic features under improved conditions.

Homogeneous catalysis and organometallic chemistry have stimulated and supported each other since the early days of hydroformylation (1938), olefin polymerization (1953), and acetaldehyde synthesis (1959) [10]. Nevertheless, heterogeneous catalysis dominates the industrial processes. Heterogeneous catalysis using the surfaces of metals or ionic platform materials may suffer from a limited number of active sites in catalysis. By contrast, homogeneous catalysis allows exquisite control on reactivity and selectivity by relatively readily analysis and facile tuning on active sites. Organometallic catalysts are said to be able to act as a bridge between heterogeneous and homogeneous catalysts. They bring together the benefits of heterogeneous catalysis (i.e., recyclability and easy removal from the reaction mixture) and homogeneous catalysis with intimate control over the transformations in the metal-ligand system.

The benefits of organometallic catalysts (activity, chemo-, regio-, and stereoselectivity) are based on a simple concept: Specific ligands keep the catalytic metal in a low nuclearity, normally monoatomic state of molecular defined stereochemistry. At the same time, the metal complexes undergo dissociation equilibria, promoting the reactivity via the coordination or addition of substrate(s) to the metal center. That is followed by a series of intramolecular transformations of the activated substrates and the decomposition of the organometallic compound to give reaction products. After the regeneration of the organometallic complex, a catalytic cycle is completed. These catalytic stoichiometric transformations are the fundamental reactions of coordination and organometallic chemistry, which are including ligand coordination, oxidative addition, insertion, isomerization, reductive elimination, etc.

Metal catalysis, especially transition metal catalysis, plays a central role in contemporary organic synthesis due to their ability to activate organic substrates and promote various interactions [11]. The ability of metal complexes to catalyze organic reactions constitutes one of the most powerful strategies to create new types of bond forming reactions. The choice of the metal combined with the design of the ligand environment provides opportunities for electronic and steric tuning of reactivity to a high degree. The main reasons why transition metals contribute so essentially in catalysis are the follows:

(i) bonding ability: the ability to form both p- and s-bonds with other moieties
(ii) wide choice of ligands: transition elements readily form chemical bonds with almost every other element and almost any organic molecule
(iii) ligand effects: a ligand can influence the behavior of a transition metal catalyst by varying the steric or electronic environment at the active site
(iv) variability of oxidation state and coordination number
(v) ability to readily interchanges between oxidation states during a catalytic reaction: transition metals can be readily involved in redox processes

A rapid progress in the study of organometallic and coordination compounds has led to the development and successful industrial application of a number of catalytic processes based on use of these compounds as catalysts. The transition metal catalysts provide an avenue for controlling selectivity and producing pure products in high yields, which has led to them widespread adoption by industry [12].

This section provides a brief overview of the selected organometallic catalysts in organic synthesis, especially for transition metal-based compounds and non-noble metal-based compounds. It highlights their potentials for the development of novel catalysis systems.

2.2.1 Transition Metal Complexes for Cross Coupling Reactions

Catalytic nucleophilic substitution reactions comprise some of the most commonly used catalytic processes in synthetic organic chemistry. When forming a chemical bond, a nucleophile is an electron-rich chemical reactant that is attracted by electron-deficient compounds. The well-known cross-coupling reactions for the formation of single C–C or carbon heteroatom C–X bonds (X = N, O, S, P, Si, B, etc.) include Kumada reaction, Negishi reaction, Stille reaction, Suzuki–Miyaura reaction, Heck reaction, Ulmann coupling, Hiyama-Denmark reaction and Buchwald–Hartwig reaction, etc. [13, 14].

Organometallic catalysts and hydrocarbons are all nucleophiles, as the C-M bonds of organometallic compounds and the C–H bonds of hydrocarbons tend to donate both of their bonding electrons to electrophiles, which will make synthetic procedures more efficient. The mechanism of various cross-coupling reactions involves a sequence of (i) oxidative addition between a metal catalyst and the electrophile or nucleophiles, (ii) transmetallation with the organoboron reagent to form metal complex, and (iii) reductive elimination of the product with the regeneration of the active metal catalyst (Fig. 2.1) [15]. The transition metal of organometallic catalyst may be compatible with different kinds of functional groups, hence making catalysis reactions accessible.

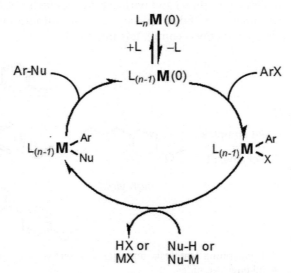

Figure 2.1 The catalysis mechanism of organometallic compound for cross coupling reaction.

16 | Structural Design for Molecular Catalysts

As aryl chlorides are more widely available substrates for the substrates of cross-coupling reactions, catalytic systems that can successfully catalyze cross-coupling reactions with these substrates were intensively studied. Metal complexes that contain electron-rich and bulky ligands may improve the catalytic activity in coupling reactions. The combination of bulky and electron-rich phosphines with different sources of palladium generate species has shown high catalytic activity in cross-coupling reactions [16–20].

Feringa et al. reported a wide range of alkyl-, aryl- and heteroaryl-lithium reagents undergo selective cross-coupling with aryl- and alkenyl-bromides in the presence of a Pd-phosphine catalyst (Fig. 2.2). The process proceeds quickly under mild conditions (room temperature) and avoids the notorious lithium halogen exchange and homocoupling [21]. The nature of the phosphine ligand is a crucial parameter to the selectivity of cross-coupling reactions [22, 23]. When using complex Pd[P(tBu)$_3$]$_2$ (P(tBu)$_3$=tri-*tert*-butylphosphine) (5 mol%) as catalysts instead of the in situ–formed catalyst of Pd$_2$(dba)$_3$ (dba=dibenzylideneacetone) and P(tBu)$_3$ or Pd$_2$(dba)$_3$ and 2-dicyclohexylphosphino-2′,6′-dimethoxybiphenyl (SPhos), cross-coupled product was obtained in high yield (96%) and with excellent selectivity. Minimal halogen exchange was taken place and the formation of the homocoupling product was completely inhibited.

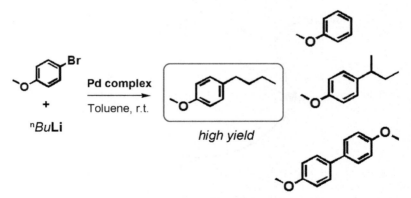

Figure 2.2 Pd-phosphine catalyst undergoes selective cross-coupling with aryl- and alkenyl-bromides.

Phosphines have attracted considerable attention and have allowed for the development of catalytic systems possessing a wide scope. Buchwald et al. developed important sets of phosphines (biaryl phosphines and cyclometallated biaryl phosphines) as excellent supporting ligands for organometallic catalyst systems [24–27]. The effectiveness of these systems was attributed to a combination of electronic and steric properties of the ligands, which favor both the oxidative addition and reductive elimination steps in the catalytic cycle. They may also be used as convenient air- and moisture-stable catalyst for aryl amination reactions. It has been suggested that the biaryl group in Buchwald's phosphines may contribute to the stabilization of ligand–metal complexes by establishing π interactions with the Pd(0) center. Based on Pd(dba)$_2$ using biaryl phosphine ligand as catalyst, Buchwald and Martin showed highly catalytic efficiency for Kumada–Corriu reactions even at temperatures as low as −65°C [28]. The Pd-catalyzed Kumada–Corriu cross-coupling reaction manifests a broad substrate scope. *Ortho-*, *meta-*, and *para-*substituted biaryls could all be efficiently prepared. In addition, a variety of functional groups were tolerated, including nitriles, amines, esters, heterocycles, and a benzylic acetal. Moreover, the process showed excellent chemoselectivity toward aryl halide substituents, like chlorides, fluorides, and even bromides making them available in many reactions, for further functionalization via conventional cross-coupling techniques.

Besides providing high turnover number catalytic efficiency, bulky electron-rich alkylphosphines may enable the use of extremely low levels of metal in the catalysis systems. Roy and Hartwig reported the first Pd-catalyzed cross-coupling of unactivated aryl tosylates with aryl Grignard reagents under mild conditions [29, 30]. With using PPF-^{t}Bu as ligand, a Josiphos-type ligand with di(*tert*-butyl)phosphino group, the catalyst loading could be as low as 0.1 mol%. The strongly electron-donating and sterical effect also makes the substrate capable of being expanded to various aryl and alkenyl tosylates.

The use of C(sp^3) nucleophiles and electrophiles for C–C bonding reactions has attracted research interests for developing new transition metal complex-based catalysis systems. Since the alkyl groups of secondary and tertiary main-group organometallic nucleophiles are prone to isomerize via β-hydride elimination/reinsertion sequence, which can result in the formation of inseparable isomeric products alongside reduced aryl products [31]. The uses of isolable, configurationally stable, optically active organometallic nucleophiles were exploited for developing the cross-coupling reactions by transition metal complex-based catalysts. Secondary alkyltin and alkylboron nucleophiles, which are stable and storable, may be activated toward transmetallation and reductive elimination by the inclusion of a sp^2-hybridized carbon atom in the α-position to the secondary alkyl centre. Researchers developed methods for the Pd-catalyzed cross-coupling of activated optically active alkylboron and alkyltin nucleophiles with aryl electrophiles, which showed the potential to alter the paradigm for asymmetric synthesis [32–35].

Biscoe et al. reported the development of a general Pd-catalyzed process for the stereoretentive cross-coupling of secondary alkyl azastannatrane nucleophiles with aryl chlorides, bromides, iodides, and triflates (Fig. 2.3) [36]. They used Pd(dba)$_2$ with ligand bis(3,5-bis(trifluoromethyl)phenyl) (2',4',6'-triisopropyl-3,6-dimethoxy- [1,1'-biphenyl]-2-yl) phosphine (JackiePhos), a bulky electron-deficient biarylphosphine, as the catalyst system. With azastannatrane backbone, secondary alkyl groups undergo transmetallation to palladium with excellent fidelity, independent of the electronic properties of the alkyl nucleophile. Coupling partners with a wide range of electronic characteristics can be tolerated well. The benchtop stability of optically active alkyl azastannatrane reagents and the ease transfer of stereochemical information via Pd-catalyzed cross-coupling reactions enable this process to be broadly applied in organic syntheses, particularly in the preparation of libraries of optically active drug candidates.

Transition metal complexes with N-heterocyclic carbenes (NHCs) are also well applied to activate aryl chlorides for cross-coupling reactions. After first demonstrated by Herrmann et al. in an imidazol-2-ylidene-palladium-catalyzed Mizoroki–Heck reaction [37], NHC complexes of various transition metals are, nowadays, privileged catalysts for a myriad of academically and commercially important processes. Some of the most important transformations mediated by NHC-metal complexes include Ir- and Ru-catalyzed hydrogenation and hydrogen transfer, Au-catalyzed activation of π-bonds, and Rh- and Pt-catalyzed hydrosilylation. While cross-coupling reaction and olefin metathesis might be the most extensively studied catalytic reactions. These two-electron donor ligands combine strong σ-donating properties with a shielding steric pattern. It thus allows for both stabilization of the metal center and enhancement of its catalytic activity. To date, NHCs are the only class of ligands that have been able to challenge the widely employed phosphines. The ratios of ligand/Pd, the steric bulk of the ligand, and the source of Pd were shown to be very important in defining the catalytic activity of the system [38].

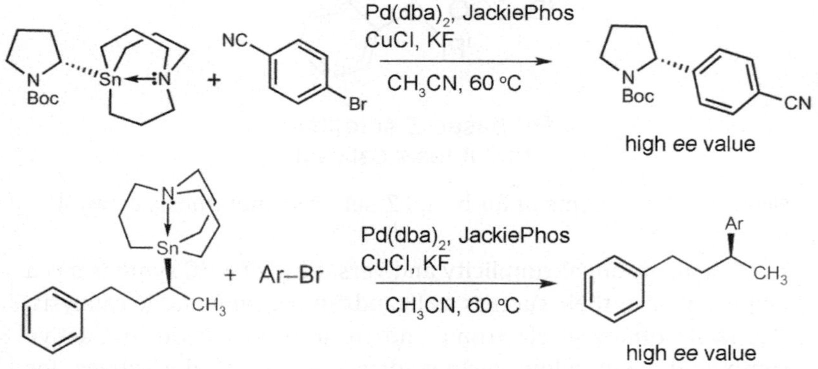

Figure 2.3 Pd-catalyzed process for the stereoretentive cross-coupling of secondary alkyl azastannatrane nucleophiles with aryl bromides.

Glorius et al. demonstrated the usefulness of bioxazoline-derived NHC ligand, IBiox, in transition metal catalysis [39]. The ligand is electron rich, sterically demanding, and has restricted flexibility. Using this ligand with palladium compound, di- and triortho-substituted biaryls can be formed in high yield at ambient temperature with nonactivated aryl chlorides as substrates. Iyer et al. described the synthesis and applications of palladacycles [40]. Due to the flexible framework and robustness of the palladacycles, good to high activity was displayed in the Heck reaction. Nolan et al. prepared NHC palladacycle complexes, which consist of a square-planar coordination around the palladium center [41]. The complex was equally active in aryl amination and α-arylation of ketones even at very low catalyst loading (0.02 mol%). Grubbs developed Ru-based metathesis catalysts with chelating NHC ligands (Fig. 2.4). The intramolecular C–H bond activation of the NHC ligand, promoted by anion ligand substitution, forms the appropriate chelate for stereo-controlled olefin metathesis. These catalysts showed remarkable Z-selectivity in cross-metathesis reactions and in a range of transformations [42, 43].

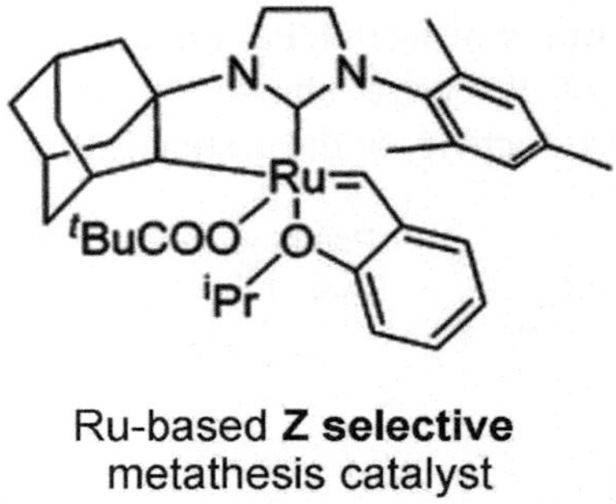

Ru-based **Z selective** metathesis catalyst

Figure 2.4 The scheme of Ru-based Z-selective metathesis catalyst.

The operational simplicity and versatility of NHC synthesis is a major factor in their success as ligands in organometallic catalysis. The facile tuning of electronic and steric properties of the active compounds will allow further development of derivatives for various specific catalysis reactions.

2.2.2 Non-Noble Metal-Based Compounds as Catalysts

In the past several decades, late and noble transition metals have been revealed to play key roles in the catalysis of chemical reactions. However, because of their expensive prices and high toxicity, efforts to replace noble metals with relatively more abundant and low-cost metals, such as Fe, Co, Ni, Cu, and Zn, are also of constant interest for research.

The use of Cu for cross-coupling reactions actually predates the discovery of Pd as a catalyst [44, 45]. Regardless of the instability of organocopper(I) species and the propensity of various Cu species to undergo radical and disproportionation reactions,

Cu salts and complexes currently rank among the most intensively investigated homogeneous catalysts for fine chemical synthesis, owing to low cost and toxicity compared with noble metals as well as to the broad variety of chemical transformations enabled by this metal center.

Cu-based catalysts may be utilized by the combination of Cu salts with organometallic reagents, such as Grignard reagents, organozinc reagents, organosilicon, and organoboron, etc. Cahiez et al. illustrated that the yield of the Cu-catalyzed coupling between alkyl Grignard reagents with alkynyl halides can be greatly improved by introducing of σ-donor ligands like N,N,N',N'-tetramethylethylenediamine (TMEDA), dimethylsulfide (Me$_2$S), triethylphosphate (OP(OEt)$_3$, and methylpyrrolidinone (NMP) [46]. In the process, the Cu species are first transmetalated with Grignard reagents to form a catalytic active complex of R$_2$CuMgX. The complex then undergoes carbocupration with alkynyl halides, and subsequently β-halogen elimination to afford the substitution product and the regeneration of catalytic Cu complex. In the case of RC≡CZ (Z=OEt, Cl), as the triple bond is stronger π acceptor than a simple terminal alkyne, a better σ-donor ligand would favor the complexation of the Cu(I) species to the triple bond by increasing the electronic density of the Cu atom and thus promote the carbocupration process. Therefore, an enhanced catalytic effect was shown with various σ-donor ligands such as TMEDA or triethyl phosphate.

Various ligands have been employed for the Cu center, to enhance its reactivity, selectivity, and stability. Several privileged ligand structures have proven to be extremely versatile and useful ligands, especially iminopyridines, which show particular effectiveness in stereoselective transformations such as nitroaldol reactions, allylic oxidations, and conjugate additions. Chelucci et al. evaluated the catalytic efficiency of Cu complexes with chiral iminopyridine ligands in carbene (cyclopropanation) and nitrene transfer reactions (aziridination, C–H amidation) [47]. They found that a better catalytic performance can be achieved for nitrene transfers, particularly in the amidation of C–H bonds.

Iron (Fe) is the most abundant transition metal on earth and plays important roles in nature. It was also one of the metals early and successfully used in chemical synthesis.

Fe-catalyzed organic transformations include nucleophilic additions, substitutions, reductions, oxidations, hydrogenations, cycloadditions, isomerizations, rearrangements, polymerizations, etc. Jiao et al. reported an Fe(OTf)$_3$/TfOH (TrO=triflate) cocatalyzed system for the coupling reaction of terminal alkynes with benzylic alcohols in the absence of base, which is a sp-sp^3 C–C bond formation process [48]. The collaboration of Fe(OTf)$_3$ and TfOH presents synergic effect on the transformation compared to the corresponding individual employment of these catalysts. As H_2O is the unique byproduct, it makes this transformation atom efficient and environmentally benign.

Sun et al. reported an unusual β-alkylation of benzylic alcohols with primary alcohols (mostly benzylic but also aliphatic) which is catalyzed by Fe-based catalyst, ferrocenecarboxaldehyde [49]. The reactions gave β-alkylated higher alcohols in high yields in the absence of any sacrificial agents (hydrogen acceptors or hydrogen donors) and nitrogen or phosphorus ligands, low yields were shown if normal Fe salts were used. In the process, Fe hydride species were generated with the oxidation of primary and secondary alcohols to the corresponding aldehyde and ketone. The base-mediated cross-aldol condensation then occurs to produce an α,β-unsaturated ketone, which undergoes transfer hydrogenation of C=O and C=C double bonds with that iron hydride species giving the desired product and regenerating the catalyst.

NHC ligands, which have been frequently applied in metal-organic synthesis, are also used as ligands for Fe-catalyzed reactions. Plietker et al. reported a regioselective alkoxy allylation of activated double bonds by Fe-catalyst of Bu$_4$N[Fe(CO)$_3$(NO)] (TBAFe) (Fig. 2.5) [50]. The system promoted the formation of sterically congested small molecules in high yields. The best regioselectivity and conversion were obtained using benzimidazole-derived N-heterocyclic carbene ligand. This transformation, in which two different bonds are formed in a decarboxylative two-component reaction with release of CO_2 as the sole byproduct, is characterized by broad functional group tolerance and mild, neutral reaction conditions. It particularly underlines the catalytic potential of TBAFe in the synthesis of complex molecules.

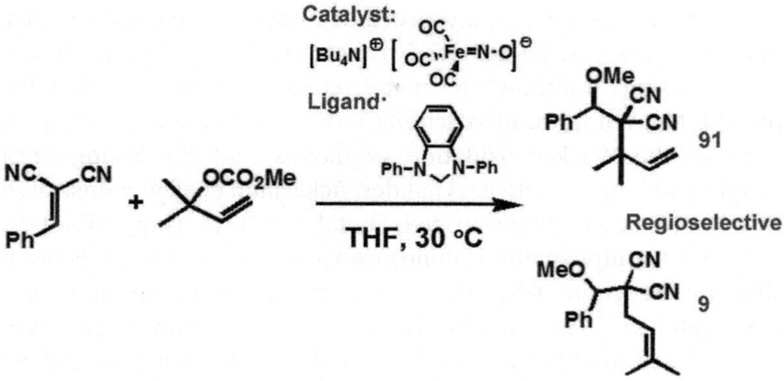

Figure 2.5 Fe-catalyst of $Bu_4N[Fe(CO)_3(NO)]$ for a regioselective alkoxy allylation of activated double bonds.

Plietker's group employed structurally defined π-allyl Fe complexes as novel, air and moisture stable precatalysts for the allylic substitution. In the presence of a NHC ligand, the aryl substituted NHC ligand SIMES (1,3-dimesitylimidazolin-2-ylidene), the activity of allyl Fe complex was greatly improved for the transformation. The better solubility of the system in organic solvents and the reduced induction period lead to an overall reduction of the catalyst concentration (down to 1 mol%). Taniguchi et al. developed an effective catalysis system with iron phthalocyanine and ethyl 2-(3,4-dichlorophenyl)hydrazinecarboxylate for the Mitsunobu reaction [51]. The hydrazinecarboxylate is oxidized in situ to an azocarboxylate by the Fe complex at aerobic conditions. The reaction between (-)-(S)-ethyl lactate and 3,5-dinitrobenzoic acid proceeded in the presence of the azocarboxylate and triphenylphosphine and afford the corresponding ester product with inversion of the stereochemistry in high enantiomeric excess (*ee*). This protocol could be applied in good to high yields for the substitution of various primary and secondary alcohols by carboxylates and N-nucleophiles.

2.2.3 Multimetallic-Based Compounds as Catalysts

Based on the studies on various catalysis systems, it can be found that a known reaction may proceed under more mild conditions or perform with better selectivity in the presence of a suitable catalyst. Many processes have proved that further improvements can be achieved by introducing additional metals or by using multinuclear metal complexes or intermetallic phases. Many important reactions rely on multimetallic catalysis, such as the Wacker oxidation of olefins and the Sonogashira coupling of alkynes with aryl halides. Ackerman et al. demonstrated that cooperativity between two metal catalysts, (bipyridine) Ni and (1,3-bis(diphenylphosphino)propane) Pd, enables a general Ullmann reaction (the cross-coupling of two different aryl electrophiles) [52]. The Pd catalyst reacts preferentially with aryl triflates to afford a persistent intermediate, while Ni catalyst preferentially reacts with aryl bromides to form a transient, reactive intermediate. The system can avoid the differentiation between multiple C–H bonds, which are required for direct arylation methods, and make the coupling reactions of aryl bromides with aryl triflates proceed in high yield without the use of arylmetal reagents.

During the past decade, tremendous efforts have been made to synthesize structurally defined heterobimetallic complexes. To synthesize useful bimetallic complexes, structural and electronic effects require very precise control in order to obtain cooperative performance. Severin et al. showed the formation of asymmetric, halogeno-bridged complexes of transition metals from simple starting materials [53]. Upon reaction of a complex having two weakly coordinating ligands (e.g. PhCN) with another complex having two terminal halogeno ligands, the corresponding asymmetric bimetallic complexes can be obtained. In the catalytic oxidation of secondary alcohols with butan-2-one, the asymmetric Rh(III)-Ru(II) complex with three chloro bridges was identified to be an exceptionally active catalyst precursor with activities surpassing that of known monomeric complexes.

Lemcoff et al. developed a comprehensive methodology to prepare nanometric size organometallic particles (ONPs) containing Ru(I), Ir(I) and Ni(0) with ROMP-derived polycycloocta-1,5-diene (pCOD) [54]. Direct exchange of the respective labile ligands of metal complexes by 1,5-hexadiene in pCOD or in situ reduction of metal ions in the presence of the polymer resulted in the formation of the polymeric complexes. These procedures can also be readily extended to the synthesis of organobimetallic nanoparticles containing two metals (Fig. 2.6), which could be added in commutative order and with specific metal ratios. The embedded metal elements are readily accessible for applications in catalysis. The close proximity of the catalytic centers leads to distinctive reactivity compared to the isolated complexes in several reactions.

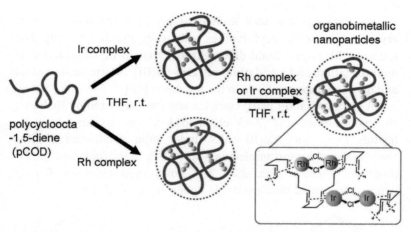

Figure 2.6 The synthesis scheme for organobimetallic nanoparticles.

2.2.4 High-Valent Organometallic Catalysts

High-valent Cu intermediates (that is, organometallic Cu(III) species) have long been proposed to have a role in Cu catalyzed cross-coupling reactions. In particular, C–C and/or C–X bond formation from an organo-Cu(III) species has been invoked as the product-releasing step for many of these transformations. However, before 2000, high-valent organometallic Cu complexes were rare and showed inert to C–C and C–X bond-forming reactions. During the past decade, tremendous developments have been achieved in this area with the detailed investigation on catalytically relevant organo-Cu(III) species in cross-coupling reactions.

Lewis basic additives such as cyanide, phosphines, pyridines, and amines have been known to improve the yield and/or rate of organocuprate conjugate addition reactions. Researchers have speculated that these Lewis bases may provide a more intimate role for the reactivity of Cu toward cross-coupling reactions. Bertz and Ogle et al. pointed out that Lewis basic ligands can drastically influence the reactivity and stability of a neutral organocopper(III) complex R$_4$Cu(III)Li [55]. Lewis basic ligands stabilize the Cu(III) center significantly more than a strongly electron donating amine and also play an important role in the elimination of side-products.

Cu catalysts are well known to promote the amination of aryl boronic acids, aryl halides, and C–H bonds [56–59]. For example, Ribas and Stahl demonstrated that in Cu-catalyzed C–H methoxylation and amidation, an aryl-Cu(III) intermediate exists and plays an important role in the reactions [60].

A stable organo-Cu(III) complex was first reported by Hedman, Hodgson, Llobet, and Stack et al. [61]. Simple triazamacrocyclic ligands react with Cu(II) salts under mild conditions to give the products of intramolecular aryl C–H activation and metal disproportionation (Fig. 2.7). The novel organometallic Cu(III) complexes formed thus are stable under protic conditions.

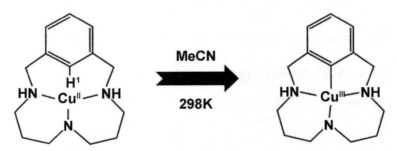

Figure 2.7 The formation of Cu(III) complex from Cu(II) complex.

Stahl and Ribas et al. synthesized a series of aryl-Cu(III)-halide complexes and represented the first direct observation of the Cu(I)/Cu(III) redox steps relevant to Ullmann-type coupling reactions [62]. The results indicated that aryl halide oxidative addition to Cu(I) may lead to the formation of an aryl-Cu(III)-X species and an aryl-Cu(III) intermediate in the turnover-limiting step of the catalytic C–N cross-coupling reaction. The work provides the basis for future studies to gain fundamental mechanistic insights into these important processes.

Although catalysis by low-valent Pd is ubiquitous and extremely synthetically useful, organo-Pd(II) complexes still have several limitations, which are the limited reactivity to form certain types of chemical bond (for example C–X and C–CF$_3$ linkages) and a high susceptibility to decomposition pathways (such as β-hydride elimination). The high-valent Pd is expected to open a potentially complementary mechanistic pathway to overcome those drawbacks [63].

It has proposed that the intermediate of high-valent Pd exists in the Pd-catalyzed reactions. To detect and isolate Pd(III) and/or Pd(IV) complexes from the reactions, electron-donating, rigid, multidentate supporting ligands (such as bidentate 2,2′-bipyridine ligand, phosphines) were employed to stabilize high-valent Pd products [64]. Arnold and Sanford et al. synthesized Pd(II) complexes containing chelating carbene ligands, which can form the stabilized Pd(IV) intermediate to promote C–X bond-forming reductive elimination [65]. The treatment of PdL(bzq) (L=OCMe$_2$CH$_2$(1-C{NCHCHNiPr}); bzq=benzo[h]quinoline) with PhICl$_2$ resulted in the oxidative addition of two chloride ligands to the Pd center and led to the formation of the first isolated Pd(IV)-NHC. The Pd(IV)-NHC complex can be stable in the solid state for at least one week. It is also the first direct observation on the C–X bond-forming reductive elimination from Pd(IV) complex (Fig. 2.8). In addition, the complex contains tunable ligands to be inert to the strongly oxidizing reaction conditions, which implies their potentials for regio- or stereocontrol over the C–H functionalization reactions.

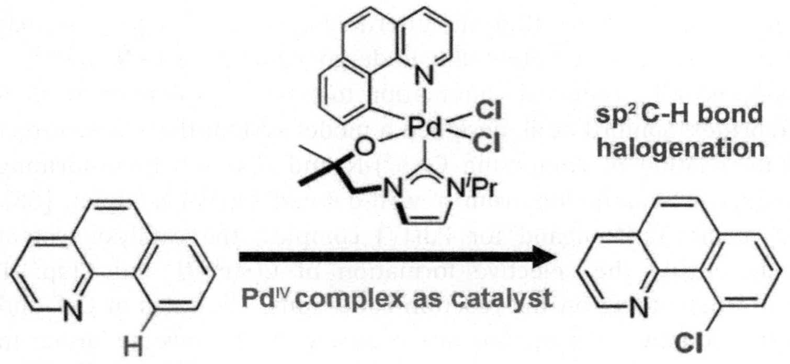

Figure 2.8 Pd(IV) complex as catalyst for C–X Bond formation.

Later, Sanford et al. demonstrated that Pd(IV)-mediated C–H activation could be achieved at an activated C(sp2)-center on a tridentate ligand featuring two N atoms [66]. The Pd(IV) complexes, which contain modular monoanionic tridentate to facially coordinate with NNN and NNC donor ligands, are stable to reductive elimination at room temperature for at lest several days. The Pd(IV) adducts may participate in both ligand substitution and C–H activation reactions.

High-valent Pd catalysis has also been exploited to facilitate the synthesis of valuable class of organic molecules. In particular, $C(sp^3)$-N bonds are ubiquitous in pharmaceuticals, agrochemicals, and commodity chemicals. Álvarez and Muñiz et al. developed intermolecular diamination reactions of alkenes employing high-oxidation-state Pd catalysis [67]. 8-Methylquinoline was chosen as a model substrate and the oxidant of N-fluorobis (phenylsulfonyl)-imide (NFSI) as a nitrogen source. The reaction proceeds with catalyst precursor [Pd(hfacac)$_2$] and NFSI, starting from the formation of palladacycle, and then the oxidation of the palladacycle with NFSI is done by a fluorinated high oxidation state Pd intermediate. It was suggested that the cationic Pd(IV) intermediate and bissulfonylimide should not combine to a neutral Pd complex but rather engage in direct nucleophilic substitution at the electrophilic carbon in the α position and install the new C–N bond. They identified NFSI as the decisive reagent for the development of a new Pd-catalyzed oxidative amidation of $C(sp^3)$–H bonds.

Since the transformations are proposed to proceed via Pd(IV) alkyl amino fluoride intermediates, it is suggested that these species can potentially undergo competing C–N and C–F bond-forming reductive elimination to yield alkyl amines or alkyl fluorides. Sanford et al. designed a model system that allow direct interrogation of competing $C(sp^3)$-N and $C(sp^3)$-F bond-forming reductive elimination from a well-defined Pd(IV) complex [68]. By using TsNH ligand for Pd(IV) complex, the catalyst system may enable the selective formation of $C(sp^3)$-N and $C(sp^3)$-F bond depending on the reaction conditions. The rates of C–C and C–F reductive elimination are expected to be inverse order in [TsNH$^-$], while that of C–N coupling is zero order in [TsNH$^-$]. As a result, the addition of NMe$_4$NHTs led to an increase in selectivity for C–N coupling. These studies provided preliminary insights into the role of reaction additives on these processes.

Trifluoromethylated arenes are essential structural motifs in a great number of pharmaceuticals, agrochemicals, and organic materials. Extensive efforts have been exerted toward the development of catalysis systems for introducing trifluoromethyl groups onto arenes. However, due to the kinetic inertness of most metal–CF_3 complexes, C–CF_3 bond formation required specialized ligands to improve the efficiency. Pd(IV) complexes containing simple bidentate, nitrogen donor ligands, have shown to enable facile C–CF_3 bond-forming reductive elimination [69, 70]. Yu et al. reported the Pd(II)-catalyzed trifluoromethylation of benzamides using an N-alkylformamide as a crucial promoter [71]. From the observation on stoichiometric oxidation of [(bzq)Pd(OAc)]$_2$ (bzq=benzo[h]quinoline) to the corresponding Pd(IV) species with CF_3^+, they suggested that the reaction with CF_3^+ may proceed through Pd(II)/Pd(IV) pathway with N-methylformamide as enabling ligand.

Other high-valent metals such as Rh and Co organometallic complexes have also been employed successfully to catalyze the addition reactions of arene C–H bonds to imines, aldehydes, Michael acceptors, and other polar electrophiles [72–76]. Kanai et al. demonstrated the utility of a cationic, high-valent Co complex for atom-economical directed addition reactions of aryl C–H bonds to polar electrophiles (Fig. 2.9). The [Cp*Co(III)(arene)](PF$_6$)$_2$ (Cp*=pentamethylcyclopentadienyl) complex possesses well activity and stability, which was shown be able to catalytically generate the nucleophilic species in situ without any additional reagents. It enables the directed C–H bond addition of 2-aryl pyridines to imines, enones, and α,β-unsaturated N-acyl pyrroles as ester and amide surrogates. Although Cp*Co(III) complexes have not attracted much attention in the field of synthetic organic chemistry, they may be promising catalysts for efficient C–H bond-functionalization processes.

Enzymes provide an exquisitely tailored chiral environment to foster high catalytic activities and selectivities, but their native structures are optimized for very specific biochemical transformations. Designing a protein to accommodate a non-native transition metal complex can broaden the scope of enzymatic transformations and enhance the activity/selectivity for small-molecule catalysis. Ward and Rovis et al. report the creation of a bifunctional artificial metalloenzyme in which a docked biotinylated Ru(III) complex is incorporated in an engineered Streptavidin with glutamate or aspartate residue [77]. The complex was shown to greatly accelereate asymmetric C–H activation. The coupling of benzamides and alkenes to access dihydroisoquinolones proceeds with up to nearly 100-folds rate acceleration.

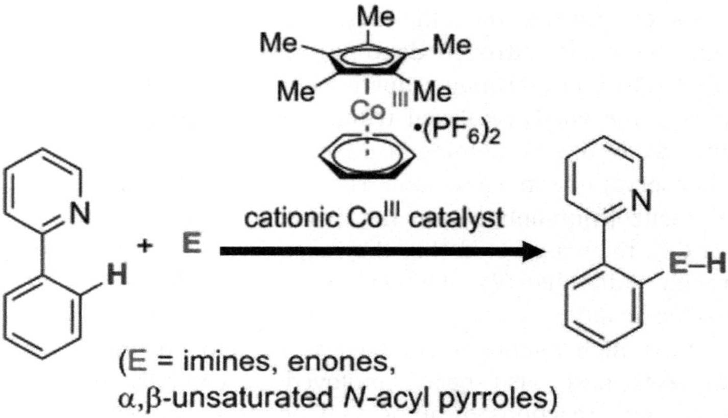

Figure 2.9 Co(III) complex as catalyst for atom-economical directed addition reactions of aryl C–H bonds.

Recently, there has been notable progress in high-valent metal catalysis, various organometallic complexes have been synthesized and a wide range of C–C and C–X bond-forming reactions can be achieved with well selectivity and reactivity. A number of them may also be competent catalysts for C–H bond halogenation, trifluoromethylation, and other reactions. Those results may stimulate more efforts on the synthetic design and utility of high-valent metal catalysts in organic synthesis.

2.3 Metal-Free Organocatalysts

Transition metal catalysts have shown important roles in the many organic transformations. However, the toxicity and high price of transition metals are of particular concerns, especially in the synthesis of pharmaceuticals, where even traces of heavy metals have to be rigorously excluded from the drug product.

The use of eco-friendly small organic molecules, called metal-free organocatalysts, has attracted considerable attention and grown quickly in chemical research [78]. In contrast to conventional metal-based catalysts and enzymes, metal-free organocatalysts are advantageous from an environmental or resource perspective, and not limited to the substrate scope like enzymes. These catalysts hold considerable promise as cheap, stable, moisture-insensitive, reproducible, and easy-to-construct alternatives to well-studied metal-based catalysts.

The scope of chemical reactions by metal-free organocatalysts has been expanded considerably during the past decade. Typical transition-metal-mediated coupling reactions, such as Suzuki, Sonogashira, Ullmann, and Heck-type coupling reactions, as well as the Tsuji–Trost reaction, can now be performed under metal-free conditions. The development of catalysts with a higher molecular weight and increased complexity often leads to a sharp improvement not only in the selectivity of the catalyst but also in its kinetic profile. There is an increasing number of asymmetric reactions, in which these catalysts meet the high standards of modern synthetic methods. In addition, organocatalysts have a great practical potential in devising multicomponent and tandem sequences. In the future, all these reactions will find their usages out of the academic environment for the synthesis of complex molecular structures.

2.3.1 Catalysis by Enamines and Imines

Although it is now widely accepted that organocatalysis is one of the main branches of enantioselective synthesis, the field of metal-free organocatalysis was formally launched not until 2000. The work of Barbas, Lerner, and List significantly proved that small organic molecule of proline could catalyze the same chemical reactions as much larger enzymes and could be extended to transformations that have a broader applicability (specifically, the intermolecular aldol reaction) [79]. Meanwhile, the report of iminium catalysis by MacMillan et al. conceptualized the new "organocatalysis" in three important ways [80, 81]:

(i) by delineating the economic, environmental and scientific benefits

(ii) by describing a general activation strategy for organocatalysis that could be applied to a broad range of reaction classes
(iii) by introducing organocatalysis to the chemical literature

Until now, the most successful catalyst for enamine-type reactions has undoubtedly been L-proline. Proline is a naturally occurring bifunctional amino acid readily available in enantiomerically pure forms. It is the only natural amino acid with a secondary amine functionality. It thus has a higher pKa value and enhanced nucleophilicity than other amino acids. Proline may react as a nucleophile with carbonyl groups or Michael acceptors to form iminium ions or enamines in the catalytic reaction. The proline-catalyzed reactions include aldol reactions, Mannich reactions, amination, aminoxylation, and Michael addition. As proline promotes the formation of highly organized transition states with extensive hydrogen-bonding networks, the high enantioselectivity of reactions can be achieved. In all proline-mediated reactions, the proton transfer from the amine or the carboxylic acid group in proline to the forming alkoxide or imide is essential for charge stabilization and C–C bond formation in the transition state [82–84]. The versatility of proline as an asymmetric catalyst is conspicuous from a broad range of C–C or C–X bond formation reactions.

The role of solvents or additives is important for the selectivity of the proline-catalyzed reactions. In the case of aldol reactions, polar aprotic organic solvents such as dimethyl sulfoxide and dimethylformamide may lead to high enantioselectivity, while polar protic solvents are better solvents for Michael reactions. Pihko et al. have observed significantly higher yields for the proline-catalyzed intermolecular aldol reaction by the addition of water [85]. The use of additives or combination of two solvents may also be a viable strategy to refine the yield and selectivity of proline-catalyzed reactions. When unmodified aldehydes were used as donors, the cross-Mannich reaction was shown to proceed faster than the competing cross-aldol reaction [86–88]. The three-component cross-Mannich reaction exhibited higher stereo- and chemoselectivity ($k_{\text{Mannich}} > k_{\text{aldol}}$) at temperatures below 0°C (Fig. 2.10). With electron-rich aromatic acceptor aldehydes, the stereodetermining proton transfer from the carboxylic acid in proline to the alkoxide is facilitated. Therefore, the aldehyde substituent occupies a pseudo-equatorial position in the aldol reaction, while in Mannich reaction the substituent is forced into a pseudoaxial orientation. These results bring insights on the importance of reaction media for the enamine formation and the stereoselectivities of the reactions.

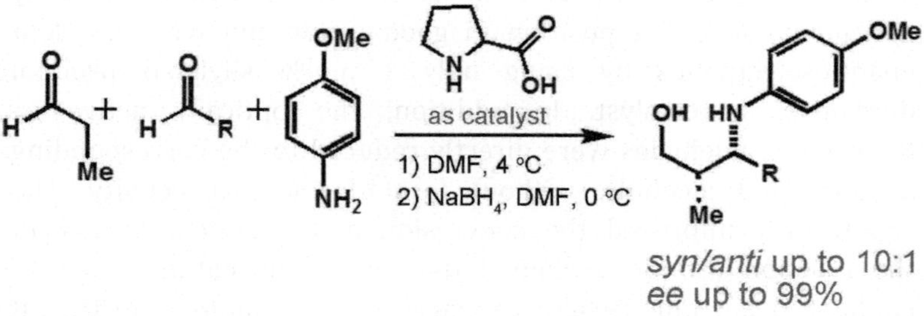

Figure 2.10 The three-component cross-Mannich reaction catalyzed by proline.

Although proline continues to play a central role in aminocatalysis, its supremacy can also be reflected by the facial synthesis for more complex proline derivatives. Both Barbas et al. and Hayashi et al. independently reported amine-derived chiral catalysts capable of inducing high enantiocontrol in the presence of large excess of water [89–90]. In their cases, highly hydrophobic proline derivatives have been used as catalysts for direct aldol reactions. Itoh et al. synthesized a series of N-substituted prolines and studied the catalytic performance for the asymmetric reduction of imines with trichlorosilane [91]. The reduction of N-aryl imines in the presence of 10 mol% N-pivaloyl-L-proline anilide was shown to give the corresponding amines in excellent yields (up to 99%) and with high enantioselectivities (up to 93% ee). Since the catalyst could interact with the substrate imine, the Lewis base proline catalyst could coordinate with trichlorosilane and activate the reducing agent in a synergistic manner. It was suggested that the coordination between Si and O atom of the amide group, and hydrogen bonding between the imine nitrogen and anilide hydrogen promotes the asymmetric reaction. It thus resulted in the attack from the Re face of the imine to give the optically active (S) amine.

The unique properties of fluorinated molecules have led to the wide interest in organic syntheses, medicinal and agricultural chemistry, as well as materials sciences. Especially, structures in which the fluorine atom is attached to a chiral center are gaining increasing attention. Jørgensen et al. presented an easy access to stereogenic C–F centers through the direct enantioselective α-fluorination of aldehydes by using silylated prolinol compound as catalyst (Fig. 2.11) [92]. A range of aldehydes can be directly fluorinated in the α position in good yields and with excellent enantioselectivities by using only 1 mol% silylated prolinol derivative as catalyst. In addition, the optically active α-fluorinated aldehydes were directly reduced to the corresponding α-fluorinated alcohols, without loss of enantioselectivity. The significantly improved the conversion and enantioselectivity of the reaction can be explained due to the formation of an E-configured enamine, where the sterically demanding substituent of the pyrrolidine ring shields the Re face of the enamine. As a consequence, the electrophilic F⁺ ion attack occurs from the Si face, providing excellent enantioselectivities.

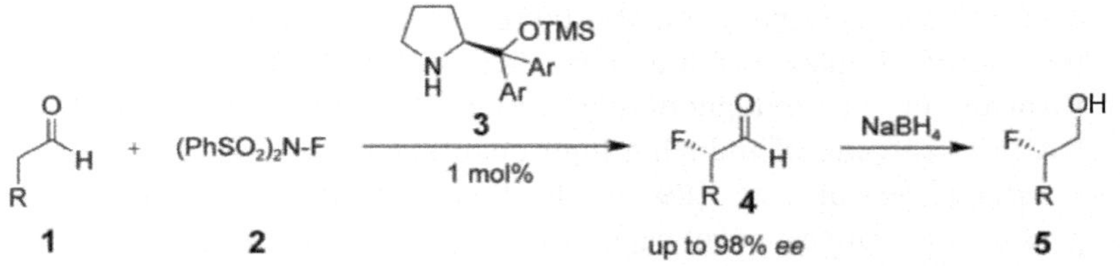

Figure 2.11 The silylated prolinol derivative as catalyst for enantioselective α-fluorination of aldehydes.

Condensation reactions between two aldehydes have been promoted by primary and secondary amine catalysis. Cross-condensation reactions between simple alkyl aldehydes and formaldehyde, in turn, are typically performed using secondary amines and acid cocatalysts under relatively drastic conditions, including high temperature, high pressure, and rapid distillation of the product from the reaction mixture. Erkkilä and Pihko recently disclosed an amine-iminium-catalyzed method for α-methylenation of aldehydes [93]. 10 mol% propionic acid salt of pyrrolidine promoted condensation of α-monosubstituted aldehydes with formaldehyde. The iminium ions with saturated α-carbons readily lose an electrophile at the α-carbon. The formaldehyde forming an iminium ion with pyrrolidine salt acts as a key intermediate for the reaction.

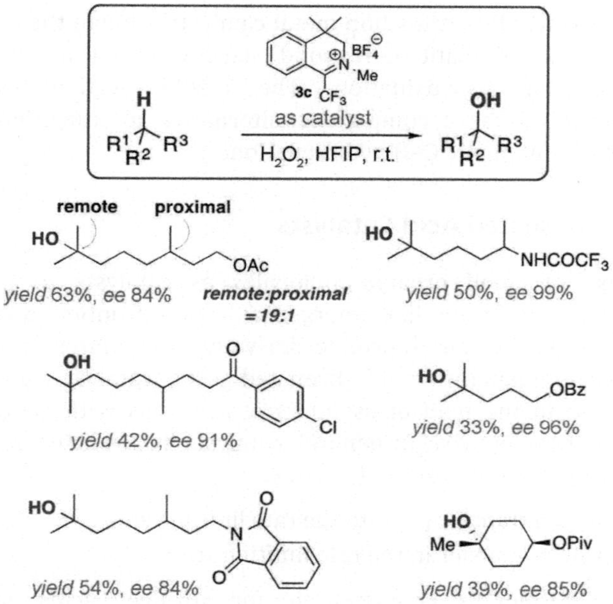

Figure 2.12 The iminium salts for C–H hydroxylation at room temperature.

Aliphatic C–H bonds are among the least reactive in organic chemistry. However, except catalysis by enzymes or bio-inspired catalysts, aliphatic C–H oxidations by molecular catalysts were always not sufficiently high in yield or predictable in site-selective manner. Powerful small-molecule catalysts thus were invented to achieve high enough yields (>50%) and predictably discriminate between aliphatic C–H bonds even within complex molecules possessing many possible sites of oxidation [94]. The development of organocatalysts has focused on harnessing the reactivity of dioxiranes and oxaziridines to hydroxylate unactivated aliphatic C–H bonds intermolecularly in a catalytic cycle. Hilinski et al. presented the first disclosure of a new class of organocatalysts: iminium salts, for C–H hydroxylation at room temperature (Fig. 2.12) [95]. The iminium salts with trifluoromethyl substitution were shown to improve the reactivity over other organocatalysts and are inclined to generate the product of hydroxylation at the site remote from the ester. Furthermore, chemoselectivity for aliphatic hydroxylation over alcohol oxidation was enhanced in compared to the existed catalysts (including transition metal catalysts), due to the ability to oxidize a 2° aliphatic C–H bond selectively but a 2° alcohol with limited over-oxidation. The results further establish organocatalysis as a competitive alternative to transition metal catalysis for aliphatic C–H hydroxylation.

2.3.2 Brønsted Acid Catalysts

The use of small organic molecules as catalysts to promote asymmetric reactions has emerged as a new frontier in reaction methodology. Besides L-proline derivatives or amine derivatives, a superior generation of Brønsted acid catalysts has been developed in the field of asymmetric reactions with good yields and enantioselectivity. In general, catalysis by Brønsted acids can operate by two types of mechanisms:

(i) proton transfer prior to the rate limiting step
(ii) proton transfer in the rate limiting step

The former is characteristic for stronger acids, such as phosphoric acids, while the latter is characteristic of weaker acids, such as thioureas. The Brønsted acid-based catalysis is a growing area with many possibilities. Through the design of Brønsted acids with a variety of structures and activation mechanisms, they can be tuned to support high yield and enantioselectivity in a variety of reactions, including Strecker and Mannich reactions, Diels–Alder reaction, hydrogenations, Michael additions, etc.

The enantioselective Mannich-type reaction of an enolate or an enolate anion equivalent with aldimines constitutes a useful method for the preparation of chiral β-amino carbonyl compounds, which are the precursors of biologically important compounds such as β-lactams and β-amino acids. Chiral Brønsted acids, in which the proton is surrounded by bulky substituents, were shown to be capable of leading to effective asymmetric induction for the reactions. Akiyama et al. reported enantioselective Mannich-type reaction of silyl enolates with aldimines catalyzed by a chiral metal-free Brønsted acid (Fig. 2.13) [96]. Aromatic groups (4-nitrophenyl groups) were introduced at the 3,3′-positions of

1,1'-Bi-2-naphthol(BINOL)-derived chiral phosphate. It is supposed that 3,3'-diaryl groups, which are not coplanar with the naphthyl groups, would effectively shield the phosphate moiety, leading to efficient asymmetric induction. The resulting chiral Brønsted acid showed catalysis with high to excellent enantioselectivities (up to 96% ee) under metal-free conditions in enantioselective Mannich-type reaction of aldimines with silyl enolates and β-aminoesters.

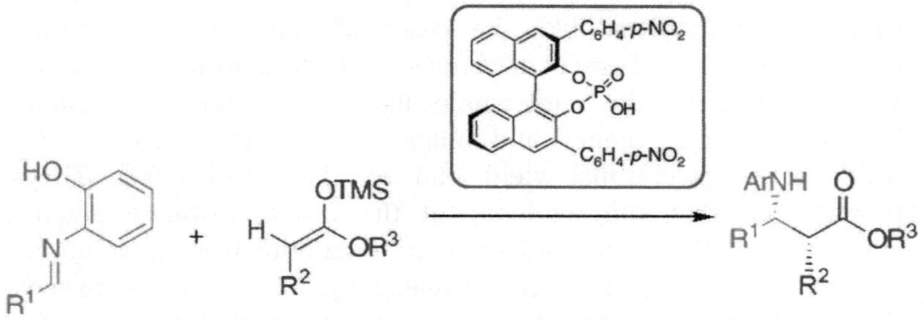

Figure 2.13 Mannich-type reaction of silyl enolates with aldimines catalyzed by a chiral metal-free Brønsted acid.

Schneider et al. developed a second-generation 2,2'-dihydroxy-1,1'-binaphthyl (BINOL)-based phosphoric acid and further optimized as an enantioselective organocatalyst of vinylogous Mannich reaction of acyclic silyl dienolates [97]. They investigated BINOL-based phosphoric acids with varied the size of the *ortho* and *para* substituents within the 3,3'-aryl groups, as these positions were of prime importance for the enantioselectivity of the reaction. Placing a ᵗBu group instead of a methyl group in the *para* position was shown to significantly increase the selectivity to 96% ee with a high yield production of the vinylogous Mannich reaction. Upon protonation of the imines, chiral contact ion pairs are generated in situ and attacked highly diastereoselectively by the nucleophile. One of the 3-aryl groups within the Brønsted acid catalyst shields the *Re* face of the imine, thereby directing the incoming nucleophile to the opposite side. They found that γ-substituted silyl dienolates are suitable substrates for this process, thus giving rise to products with two new chiral centers in good diastereo- and enantiocontrol.

Although a number of structurally diverse strong Brønsted acid catalysts have been developed for the addition of alcohols and other protic nucleophiles to simple alkenes, the highly enantioselective reactions reported are generally restricted to the activation of an electrophilic C–X or X–X multiple bond. By conjugation a nucleophilic base to a chiral Brønsted acid, it was revealed that the chiral catalyst can be directly bound to the substrate in the nucleophilic addition step, and may facilitate a highly enantioselective transformation based on "covalent catalysis" mechanisms like the cases of enamine or iminium catalysis [98]. Dean Toste et al. showed that chiral dithiophosphoric acids catalyze the intramolecular hydroamination and hydroarylation of dienes and allenes to generate heterocyclic products in exceptional yield and *ee* (Fig. 2.14) [99]. They pointed out that the tendency of the dithiophosphate added covalently to the diene rather than remaining free in solution may result in surprising chemoselectivity. The tether between the nucleophile and the diene can also be varied to generate spirocyclic products. Besides serving as a useful means for valuable chiral hetero- and carbo-cyclic products, the hydroamination and hydroarylation reactions also present a fundamentally distinct catalytic pathway from those reactions using conventional chiral organocatalysts.

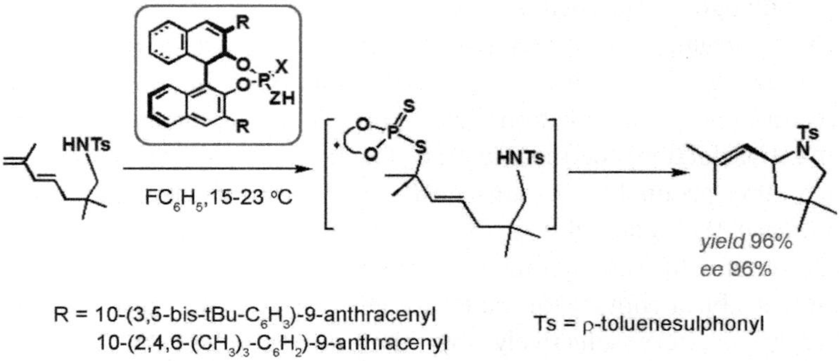

Figure 2.14 The chiral dithiophosphoric acids for the intramolecular hydroamination and hydroarylation of dienes.

Azlactones have been attractive since they may be served as protecting amino acids in the synthesis of natural or synthetic bioactive molecules. The multi-functional heterocyclic azlactone structure confers to such molecules a diverse chemistry, acting as electrophiles via two distinct sites, while also suffering from nucleophile attack. A successful catalyst must bring about a selective kinetic resolution (i.e., $k_{fast} \gg k_{slow}$). It must also be capable of catalyzing the epimerization equilibrium of the azlactone with an efficiency, which ensures that racemization occurs at a considerably faster rate than the addition of the alcohol to the less preferred substrate enantiomer. Connon and Palacio designed a series of C-5′-hydroxylated cinchona alkaloid catalysts for the dynamic kinetic resolution (DKR) of azlactones by thiolysis [100]. They achieved the thiolytic DKR of azlactones with excellent enantioselectivity (84–92% ee) for the first time, and surpassed the levels for all substrates previously evaluated. Eberlin and Amarante et al. showed camphorsulfonic acid (CSA) acting as an effective Brønsted acid catalyst for the azlactone ring (Fig. 2.15) [101]. With nucleophiles applied in the reaction system, protected amino esters and amides derivatives may be produced in good to excellent yields. The azlactone activation proceeds by forming an ion-pairing intermediate with CSA, and following by nucleophilic attack. CSA is responsible for the protonation and de-protonation step.

Heteroatom-substituted alkynes probably represent the most versatile class of alkynes. Especially ynamines, with their uniquely polarized triple bond, allow for an exceptionally high level of reactivity together with a strong differentiation of the two sp-hybridized carbon atoms. They are emerging as especially useful and versatile building blocks for organic synthesis. The development of new reactions or synthetic sequences starting from ynamides has attracted great interest, resulting in an ever-increasing number of publications [102]. Most of reactions of ynamines involve one of the multiple modes of cycloaddition and rely on soft, alkynophilic transition metal catalysts or promoters. While the metal-free catalysis may allow the functionalization of C–H bonds through mechanistically novel pathways. Maulide et al. reported the use of a metal-free Brønsted acid (triflic acid) as catalyst for redox arylation reaction of ynamides through a [3,3]-sigmatropic rearrangement [103]. This redox arylation reaction is shown to be compatible with a broad range of aryl sulfoxides and delivers both electron-donating and -withdrawing substituted α-arylated oxazolidinones in good to excellent yields under mild conditions.

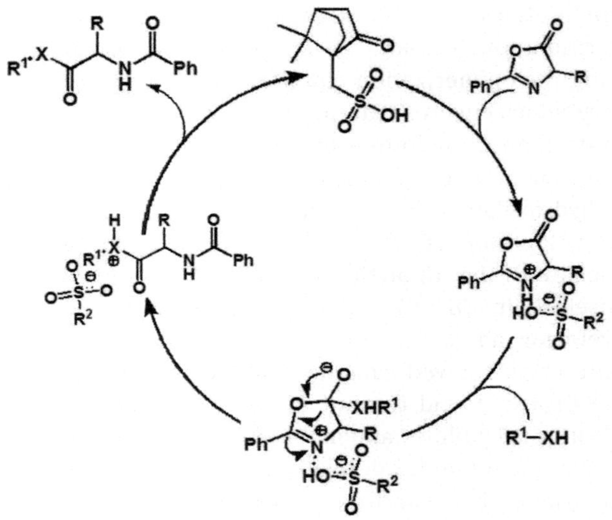

Figure 2.15 The scheme for camphor sulfonic acid as an effective catalyst via nucleophile attack.

The current global economic and environmental landscape has accelerated research in carbon dioxide (CO_2) capture and storage (CCS) technology across a broad range of chemical disciplines. Transformations employing CO_2 as a reagent are typically used either Lewis basic substrates with sufficient nucleophilicity to react with the poorly electrophilic CO_2 or metal-based reagents to increase the rate of CO_2 incorporation. The metal-free catalysts are founded capable of acting as more ideal catalyst, to stabilize the adduct of a weak nucleophile with CO_2, and also effectively guide enantioselective bond formation. Johnston et al. designed a metal-free organocatalyst based on a properly balanced Brønsted acid/base bifunctional catalyst, which may bring the virtues of minimalism (symmetrical catalyst, low temperature, atmospheric pressure, near-neutral pH) for the formation of C–O bonds (Fig. 2.16) [104]. They used trans-stilbene diamine incorporated pyrrolidine-substituted bis(amidine) (StilbPBAM) as the catalyst. When complexed with triflimide ($HNTf_2$), CO_2 capture reactions with homoallylic alcohols showed high enantioselectivity (80–95%

ee) and high chemical yield for the production of cyclic carbonates. The dual Brønsted acid/base catalyst possesses hydrogen-bond donor and acceptor functionality to activate and orient substrates in the enantioselective reaction. Although employed by relatively weak nucleophiles, this CO_2 fixation offers a simple equivalent to metal-free oxidations of homoallylic alcohols by generating transient acids to an alkene in combination with N-iodosuccinimide. The virtues of Brønsted acid/base activation alone make the synthesis of highly enantioselective carbonate possible.

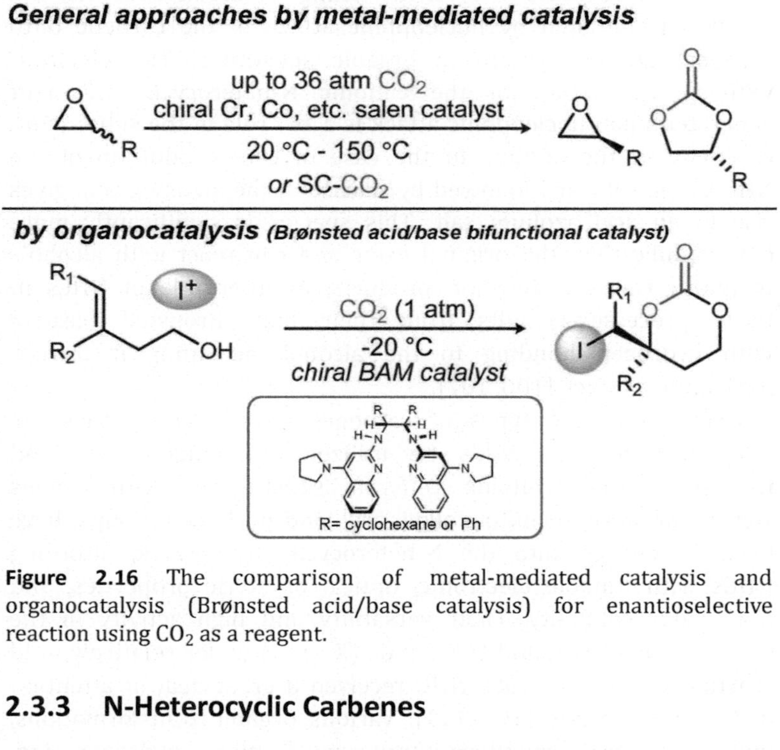

General approaches by metal-mediated catalysis

by organocatalysis (Brønsted acid/base bifunctional catalyst)

Figure 2.16 The comparison of metal-mediated catalysis and organocatalysis (Brønsted acid/base catalysis) for enantioselective reaction using CO_2 as a reagent.

2.3.3 N-Heterocyclic Carbenes

The successful isolation and characterization of an N-heterocyclic carbene in 1991 opened up a new class of organic compounds for investigation. N-heterocyclic carbenes (NHCs) today rank among the most powerful tools in organic chemistry, with numerous applications in commercially important processes [105]. Defined as neutral compounds containing a divalent carbon atom with a six-electron valence shell, carbenes are an intriguing class of carbon-containing compounds. Their incomplete electron octet and coordinative unsaturation, render free carbenes inherently unstable and can form highly reactive transient intermediates in organic transformations. As excellent ligands for transition metals, NHCs have found multiple applications in some of the most important catalytic reactions in the chemical industry. With intensive exploration on the rich chemistry of these compounds, they have become new major class of organocatalysts for research.

In the processes with NHCs acting as organocatalysts, the reaction is initiated by nucleophilic attack of the carbene onto carbonyl groups present in organic substrates. The electron-withdrawing nature of the cationic N-heterocyclic fragment generated upon nucleophilic attack is a key role in the subsequent reactivity of the adduct. In the case of esters, addition of the NHC to the carbonyl followed by release of the alkoxy group gives rise to an acyl azolium salt. This species is significantly more electrophilic than the original ester and can react with alcohols to afford transesterification products. Another role of NHCs in these processes results from their high Brønsted basicity with hydrogen bonding to the alcohol, activating it toward nucleophilic attack [106, 107].

Many different types of carbenes have been synthesized. The most popular NHCs be utilized are imidazolium- and triazolium-based scaffolds. Different substituents, with various steric hindrance, including alkyl, aryl, and perfluoro groups, have been introduced into the N-heterocyclic framework, affording NHCs with tunable electronic, optical or steric properties, and also water solubility. Their versatility and high activity in the construction of C–C and C–X bonds (X=O, N) under relatively mild reaction conditions make NHC received a great deal of attention in the past decade [108–111]. Various organic transformations, including cross couplings, transesterification, hydrogenation, cyclopropanations, hydrosilylations, aryl aminations, olefin metathesis, polymerization and ring-opening reactions, can be efficiently catalyzed by NHCs.

Sun and coworkers reported an efficient asymmetric α-fluorination of azolium enolates from simple aliphatic aldehydes and α-chloro aldehydes under NHC catalysis [112].

With a suitable combination of precatalyst, oxidant, base, and fluorination reagent, the reaction proceeded smoothly to yield a wide range of α-fluoro esters, amides, and thioesters with excellent enantioselectivity. Huang et al. reported the enantioselective C–C and C–S forming reaction using chiral NHCs (Fig. 2.17) [113, 114]. They suggested that the NHC might serve as a dual functional catalyst through noncovalent interactions: hydrogen bonding and π–π stacking, unlike classical NHCs. As catalytic amounts of hexafluoroisopropanol are the critical proton shuttle that facilitates hydrogen transfer to provide high-reaction rates and high enantioselectivity, successful asymmetric reactions by NHC catalysts can be accomplished through a rational design from balancing the pKa values of the substrate, the carbene precursor and the product.

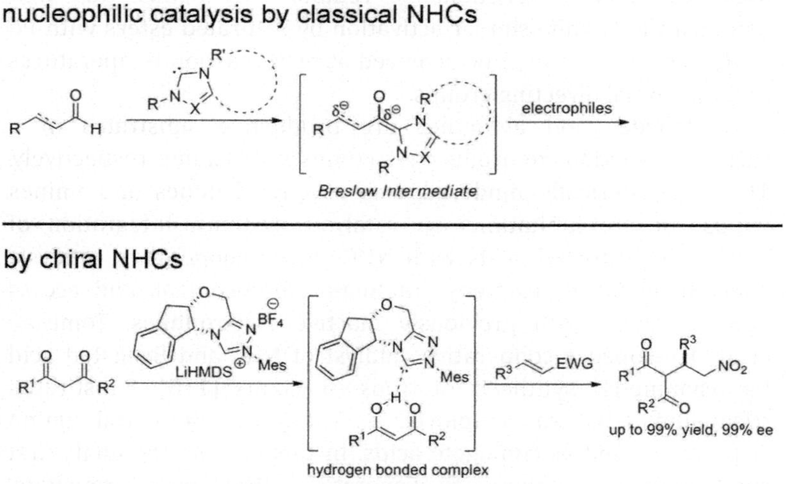

Figure 2.17 NHCs as non-covalent chiral catalyst for asymmetric reaction.

The horizon of chemical synthesis is suggested capable greatly expanded if the typically inert β-carbons of saturated esters could be used as nucleophiles in organic synthesis. Chi et al. reported the catalytic activation of simple saturated ester β-carbons as nucleophiles (β-carbon activation) using NHC as organocatalyst [115]. The addition of NHC catalyst (imidazolium compounds) to ester substrate yields an NHC-bounded ester intermediate II. The reactive ester intermediate bearing acidic α-CHs can undergo deprotonation to form an enolate intermediate containing a nucleophilic α-carbon. Because of the electron-withdrawing ability of the triazolium moiety and the conjugated nature of the triazolium-bounded enolate intermediate, the formal ester β-CH protons could become acidic. Deprotonation of the β-CHs of enolate intermediate could afford the intermediate with β-carbon as a nucleophilic centre. The catalytically generated nucleophilic β-carbons undergo enantioselective reactions with electrophiles such as enones, trifluoroketones and hydrazones to afford cyclopentenes, γ-lactones and γ-lactams, respectively. This catalytic activation mode for saturated ester β-carbons under mild conditions opens a valuable new arena for chemical reactions and concise synthetic strategies. The capability of metal-free organocatalysis to functionalize remote C–H bonds was well demonstrated, while similar activation by saturated esters with Pd or Rh catalysis can only be achieved at high reaction temperatures or with special directing groups.

Aldehydes and aldimines are productive substrates in a variety of reactions to produce γ-lactones and lactams, respectively. The more sterically hindered, less reactive ketones and imines impose severe limitations on catalysis systems. Integration of Lewis and Brønsted acids with NHCs bring cooperative catalysis effect to enhance reactivity, modulate stereocontrol, and access new reactivity with previously inactive electrophiles. Tomislav et al. developed a cooperative catalyst of NHC and Brønsted acid for asymmetric synthesis of trans-γ-Lactams [116]. Substituted triazolium salts was employed as catalysts, and chiral amino acids were used as conjugate acids. In this system, the enals first react with the carbene to generate a Breslow intermediate, which attacks the acid-activated imine via hydrogen-bonding intermediate. Steric hindrance leads to anti-orientation of R3 and the alkenyl. Proton transfer then results in the formation of acyl carboxylate. The nitrogen species replaces the carbene to afford the product and the free carbene. Through this cooperation of Brønsted acid with NHC, an efficient organocatalysis asymmetric approach for the synthesis of trans-γ-lactams in high yields (up to 99%), high enantioselectivities (93% ee), and complete diastereoselectivity (>20/1 dr) can be achieved.

To achieve new advances in rational design of NHC structure with anticipated reactivity, computationally guided rational catalyst design has been utilized to improve selectivity for some catalysis processes, including of NHC catalysis. Sigman's group demonstrated the use of correlative methods to dictate improved NHC catalyst scaffolds for epoxidation, hetero-Diels–Alder, and other catalytic asymmetric reactions [117]. Enhanced catalyst performance can be achieved. The relationship between catalyst structure and enantioselectivity can also be constructed. Cheong and Scheidt et al. designed a new tailored C1-symmetric biaryl-saturated imidazolium-derived NHC catalyst, based on a rational design strategy guided by computational modeling of competing transition states (Fig. 2.18) [118]. They investigated a computational model of the stereodetermining homoaldol step with density functional theory (DFT), B3LYP/6-31G*. For the addition of α,β-unsaturated aldehydes to α-ketophosphonates, they found that nucleophilic attack of the homoenol to the acyl phosphonate carbonyl is concomitant with deprotonation of the enol proton by the forming alkoxide. Stereoselectivity arises from differential stabilization of the phosphonyl oxygen by aryl protons of the catalyst and through nonclassical hydrogen bonds (NCHBs). Based on the computational prediction, new C1-symmetric biaryl-saturated imidazolium catalysts can be created. With more substituents at sterical 3,5-positions in the aryl ring, the catalysts result in the production of the γ-butyrolactones in good to excellent yield and enantioselectivity. In addition, this new enantioselective platform provides a distinct approach for 1,4-carbonyl compounds, which are difficult to access through traditional methods. Various substituents can be incorporated through the appropriate choice of the acyl phosphonate and aldehyde starting materials.

Advances in polymer chemistry in the 20th century have been dominated by metal-based catalysis. Except for polymers obtained by free radical polymerization, most of the industrial synthetic polymers used organometallic species to catalyze or initiate their synthesis. Concepts of organocatalysis have been applied to some polymerization reactions from the past decade, in an attempt to compete with metallic catalysts [119]. Among organocatalysts for polymerizations, NHCs have been extensively investigated, where they offer an alternative to traditional organometallic catalysts and initiators. Based on the pioneering works of Hedrick et al. [120, 121], NHCs organocatalysts have been expanded to polymerizations, such as the ring-opening polymerization (ROP) of heterocyclic monomers, step-growth polymerization of bifunctional monomers, and the group transfer polymerization of acrylic monomers. Proton-transfer polymerization (HTP) describes a polymerization in which each propagation step involves a proton transfer, a process to (re)generate the active propagating species or activate the monomer and the nucleophile. Matsuoka et al. reported the polymerization of hydroxyalkyl acrylates by an NHC catalyst, producing low MW poly(ester-ether)s with M_n up to 2400 g/mol. They proposed that this NHC-catalyzed HTP process is proceeded via an initial zwitterionic intermediate, which is derived from Michael addition of NHC to the monomer, then a proton transfer from OH-carrying moiety to generate alkoxide. The alkoxide undergoes oxa-Michael addition to monomers and after another proton transfer the propagation cycle is completed [122]. In most of NHC catalyzed HTP, monomers bearing acidic protons are required. Through the proper design and select of NHCs, polymerization of common acyclic monomers containing no such protic groups, like typical dimethacrylates, can form unsaturated polyesters uniquely. Such unsaturated polyesters are of especially scientific and technological interest for producing tailor-made polyester materials through post-functionalization and cross-linking. Cavallo and Chen et al. synthesized new triazolylidene carbenes for HTP of common dimethacrylates (DMAs) containing no protic groups into unsaturated polyesters (Fig. 2.19) [123]. They found that triazolylidene carbene of TPT (1,3,4-triphenyl-4,5-dihydro-1H-1,2,4-triazol-5-ylidene) with two MeO-substitutent (OMe$_2$TPT) exhibited the highest HTP activity and also produced the polyester with the highest molecular weight. The TPT derivative with two MeO substituents makes this NHC catalyst bearing both a strong nucleophile and a good leaving group. They revealed that the subsequent key H-transfer step follows an intramolecular mechanism via a five-membered transition state (thus formally 1,4-H shift). The zwitterionic intermediates along the dimerization pathway can adopt the more stable closed spirocyclic structures.

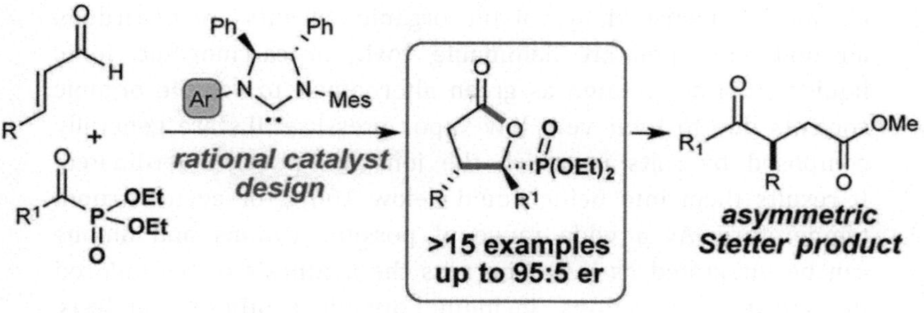

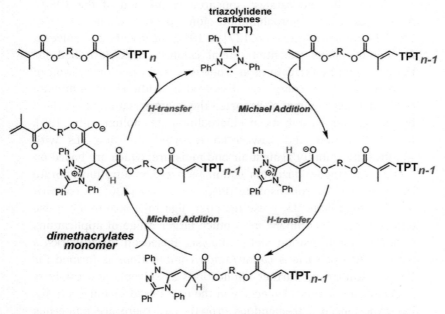

Figure 2.18 Tailored C1-symmetric biaryl-saturated imidazolium-derived NHC catalyst.

Figure 2.19 N-heterocyclic carbene (NHC)-catalyzed proton-transfer polymerization.

2.4 Ionic Liquids for Catalysts

The design of environmentally benign solvents or solventless systems has been also one of the most active areas of green chemistry. Organic solvents pose a particular concern to the chemical industry. Many volatile organic solvents are hazardous air pollutants, and are flammable, toxic, or carcinogenic. Ionic liquids (ILs) are treated as green alternatives to volatile organic solvents due to their very low vapor pressure. ILs are generally composed by salts in which the ions are poorly coordinated. It results them into being liquid below 100°C, or even at room temperature. As a wide range of possible cations and anions can be integrated as ILs, it permits the synthesis of ILs tailored for specific applications, including organic synthesis, catalysis, electrochemical devices, solvent extraction, etc. [124, 125].

The first publications in which ILs were described as new reaction media and catalyst appeared at the end of the 1980s. Acidic ILs with chloroaluminate ions proved to be effective Friedel–Crafts catalysts [126, 127]. In 1990, the use of ILs as solvents for homogeneous transition metal catalysts was described for the first time by Chauvin et al. They reported the dimerization of propene by nickel complexes dissolved in acidic chloroaluminate IL [128]. Osteryoung et al. reported the polymerization of ethylene by Ziegler-Natta catalysts in chloroaluminate molten salts [129]. In 1992, Wilkes and Zaworotko reported the first ILs with significantly stability against air and moisture, which are based on 1-ethyl-3-methylimidazolium cation with either tetrafluoroborate $[BF_4]^-$ or hexafluorophosphate $[PF_6]^-$ as anions [130]. In contrast to chloroaluminate ILs, these ILs offer high tolerance on both use and storage, which opens up a much larger range of applications especially for transition metal catalysis. Based on these works, a whole range of various cation/anion combinations is formed for ILs and studied for further applications as solvents and catalysts in chemical reactions. Especially in the past decade, catalysis in ILs has experienced a tremendous growth, and there are numerous examples of a variety of catalytic reactions that have been successfully carried out in these new media. Acting as solvents or catalysts themselves, many reactions show enhanced reaction rates, improved selectivity, or easier reuse of catalysts when carried out in ILs. Therefore, the design of next-generation ILs holds significant promise for improved green chemistry and environmental benefits.

2.4.1 Ionic Liquids and Their Physical Properties

The interests in ionic liquids was largely because of their advantageous properties, i.e., negligible vapor pressure, high thermal and chemical stability, selective dissolvability to many organic/inorganic materials, good electrical conductivity and ionic mobility. Ionic liquids are composed of cations and anions (Fig. 2.20). Their properties are designable according to the nature of the cations and anions composed. Since there are numerous combination possibilities on cations and anions, ionic liquids with diverse miscibility, viscosity, density, conductivity, and polarity can be adjusted. Here, we mainly describe the properties that may be the key factors that affect their applications in chemical reactions.

Viscosity is an important parameter of ionic liquids. The value of viscosity varies tremendously with chemical structure, composition, temperature. A high viscosity reduce the reaction rate or the diffusion rate of the redox species. The viscosity of an IL is determined by both the cation and anion. The more asymmetrical cations and less interactions (van der Waals force, hydrogen bonding, electrostatic forces) between ions may result in a lower viscosity. It has been shown that the viscosity of imidazolium-based ILs can be decreased by using highly branched and compact alkyl chains [131]. Phosphonium ILs tend to have higher viscosity than imidazolium-based ILs. ILs based on the dicyanamide anions have the lowest viscosity of all the ILs. For the same cation, the viscosity decreases as follows with change in anion:

$$[Cl]^- > [PF_6]^- > [BF_4]^- > [NO_3]^- > [N(SO_2CF_3)_2]^-$$

The solvation property of ILs is a necessary character for the dissolution of reactants and the recovery of products. Particularly the miscibility of ILs with water is interesting, which is an important factor for the industrial application of these solvents. The water content of ILs can affect the rates and selectivity of reactions. The miscibility of ILs with water depends on the nature of the anion, temperature and the length of the alkyl chain on the cation. It is because that strong hydrogen bonds can form between the water and the anion of the IL. The anions like [PF$_6$]$^-$, [BF$_4$]$^-$, [ClO$_4$]$^-$, [NO$_3$]$^-$ and [CF$_3$SO$_3$]$^-$ can form hydrogen bondings between anions and water molecules by symmetrical 1:2-type hydrogen-bonded complexes: anion—HOH—anion. The strength of bonding increases in the order [PF$_6$]$^-$ < [BF$_4$]$^-$ < [ClO$_4$]$^-$ <[NO$_3$]$^-$ < [CF$_3$CO$_2$]$^-$. The solubility of water in ionic liquids can be manipulated by adding short-chain alcohols to biphasic systems or by increasing temperature. This behavior has been used to generate a thermally controlled reaction system, where the reaction mixture is monophasic under reaction temperature for maximum mixing of the starting materials, but biphasic at room temperature, allowing facile separation of the products and catalyst [132].

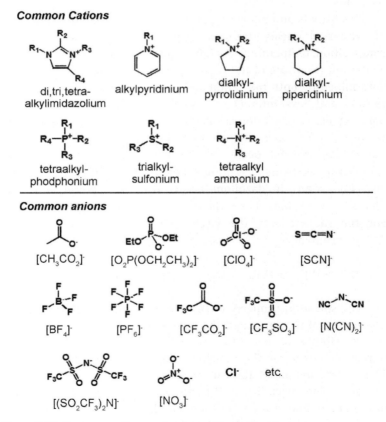

Figure 2.20 Commoncations and anions of ionic liquids.

It is well known that the solvent polarity could dramatically affect the rates and selectivity of chemical reactions [133–136]. As a "green" alternative of solvents for chemical reactions, the polarity of ILs is another important property. In chemistry, polarity is a separation of electric charge which leads to a molecule or its chemical groups having an electric dipole or multipole moment. For solvents, polarity means the sum of all possible specific and non-specific intermolecular interactions between the solvent and any solute. In contrast to molecular solvents, where the solute cation and anion stay close to each other to preserve charge neutrality, ionic liquids solvate ions with completely divorcing the cations and anions from each other. Although ILs are considered inhomogeneous solvents, the ionic liquid itself is capable of preserving the charge neutrality during the solvation. Due to the presence of both cations and anions, there is a much wider range of solvent–solute interactions in ILs than conventional organic solvents. Kamlet and Taft introduced multi-parameter system based on linear solvation energy relationships, which are composed of the complimentary scales of hydrogen bond acidity (α), hydrogen bond basicity (β), and dipolarity/polarizability effects (π^*) [137–145]. The π^* value lies at the heart of the Kamlet–Taft system, which resembles the most to the polarity, ignoring the effects from hydrogen bonding. It revealed that the π^* values for the commonly used ionic liquids are high in comparison with most non-aqueous molecular solvents like acetone, dimethyl sulfoxide, etc., but less than water or short-chain alcohols. A comprehensive evaluation of the polarity of ILs is also reported using solvatochromic dyes such as Reichardt's dye. ILs show interesting synergistic effects in their binary mixtures with water or organic cosolvents. Dramatic changes on the polarity can be observed, which were termed as "hyperpolarity" [146].

2.4.2 Ionic Liquids for Catalytic Reactions

Over the past decade, ILs have been applied to a wide range of organic reactions, including hydrogenation, oxygenation, cross coupling, Diels–Alder, carbonyl protection, epoxidation, ring closing metathesis, Knoevenagel condensation, esterification hydroformylation and Friedel–Crafts alkylation, etc. There seems no limit of ILs to be employed to the types of catalytic reactions.

Table 2.1 presents the list of the representative catalytic reactions using ILs.

Although many studies have found that the reaction rate, conversion and selectivity are enhanced to different extents, or even have significant commercial applications due to the conduction of ILs in the reaction systems, the role of ILs and the relationship between the properties of ILs and catalytic performance are still being unrevealed, which motivate chemists or scientists to keep studying reactions in this fascinating new environment. New methods and strategies are expected to be developed to manipulate and study ILs in catalysis, which may further promote the use of ILs in the green chemistry.

Table 2.1 The survey of catalytic reactions using various ionic liquids

Reaction	Reaction scheme	Representative ILs used	Ref.
Hydrogenation	Hydrogenation of 1-pentene with isomerization to 2-pentene using	[bmin]BF$_4$/PF$_6$	[147]
	Hydrogenation of cyclohexene catalyzed by rhodium complexes	[bmin]BF$_4$/PF$_6$	[148]
C–C and C–O cleavage	Acylative cleavage of a series of cyclic and acyclic ethers	[emim]I/AlCl$_3$	[149]
	Asymmetric ring opening reaction	[bmin]BF$_4$/PF$_6$/SbF$_6$	[150]
C–C coupling	Friedel–Crafts reaction	[bmin]BF$_4$/PF$_6$; [emin]Cl;	[151, 152]
	Diels–Alder reaction	[emin]Cl; [emin]BF$_4$/ClO$_4$/ CF$_3$SO$_3$/NO$_3$/PF$_6$; [bmin]PF$_6$	[153–155]
	Oligomerization/ polymerization	[emin]Cl; [bmin]Cl	[156–158]
	Alkylation	[bmin]BF$_4$/PF$_6$	[159]
	Allylation	[bmin]BF$_4$/PF$_6$	[160–162]
	Heck reaction	[bmin]Br/BF$_4$; [Nbu$_4$]Br	[163, 164]
	Suzuki cross coupling	[bmin]BF$_4$	[165]

Reaction	Reaction scheme	Representative ILs used	Ref.
	Heck reaction	[bmin]Br/BF$_4$; [Nbu$_4$]Br	[163, 164]
	Suzuki cross coupling	[bmin]BF$_4$	[165]
	Hydroformylation	[bmin] PF$_6$ Dialkylimidazolium, Trialkylimidazolium, N,N dialkylpyrrolidinium with BF$_4^-$, PF$_6^-$, CF$_3$SO$_3^-$, N(CF$_3$SO$_2$)$_2^-$	[166–168]
	Oxidation	[bmin]PF$_6$	[169]
	Nucleophilic displacement	[bmin]PF$_6$	[170]
	Nitration of aromatics	[emin]OTf/CF$_3$COO/NO$_3$; [HNEtPr^{i}2]CF$_3$COO	[171]
	Reduction of aldehydes	[bmin]BF$_4$, [emin] BF$_4$/PF$_6$	[172]
	Witting reaction	[bmin]BF$_4$	[173]
	Cycloaddition	[bmin]/[bpy]Cl/BF$_4$/PF$_6$	[174]
	Stille coupling	[bmin]BF$_4$	[175]
	Transesterification	[bmin]PF$_6$; [emin]BF$_4$	[176]
	Alcoholysis, ammoniolysis	[bmin]BF$_4$/PF$_6$	[177]

2.4.3 Biphasic Ionic Liquids

ILs have obvious potential as reaction media, and one important challenge is to use their unique solvent properties to develop efficient methods for product separation and IL recycling. Due to the polar nature of ionic liquids, biphasic systems tend to form with either unpolar organic solvents or aqueous media. The mixing of ILs with other solvents is also a much easier pathway for tuning their properties in contrast to the structural design of pure ILs by synthesis. For an ideal liquid–liquid biphasic catalysis system, the ionic liquid may be able to dissolve the active species while also be partially miscible with the substrate, and limited solubility for the products. The process will allow simple removal of products without affecting the activity of catalyst. These particular features make biphasic ILs very attractive liquid media for catalysis.

IL–aqueous biphasic system has attracted more interests, as water is a fundamental media and plays an important role in many biological and chemical systems. IL–water combinations were superior compared to conventional organic solvents and biphasic IL/organic co-solvent media with respect to catalytic performance as well as to catalyst separation and recycling [178]. The properties of ILs can be significantly influenced by the addition of water. The reaction rates and selectivity of the catalytic process may also be affected.

Lee et al. demonstrated the oxidation reaction of benzylic alcohols with trichloroisocyanuric acid in 1-butyl-3-methylimidazolium tetrafluoroborate [bmim][BF$_4$]–water media at room temperature. The reaction system to the corresponding carbonyl compounds showed good yields under mild conditions and simple operation compared with previously reported methods [179]. Zlotin and coworkers designed and synthesized a novel chiral ionic liquid as an efficient recoverable organocatalyst for the direct asymmetric aldol reaction between cycloalkanones and aromatic aldehydes in the presence of water [180]. The catalyst retains its activity and selectivity over at least five reaction cycles [181]. Zlotin and coworkers further developed the chiral IL–water media for asymmetric aldol reactions. The cyclic ketones and methylketones react with aromatic aldehydes afforded respective aldols in high yields and with excellent regio-, diastereo- and enantioselectivities in this biphasic system.

Supercritical carbon dioxide (SC-CO$_2$) has been also considered as solvent for green processes in recent decades. Recent research has verified the possibility to carry out integral green catalytic processes by combining SC-CO$_2$ and ILs. Their different miscibilities make a two-phase system, which showed an exceptional ability to perform both the transformation and the products extraction simultaneously, even under extreme conditions (such as 150°C and 100 bar) [182–185]. IL/SC-CO$_2$ systems may have a stabilizing effect against enzyme deactivation caused by SC-CO$_2$. Some ILs have proved to be by far the best nonaqueous media for enzyme-catalyzed reactions, displaying a high level of activity and stereoselectivity in chemical transformations [186, 187]. An overstabilization effect on the biocatalysts can also be exerted (over 2,300-folds half-life time with respect to classical organic solvents) [188, 189]. Another advantage of ILs/SC-CO$_2$ system is the much improved performance on product separation and catalyst recycling. SC-CO$_2$ may efficiently isolate organic products, especially hydrophobic molecules. As the solvent properties can be adjusted by changing either the pressure or the temperature, it has been described as an excellent solvent for the transport of hydrophobic compounds [190]. Leitner et al. designed a continuous-flow process based on a chiral transition metal complex in a supported ionic liquid phase (SILP) with SC-CO$_2$) as the mobile phase for asymmetric catalytic transformations of low-volatility organic substrates at mild reaction temperatures [191]. Enantioselectivity of >99% ee and quantitative conversion were achieved in the hydrogenation of dimethylitaconate, achieving turnover numbers beyond 100,000 for the chiral quinaphos-Rh complex. These results demonstrate that the SILP/SC-CO$_2$ approach to continuous-flow asymmetric catalysis can provide a viable option for the production of chiral fine chemicals and pharmaceuticals, in which process the selectivity, product purity, production efficiency, and flexibility are especially important.

2.4.4 Task-Specific Ionic Liquids for Catalysis

The application of ionic liquids in catalysis has problems, too. Purity is a major concern for using ILs. In most synthetic processes, ILs were neither distilled nor recrystallized for purification. The main impurities in ILs are halide anions or trace organic compounds of low volatility which are generally from unreacted starting materials and water [192]. Therefore, besides optimizing the synthesis process for highly pure ionic liquids, the strategies for developing task-specific ionic liquids (TSILs) were prospected [193–195].

The prospect of designing ionic liquids that exhibit specific properties to enhance catalytic activity opens up new potentials and possibilities in catalysis and the design of so-called task specific ionic liquids (i.e., ionic liquids containing functional groups) may favor the breakthrough of the boundaries in ionic liquid catalysis. A selection of functionalized ionic liquid cations is shown in Fig. 2.21, many of which have been developed for applications in catalysis. TSILs with functional groups behave not only as a reaction medium but also as a reagent or catalyst in some reactions or processes.

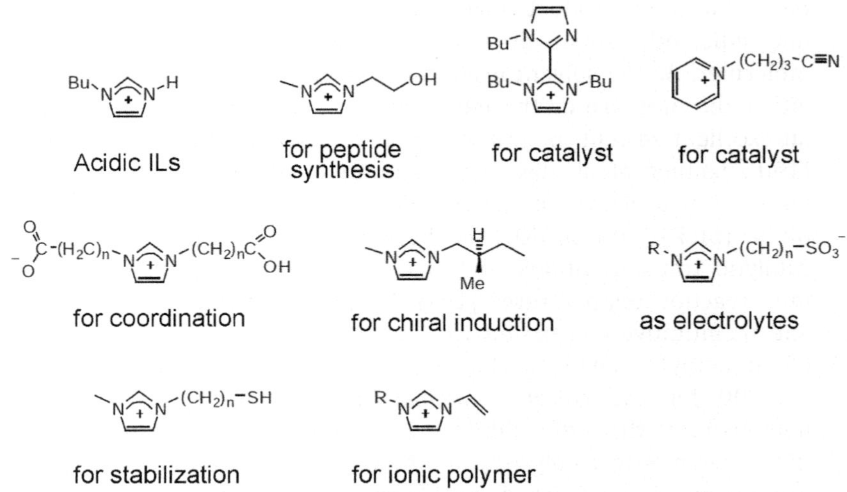

Figure 2.21 Some functionalized ionic liquids.

In some cases of transition metal catalysis using ILs as immobilization media, the uncharged catalysts may not be well retained during product separation or for extensive reuse. The functionalized ligands are required to prevent the loss or leaching of the catalysts. Functional groups, such as amines, amides, ethers, alcohols, acids, urea, and thiourea tethered to cations have been known for the appropriate coordinating groups in ILs to improve the immobilization of metal catalysts effectively. Dyson et al. designed N-butyronitrile pyridinium cation-based ILs to improve catalyst's retention [196]. It was envisaged that the nitrile group could weakly coordinate to the metal center thereby anchoring it in the ionic liquid phase. The catalytic activity of the palladium complexes in such ILs was evaluated in Suzuki and Stille coupling reactions. Excellent recycling properties relative to other ILs are exhibited. The specific functionalized ILs can improve the solubility of metal ions but do not block the activity or even act as a stabilizer by coordinating [197, 198].

Chiral ionic liquids (chiral ILs) have been recognized for years as an alternative strategy to carry out chiral transformation, especially for asymmetric synthesis. Functionalized ionic liquids may allow structural variability for the design of tailor-made solvents. Nowadays a large pool of chiral ionic liquids bearing chiral cation, anion, or both functionalities is available. Chiral ILs may improve the outcome of asymmetric reactions or facilitate separation. Tran et al. synthesized a series of structurally novel chiral ILs in which cation is an imidazolium group, while anion is are based on a borate ion with spiral structure and chiral substituents [199]. The chiral recognition was dependent on solvent dielectric constant, concentration, and structure of the ILs. Stronger enantiomeric recognition was found at higher concentration of ILs and for anions with a relatively larger substituent group (e.g., chiral anion with a phenylmethyl group exhibits stronger chiral recognition compared to that with a phenyl group, and an anion with an isobutyl group has the weakest chiral recognition). Structurally novel and strong intra- and intermolecular chiral recognition ability of these chiral ILs indicate that they can be used in a variety of applications including as chiral solvent for asymmetric synthesis and as chiral stationary phase for chromatographic separations.

TSILs may also act as liquid supports for certain combinatorial synthesis. In peptide synthesis, the molecules are bound to a solid phase throughout the synthesis and only cleaved until the last step. TSILs can be a linker to the liquid phase in the synthesis processes. Miao and Chan used a hydroxy-functionalized imidazolium IL for the synthesis of oligopeptides [200]. NMR studies showed that these chiral ILs exhibit intramolecular as well as intermolecular enantiomeric recognition. Good yield of a pentapeptide without the need for chromatography can be achieved. Same approach was also utilized for multi-step synthesis of drugs [201, 202]. Compared with other supports, these liquid supports may have higher loading capacity, facial analysis on the intermediates.

Through the functionalization of ILs, the chemical/physical properties of ILs can adapt for various applications. Lütz et al. described ionic liquids as "performance additives" in electroenzymatic reactions [203]. In oxidoreductase-catalyzed reactions, the addition of small amounts of an IL led to enhanced conductivity of the reaction medium and to the higher stability of biocatalysts and cofactors. Through the incorporation of active pharmaceutical component into the anion or the cation of the IL, a multitude of active-agent combinations is imaginable. An IL derived from didecyldimethylammonium bromide and sodium ibuprofen ("didecyldimethylammonium ibuprofen") would show both antibacterial and anti-inflammatory activity [204].

2.4.5 Supported Ionic Liquids for Catalysis

Supported ionic liquid catalysis is a concept that combines the advantages of ionic liquids with those of heterogeneous support materials. The applying solid supports for the immobilization of ionic liquids exhibit similar or advanced chemical behaviors and facilitate separation of catalysts from reaction mixtures [205–207]. Various types of immobilized ILs materials have been proposed in the literatures with acronyms like SCILs (solid catalysts with ILs), SILPC (supported ionic liquid phase catalyst), SILFs (supported ionic liquid films), PSILs (polymer-supported ILs), SILNPs (supported IL nanoparticles), etc.

Various types of supports have been applied for the immobilization of ILs, such as zeolites [208, 209], mesoporous silica [210, 211], alumina derivatives [212, 213], polymers [214, 215], carbon nanostructures [216], and magnetic nanoparticles [217].

Due to the unique characteristics (controllable pore nanostructures, high chemical/physical stability) of mesoporous materials, they are widely studied and employed as supports for ILs. The degree of IL immobilization in the mesoporous materials may have profound influence on their catalytic performances. The strategies for the immobilization of ILs include wet impregnation [218, 219], grafting [220–223], physical adsorption [224–226], ion-exchange [227], sol-gel polymerization [228–230]. However, it is not easy to maintain the open porous framework and keep stable immobilization of ILs at the same time. The combination design on the functional ILs with the formation of mesoporous framework is needed to fully exert the synergistic effects as a novel organic/inorganic supported catalyst. Brunel et al. synthesized a trialkoxysilylated ionic liquid based on disilylated guanidinium and monosilylated sulfonimide species. This IL allowed the successful fabrication of periodic mesoporous organosilicas containing covalently anchored ion-pairs through both organo-cationic and organo-anionic moieties [231]. Yin et al. reported the formation of novel periodic mesoporous organosilica materials incorporating Lewis acidic chloroindate(III) ionic liquid moieties [232]. N-(3-triethoxysilylpropyl)-N(3)-(3-trimethoxysilylpropyl-

4,5-dihydroimidazolium chloride was first synthesized and used as bridging organosilica precursor to co-condense with tetraethoxysilane (TEOS) in the presence of P123. After the removal of the template, the addition of $InCl_3$ gave rise to mesoporous materials incorporating chloroindate(III) ionic liquid moieties with Lewis acidity. For this supported catalyst, ILs located within the channel walls without blocking mesopores, which will allow reactant molecules to diffuse easily into channels and contact conveniently with the internal catalytically active sites. Their application in the Friedel–Crafts catalytic reaction showed that this catalyst possesses both high catalytic efficiency and higher hydrolytic stability.

Polymers are also main support materials for catalysts, which were developed quite in time followed with the catalysts. Extensive work has been reported on the polymer-supported ILs. Various polymeric substrates, such as polystyrene (PS) [233], resin [234], PEG [235], Nafion [236], and chitosan [237], are proven to be efficient and recoverable catalysts for a number of catalytic reactions, such as nucleophilic substitution, esterifications cylcoaddition of CO_2 to epoxides, and allylic substitution. Compared with inorganic materials, polymers might be considered to possess relatively weaker mechanical, thermal and chemical stabilities. However, the discovery of the FDU-type periodic mesoporous polymers with rigid phenolic resin walls opens up new possibilities for polymer-supported catalysts. FDU mesopolymers contain a three-dimensionally connected benzene ring framework and plenteous phenolic hydroxyl groups. They exhibit attractive merits of tunable mesostructures, high surface area, uniform pore sizes as well as high thermal and chemical stability [238, 239]. Wu and He et al. developed a series of novel heterogeneous catalysts by immobilizing imidazolium-based functionalized ionic liquids on FDU-type mesoporous polymer (Fig. 2.22) [240]. This system served as an efficient catalyst in the solvent-free cycloaddition reaction of CO_2 with epoxides without the use of any co-catalyst. Compared with polystyrene (PS) and mesosilica SBA-15 supports, this catalyst can be easily recovered and reused without a significant loss of activity. The mesoporous organic framework and phenolic hydroxyl groups are proved to be key factors for improving the catalytic performance of supported ILs.

60 | *Structural Design for Molecular Catalysts*

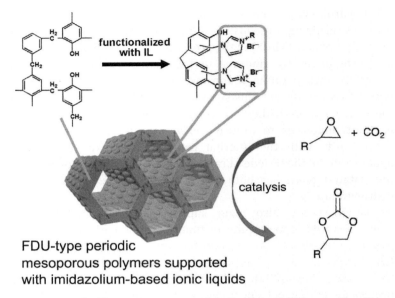

Figure 2.22 The heterogeneous catalyst by immobilizing imidazolium-based ionic liquids on FDU-type mesoporous polymer.

Carbon nanotubes (CNTs) are another class of interesting supports for IL-based materials. Lee et al. covalently modified CNTs with imidazolium polymer salts with different anions, Pd/polyIL(X)-CNTs (X=Cl, Br, I, ClO_4, BF_4, PF_6) [241]. The anions of IL can significantly impact the kinetics of the oxygen reduction reaction. The subtly varied structures of the IL moiety may profoundly influence the interfacial interactions between the support material, catalysts, and electrolytes, which is essential to the performance of IL-CNT hybrid catalysts, especially for fuel cells. Fullerene spheres can also be used as supports for ILs. Giacalone and Gruttadauria et al. synthesized a series of fullerene-ionic liquid (IL) hybrids, in which the number of IL moieties (two or twelve), anions and cations have been varied (Fig. 2.23) [242]. The combination of C_{60} and IL gives rise to new unique properties in the conjugates. It also allows the further loading of catalytic nanoparticles in the systems. The C_{60}-IL hybrids were successfully applied as catalyst in Suzuki and Mizoroki–Heck reactions in a concentration of just 0.2 mol% and showed recyclability for five runs with no loss in activity.

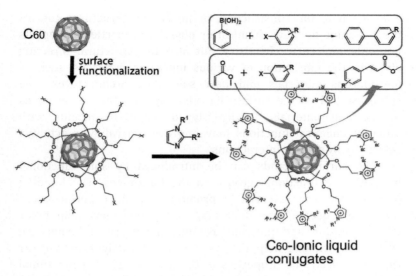

Figure 2.23 C_{60}–ionic liquid conjugates applied as catalyst in Suzuki and Mizoroki–Heck reactions.

Magnetic nanoparticles are attractive supports for ILs, since magnetic properties may facilitate the recovery and recycle of the catalysts. Cheng et al. prepared the magnetic nanoparticle supported IL (MNP-IL) catalyst via the surface modification of Fe_3O_4 nanoparticles with N-alkyl imidazole groups [243]. The resulting MNP-IL catalyst demonstrated high activity in CO_2 cycloaddition reactions at a lower CO_2 pressure. The catalysts could be easily recycled by an external magnetic field and reused for up to 11 times with essentially no loss of activity. This kind of MNP-ILs has also been used as adsorbents for the preconcentration and separation of polycyclic aromatic hydrocarbons (PAHs) from environmental samples [244]. It was shown that only a small amount of ILs and Fe_3O_4 NPs was needed to obtain the satisfactory extraction recoveries.

2.5 Surface Molecular Ligands for Catalysts

Owing to the continuous developments in synthetic organic methodologies and increasing attention on nanomaterials for various applications, including catalysis, ligand design has become a major part for construction of novel nanostructures toward the searching for new chemical or physical properties. It impacts not only molecular chemistry but also the structural directing effect in the fabrication of various inorganic materials, such as mesoporous materials and new solid-state architectures. The subtle control on the surface ligands may vary the nanofeatures of the nanomaterials. Particularly for catalysts, ligands with different chemical functions may improve the catalytic activity/selectivity by surface interactions.

The surface ligands can be introduced by sole functional molecules or functional groups in the backbones. The bonding patterns are also essential in promoting catalytic performance. The emergence of new molecular interactions may help tailor the orientation and possible dynamic behaviors of ligands on surface, which can help the fine-tuning of catalytic activity or the stereoelectronic properties of the active sites. The functional ligands may also promote the development of new mesoporous materials containing binding sites for the specific incorporation of catalysts. The ligand designs may open up new avenues for the bottom-up fabrication of nanomaterials and novel organic-inorganic hybrid materials, which will lead to their further applications in various fields, not only as catalysts but also in biology, molecular devices, sensors, etc.

2.5.1 Surface Modification of Nanoparticles and Nanoclusters

Nanocatalysts are usually metal nanoparticles or nanoclusters. As metal nanoparticles have a higher surface area, they can increase the catalytic activity and speed up the catalytic process. Not only particle size, shape, crystalline structure, but also the surface functional groups are the characteristic of nanoparticles for catalysis, which are vital factors to influence the catalytic properties of metal nanoparticles. To avoid the aggregation of nanoparticles and keep the monodispersion state, the surface stabilizers are used for the preparation of metal nanoparticles. For homogeneous catalysis, it is known that the ligand can positively contribute to catalytic reactions and enable the control of selectivity by interacting with the adsorbed reactant. However, in heterogeneous catalysis, it has been considered that the presence of surface ligands can hinder the catalytic activity of metal nanoparticles in most cases. As only the surface atoms of the catalytic nanoparticles contribute to the reaction, ligands may block the free coordination sites of the surface atoms, and thus the ability to adsorb and catalytically convert reactants may eventually become inhibited. However, recent studies have highlighted the importance of surface modification to enhance the catalysis capability of metal nanoparticles.

Control selectivity of reactions is one of the major research goals and fundamental challenge in catalysis for the following decades. The use of ligands becomes a breakthrough to open new strategies for catalytic properties. It has been demonstrated that the chemo- and stereoselectivity of nano-catalysts for heterocatalysis can be greatly improved by the surface ligands. Medlin et al. prepared highly selective Pd catalysts with the deposition of n-alkanethiol self-assembled monolayer (SAM) on the surface [245]. These coating layers were shown to improve the selectivity of 1-epoxybutane formation from 1-epoxy-3-butene on Pd catalysts from 11 to 94% at equivalent reaction conditions and conversions, although sulfur species are generally considered to be indiscriminate catalyst poisons. The comparison of SAM-coated catalysts with catalysts modified by carbon monoxide, hydrocarbons or sulfur atoms indicated that SAMs restrict sulfur coverage to enhance selectivity without significantly poisoning the activity of the desired reaction.

To design the surface of nanocatalysts and introduce the functional ligands on the surface, ligand-free metal nanoparticles or nanoparticles with replaceable ligands might be required to carry out further modifications. The chemical preparation of nanoparticles always begins with molecular precursors, which are decomposed by either thermal/optical excitation or a reducing agent to generate individual metal atoms with an organic residue, and finally nucleate forming nanoparticles. Wang et al. prepared stable Pt, Rh, and Ru metal nanoclusters with small particle size (1–2 nm) in the absence of any usual protective agent [246]. By heating corresponding metal hydroxide colloids in ethylene glycol containing with NaOH, the "unprotected" nanoclusters can be obtained. The "unprotected" Pt nanoclusters can also be easily transformed to various protected Pt nanoclusters with the same Pt cores. Kunz et al. modified this kind of "unprotected" Pt nanoclusters with hydrophilic organic ligand (L-proline) for heterogeneous catalysts [247]. In contrast to the supported or "unprotected" Pt nanoclusters, the proline-functionalized Pt nanoclusters are highly chemoselective, which may achieve 100% conversion even for the selective hydrogenation of acetophenone. Experiments under kinetically controlled conditions also revealed that this high chemoselectivity is not accompanied by a loss of catalytic activity. An enhanced rate toward the desired product was found for proline-functionalized Pt nanocatalysts.

Surface functionalization with organic molecules has been utilized as a powerful tool to further manipulate the properties of noble metal nanoparticles. Among those organic molecules, mercapto derivatives have been used extensively due to the strong affinity of thiol moieties to transition metal surfaces. However, although the stability of nanoparticles can be greatly improved, the surface metal–thiolate linkages may completely destroy the catalytic activity of metal nanoparticles [248]. Inspired from organometallic chemistry, recently a wide variety of metal–ligand bonds have been explored to functionalize metal nanoparticles, beyond the conventional metal–thiolate linkages [249]. Tsukuda et al. functionalized Au nanoclusters by terminal alkynes. 1-Octyne (OC-H), phenylacetylene (PA-H), and 9-ethynyl-phenanthrene (EPT-H) were bound on the Au surface by the ligand exchange of small Au clusters (diameter < 2 nm) stabilized by polyvinyl-pyrrolidone (PVP) [250]. Because of intimate coupling between the alkynyl group (RC≡C–) and Au clusters via Au–C covalent bonds, novel photophysical and electronic properties may be exhibited. Chen et al. prepared a series of AuPd alloy nanoparticles with surface alkyne protected (Fig. 2.24) [251]. Owing to the unique metal–ligand interfacial interactions to impact the binding with reaction intermediates, the resulting alkyne-protected nanoparticles exhibited apparent electrocatalytic activity in the oxygen reduction reactions (ORR). The mass activity was over eight times better and the specific activity was almost twice in contrast to commercial Pd black. The metal–carbon interfacial bonds can modify the electronic surface structure of the catalyst, which may strongly influence the binding to oxygenated intermediates and result enhanced electrocatalytic activity.

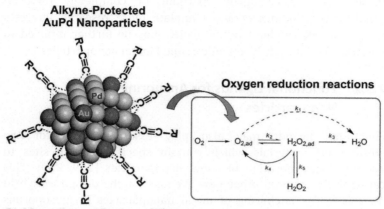

Figure 2.24 Alkyne-protected AuPd nanoparticles for electrocatalytic reduction of oxygen.

An increasing number of recent studies have presented more complicated surface functionalization of nanoparticles, which can be manipulated through diverse metal–ligands interactions. However, the roles of surface ligands in modulating the catalysis of metal nanoparticles are awaiting fundamental understanding. As a well-defined composition and structure can be expected in the monodispersed state, atomical metal nanoclusters may be served as an ideal system to gain insight into the influence of surface ligands on the catalysis of metal nanoparticles [252–256]. It is highly desirable to unambiguously identify the roles of surface ligands on catalysis from nanoclusters with removable surface ligands. However, in some cases, the strong interaction between ligands and metals may destroy the well-defined structures of nanoparticles and lead to heavy aggregation during the removal of surface ligands. Zeng and Wang et al. developed alkynyl-stabilized metal nanoclusters, which are labile enough to be readily removed under mild conditions while maintaining their overall structures [257]. They used phenylalkynyl-capped intermetallic 62-metal-atom $Au_{34}Ag_{28}$ cluster as a model system to evaluate the importance of surface ligands in catalysis of metal nanoparticles. They proved that the nanoclusters fully capped by surface phenylalkynyls are more catalytic active than those with surface ligands partially or completely removed [258]. These findings should stimulate more research interests in fundamental understanding, which may be further ultilized to optimize the catalysis by organic-capped metal nanoparticles.

2.5.2 Functional Ligands for Anchoring of Nanoparticles

Functional ligands may also play essential roles for the mesoporous materials which contain specific binding sites to allow the incorporation of metal nanoparticles [259, 260]. This can thereby open up other new avenues for the design of hybrid materials. The anchoring of metal nanoparticles in mesoporous materials may overcome the drawbacks of nanoparticles on easy aggregation and losing activity due to surface stabilizers or recovery from usage, and thus promote their applications in industry catalytic productions. Compared with the precise design on the surface of nanoparticles, the functionalization of mesopores is a relatively easy task and various catalysts can be tethered in such structures.

Especially for mesoporous silica nanoparticles (MSN), they have received much attention as catalyst scaffolds in the area of mesoporous supports. Their applications for the supported catalysts include olefin hydrogenation, Heck and Sonogashira reactions, ring opening of epoxides, hydroamination of olefins, and hydrocarboxylation of aryl olefins and alcohols. MSN-supported catalysts not only exhibit increased stability but also result in an increase in the selectivity versus its homogeneous analog. Zhou et al. synthesized Pd(0) complex catalyst by supported in amino-modified MCM-41 materials [261]. MCM-41 was post-modified with aminopropyltriethoxysilane and followed the mixing with $PdCl_2$. After reduction by $NaBH_4$, supported Pd(0) complex catalyst was obtained by a rather simple process. This complex showed high activity and steroselectivity for Heck reaction at 70°C. In addition, even after the catalyst had been exposed to air for 50 days, its activity did not decrease. Based on a rather simple and convenient process, this system offers excellent practical advantages of high activity, easy handling, separation from the product, and recycling.

Functionalized mesoporous silica materials may be also directly employed as effective catalysts. Mesoporous materials containing acid groups have been used widely as acid catalysts because the acidic functional sites are capable of acting as active sites for a range of reactions. For example, sulfonated periodic mesoporous organosilica (PMO) materials have been synthesized using co-condensation, direct sulfonation and postfunctionalization methods [262, 263]. The prepared sulfonic acid–containing mesoporous materials exhibited superior catalytic performance in esterification, acetylation, and condensation reactions. Wang et al. modified mesoporous SBA-15 with zinc carboxylate (ZC/SBA-15), and used the resulted mesoporous materials as a heterogeneous catalyst for the selective synthesis of methyl-4,40-di(phenylcarbamate) (MDC) from dimethyl carbonate (DMC) and 4,4'-methylenedianiline (MDA) (Fig. 2.25) [264]. They used a three-step synthesis process — that is, grafting SBA-15 with –CN group; converting into –COOH group; ion-exchange by Zn^{2+} ions to graft zinc carboxylate groups –COO(Zn) onto the pore surface of mesoporous SBA-15. The catalyst ZC/SBA-15 exhibits high MDA conversion (near 100%) and MDC selectivity (about 86.5%) with good recyclability. By control of surface ligands on mesoporous materials, this work presented a new approach for non-phosgene synthesis of isocyanates.

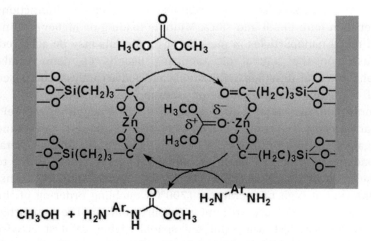

Figure 2.25 Zinc carboxylate functionalized mesoporous material as catalyst for selective synthesis of methyl-4,4'-di(phenylcarbamate).

Metal complexes with phosphorous-type ligands have been employed in various transition metal–catalyzed organic reactions. It has been found that the electron-withdrawing groups of phosphinites increase the strength of the metal phosphorous bond and act as a better acceptor. Phosphinite complexes with different metals (for example Rh, Pd, etc.) have shown high catalytic activity in different homogenous or heterogeneous reactions. Farjadian and coworkers modified mesoporous silica with a series of mono-, di-, and tri-phosphinite ligands for the preparation of supported Pd phosphinite complexes [265]. High efficiency and stability of Pd catalysts in the phosphinite-functionalized silica supports were shown in the Heck coupling reaction. The high yields of the resulting products indicated the good reactivity of diverse (electron-neutral, -rich, and -poor) haloarenes in these heterogeneous systems. Through simple filtration, the catalyst can also be recycled for repeated usage.

Besides anchoring nanoparticles by introducing suitable ligands on the surface of mesoporous supports, the successful manipulation on the loading of metal nanoparticles in mesoporous supports can also be exerted from the design of stabilizing ligands on metal particles. As the post-modification on the mesoporous particles may partially block the pores, the loading efficiency of metal nanoparticles or precursors may be decreased. Proper stabilizers on the nanoparticles may strengthen short-ranged interactions between bare metal and silica. Multiple loading or higher loading of metal nanoparticles in mesoporous materials may be achieved. Based on the control of stabilizing ligands on nanocrystals, Johnston et al. designed highly stable and active catalysts composed of bimetallic nanocrystals on mesoporous supports with a controlled intermetallic alloy structure [266]. Presynthesized FePt nanocrystals (<4 nm) coated with oleic acid and oleylamine ligands are infused in the mesoporous silica pores at high metal loadings (>10 wt%). The strong metal-support interactions due to the weak binding ligands on nanocrystals impart stability against sintering at high temperatures (700°C), enabling ordering of the nanocrystal structure without size perturbation. The supported FePt nanocatalyst was exhibited sixfold higher catalyst activity than commercial Pd-Al catalyst for liquid 1-decene hydrogenation and was stable for multiple reactions. In contrast to the reduction of multiple metal precursors within porous supports, more precise control over size, composition, and crystalline structure can be achieved. With the high loading into the mesoporous materials, it offers novel bimetallic or multi-metallic catalysts with enhanced catalysis activity, selectivity and stability.

2.6 Summary and Perspectives

With the increasing development in chemical synthesis, nanotechnology, and theoretical approaches, the design of molecular structures may be realized using qualitative ideas and become more imaginable. A broad impact of molecular-based catalysts is witnessed in the field of organic synthesis, which may allow for the mild construction of complex molecules from simple starting materials. With the use of nanoarchitectured supports for molecular catalysts, novel catalyst systems can also be developed, which may combine the advantages of homogeneous and heterogeneous catalysis. The advances for this kind catalytic system have shown tremendous potential for commercial processes and will inspire more novel applications from molecular catalysts.

Molecular catalysts are being successfully used in industrial settings and for natural product syntheses. A wide variety of possibilities may also be offered for new transformations by molecular designs. New challenges may no doubt continue to appear, where designing efficient molecular structures and functions may unfold a large stage for more reactive, selective, and economical molecular catalysts.

References

1. Macmillan, D. W. C. (2008). The advent and development of organocatalysis, *Nature*, **455**, pp. 304–308.
2. Dalko, P. I., and Moisan, L. (2004). In the golden age of organocatalysis, *Angew. Chem. Int. Ed.*, **43**, pp. 5138–5175.
3. Dondoni, A., and Massi, A. (2008). Asymmetric organocatalysis: From infancy to adolescence, *Angew. Chem. Int. Ed.*, **47**, pp. 4638–4660.

4. Long, N. J. (1995). Organometallic compounds for nonlinear optics—the search for Enlightenment!, *Angew. Chem. Int. Ed. Engl.*, **34**, pp. 21–38.
5. Jones, R. G., and Gilman, H. (1954). Methods of preparation of organometallic compounds, *Chem. Rev.*, **54**, pp. 835–890.
6. Patra, M., and Gasser, G. (2012). Organometallic compounds: An opportunity for chemical biology?, *ChemBioChem*, **13**, pp. 1232–1252.
7. Li, X., and Hou, Z. (2008). Organometallic catalysts for copolymerization of cyclic olefins, *Coord. Chem. Rev.*, **252**, pp. 1842–1869.
8. Coperet, C., Comasvives, A., Conley, M. P., Estes, D. P., Fedorov, A., Mougel, V., Nagae, H., Nunezzarur, F., and Zhizhko, P. A. (2016). Surface organometallic and coordination chemistry toward single-site heterogeneous catalysts: Strategies, methods, structures, and activities, *Chem. Rev.*, **116**, pp. 323–421.
9. Hartwig, J. F. (2008). Carbon-heteroatom bond formation catalysed by organometallic complexes, *Nature*, **455**, pp. 314–322.
10. Herrmann, W. A., and Cornils, B. (1997). Organometallic homogeneous catalysis-Quo Vadis? *Angew. Chem. Int. Ed. Engl.*, **36**, pp. 1048–1067.
11. Crabtree, R. H. (2014). *The Organometallic Chemistry of the Transition Metals*, 6th ed. (John Wiley & Sons, Inc.).
12. Beller, M., and Blaser, H.-U. (2012). *Organometallics as Catalysts in the Fine Chemical Industry, Topics in Organometallic Chemistry*, vol. 42. (Springer, Berlin, Heidelberg, Germany).
13. Cahiez, G., and Moyeux, A. (2010). Cobalt-catalyzed cross-coupling reactions, *Chem. Rev.*, **110**, pp. 1435–1462.
14. Nicolaou, K. C., Bulger, P. G., and Sarlah, D. (2005). Catalyzed cross-coupling reactions in total synthesis, *Angew. Chem. Int. Ed.*, **44**, pp. 4442–4489.
15. Miyaura, N., and Suzuki, A. (1995). Palladium-catalyzed cross-coupling reactions of organoboron compounds, *Chem. Rev.*, **95**, pp. 2457–2483.
16. Yamamoto, T., Nishiyama, M., and Koie, Y. (1998). Palladium-catalyzed synthesis of triarylamines from aryl halides and diarylamines, *Tetrahedron Lett.*, **39**, pp. 2367–2370.
17. Littke, A. F., Dai, C., and Fu, G. C. (2000). Versatile catalysts for the Suzuki cross-coupling of arylboronic acids with aryl and vinyl halides and triflates under mild conditions, *J. Am. Chem. Soc.*, **122**, pp. 4020–4028.

18. Kawatsura, M., and Hartwig, J. F. (1999). Simple, highly active palladium catalysts for ketone and malonate arylation: Dissecting the importance of chelation and steric hindrance, *J. Am. Chem. Soc.*, **121,** pp. 1473–1478.
19. Zapf, A., Ehrentraut, A., and Beller, M. (2000). Versatile catalysts for the Suzuki cross-coupling of arylboronic acids with aryl and vinyl halides and triflates under mild conditions, *Angew. Chem. Int. Ed.*, **39**, pp. 4153–4155.
20. Schroeter, F., Soellner, J., and Strassner, T. (2017). Cross-coupling catalysis by an anionic palladium complex, *ACS Catal.*, **7**, pp. 3004–3009.
21. Giannerini, M., Fañanás-Mastral, M., and Feringa, B. L. (2013). Direct catalytic cross-coupling of organolithium compounds, *Nat. Chem.*, **5**, pp. 667–672.
22. Martin, R., and Buchwald, S. L. (2008). Palladium-catalyzed Suzuki-Miyaura cross-coupling reactions employing dialkylbiaryl phosphine ligands, *Acc. Chem. Res.*, **41**, pp. 1461–1473.
23. Fu, G. C. (2008). The development of versatile methods for palladium-catalyzed coupling reactions of aryl electrophiles through the use of P(^{t}Bu)$_3$ and PCy$_3$ as ligands, *Acc. Chem. Res.*, **41**, pp. 1555–1564.
24. Aranyos, A., Old, D. W., Kiyomori, A., Wolfe, J. P., Sadighi, J. P., and Buchwald, S. L. (1999). Novel electron-rich bulky phosphine ligands facilitate the palladium-catalyzed preparation of diaryl ethers, *J. Am. Chem. Soc.*, **121**, pp. 4369–4378.
25. Wolfe, J. P., and Buchwald, S. L. (1999). A highly active catalyst for the room-temperature amination and Suzuki coupling of aryl chlorides, *Angew. Chem. Int. Ed.*, **38,** pp. 2413–2416.
26. Huang, X., Anderson, K. W., Zim, D., Jiang, L., Klapars, A. and Buchwald, S. L. (2003). Expanding Pd-catalyzed C-N bond-forming processes: The first amidation of aryl sulfonates, aqueous amination, and complementarity with Cu-catalyzed reactions, *J. Am. Chem. Soc.*, **125**, pp. 6653–6655.
27. Huang, X., Anderson, K. W., Zim, D., Jiang, L., Klapars, A., and Buchwald, S. L. (2010). A multiligand based Pd catalyst for C-N cross-coupling reactions, *J. Am. Chem. Soc.*, **132**, pp. 15914–15917.
28. Martin, R., and Buchwald, S. L. (2007). Pd-catalyzed Kumada-Corriu cross-coupling reactions at low temperatures allow the use of Knochel-type Grignard reagents, *J. Am. Chem. Soc.*, **129**, pp. 3844–3845.

29. Roy, A. H., and Hartwig, J. F. (2004). Oxidative addition of aryl sulfonates to Palladium (0) complexes of mono-and bidentate phosphines. Mild addition of aryl tosylates and the effects of anions on rate and mechanism, *Organometallics*, **23,** pp. 194–202.
30. Limmert, M. E., Roy, A. H., and Hartwig, J. F. (2005). Kumada coupling of aryl and vinyl tosylates under mild conditions, *J. Org. Chem.*, **70**, pp. 9364–9370.
31. Rudolph, A., and Lautens, M. (2009). Secondary alkyl halides in transition-metal-catalyzed cross-coupling reactions, *Angew. Chem. Int. Ed.*, **48**, pp. 2656–2670.
32. Sandrock, D. L., Jean-Gerard, L., Chen, C.-Y., Dreher, S. D., and Molander, G. A. (2010). Stereospecific cross-coupling of secondary alkyl β-trifluoroboratoamides, *J. Am. Chem. Soc.*, **132,** pp. 17108–17110.
33. Ohmura, T., Awano, T., and Suginome, M. (2010). Stereospecific Suzuki-Miyaura coupling of chiral α-(acylamino) benzylboronic esters with inversion of configuration, *J. Am. Chem. Soc.*, **132**, pp. 13191–13193.
34. Lee, J. C. H., McDonald, R., and Hall, D. G. (2011). Enantioselective preparation and chemoselective cross-coupling of 1,1-diboron compounds, *Nat. Chem.*, **3**, pp. 894–899.
35. Molander, G. A., and Wisniewski, S. R. (2012). Stereospecific cross-coupling of secondary organotrifluoroborates: Potassium 1-(benzyloxy) alkyltrifluoroborates, *J. Am. Chem. Soc.*, **134**, pp. 16856–16868.
36. Li, L., Wang, C.-Y., Huang, R., and Biscoe, M. R. (2013). Stereoretentive Pd-catalysed Stille cross-coupling reactions of secondary alkyl azastannatranes and aryl halides, *Nat. Chem.*, **5**, pp. 607–612.
37. Herrmann, W. A., Elison, M., Fischer, J., Köcher, C., and Artus, G. R. J. (1995). Metal complexes of N-heterocyclic carbenes: A new structural principle for catalysts in homogeneous catalysis, *Angew. Chem. Int. Ed. Engl.*, **34**, pp. 2371–2374.
38. Marion, N., and Nolan, S. P. (2008). Well-defined N-heterocyclic carbenes-palladium (II) precatalysts for cross-coupling reactions, *Acc. Chem. Res.*, **41**, pp. 1440–1449.
39. Altenhoff, G., Goddard, R., Lehmann, C. W., and Glorius, F. (2004). Sterically demanding, bioxazoline-derived N-heterocyclic carbene ligands with restricted flexibility for catalysis, *J. Am. Chem. Soc.*, **126**, pp. 15195–15201.

40. Iyer, S., and Jayanthi, A. (2003). Saturated N-heterocyclic carbene oxime and amine palladacycle catalysis of the Mizoroki-Heck and the Suzuki reactions, *Synlett,* **8**, pp. 1125–1128.
41. Viciu, M. S., Kelly, R. A., Stevens, E. D., Naud, F., Studer, M., and Nolan, S. P. (2003). Synthesis, characterization, and catalytic activity of N-heterocyclic carbene (NHC) palladacycle complexes, *Org. Lett.*, **5**, pp. 1479–1482.
42. Endo, K., and Grubbs, R. H. (2011). Chelated ruthenium catalysts for Z-selective olefin metathesis, *J. Am. Chem. Soc.*, **133**, pp. 8525–8527.
43. Keitz, B. K., Endo, K., Patel, P. R., Herbert, M. B., and Grubbs, R. H. (2012). Improved ruthenium catalysts for Z-selective olefin metathesis, *J. Am. Chem. Soc.*, **134**, pp. 693–699.
44. Negishi, E.-I. (2011). Magical power of transition metals: Past, present, and future (Nobel lecture), *Angew. Chem. Int. Ed.*, **50**, pp. 6738–6764.
45. Johansson Seechurn, C. C., Kitching, M. O., Colacot, T. J., and Snieckus, V. (2012). Palladium-catalyzed cross-coupling: A historical contextual perspective to the 2010 Nobel prize, *Angew. Chem. Int. Ed.*, **51**, pp. 5062–5085.
46. Cahiez, G., Gager, O., and Buendia, J. (2010). Copper-catalyzed cross-coupling of alkyl and aryl Grignard reagents with alkynyl halides, *Angew. Chem. Int. Ed.*, **49**, pp. 1278–1281.
47. Abedi, Y., Biffis, A., Gava, R., Tubaro, C., Chelucci, G., and Stoccoro, S. (2014). Cu-iminopyridine complexes as catalysts for carbene and nitrene transfer reactions, *Appl. Organometal. Chem.*, **28**, pp. 512–516.
48. Xiang, S.-K., Zhang, L.-H., and Jiao, N. (2009). Sp-sp^3 C-C bond formation via Fe(OTf)$_3$/TfOH cocatalyzed coupling reaction of terminal alkynes with benzylic alcohols, *Chem. Commun.*, **42**, pp. 6487–6489.
49. Yang, J., Liu, X., Meng, D.-L., Chen, H.-Y., Zong, Z.-H., Feng, T.-T., and Sun, K. (2012). Efficient iron-catalyzed direct β-alkylation of secondary alcohols with primary alcohols, *Adv. Synth. Catal.*, **354**, pp. 328–334.
50. Dieskau, A. P., Holzwarth, M. S., and Plietker, B. (2012). Fe-catalyzed multicomponent reactions: The regioselective alkoxy allylation of activated olefins and its application in sequential Fe catalysis, *Chem. Eur. J.*, **18**, pp. 2423–2429.
51. Hirose, D., Taniguchi, T., and Ishibashi, H. (2013). Recyclable Mitsunobu reagents: Catalytic Mitsunobu reactions with an

iron catalyst and atmospheric oxygen, *Angew. Chem. Int. Ed.*, **52**, pp. 4613–4617.

52. Ackerman, L. K. G., Lovell, M. M., and Weix, D. J. (2015). Multimetallic catalysed cross-coupling of aryl bromides with aryl triflates, *Nature*, **524**, pp. 454–457.

53. Severin, K. (2002). Asymmetric halogeno-bridged complexes: New reagents in organometallic synthesis and catalysis, *Chem. Eur. J.*, **8**, pp. 1515–1518.

54. Mavila, S., Rozenberg, I., and Lemcoff, N. G. (2014). A general approach to mono-and bimetallic organometallic nanoparticles, *Chem. Sci.*, **5**, pp. 4196–4203.

55. Bartholomew, E. R., Bertz, S. H., Cope, S., Dorton, D. C., Murphy, M., and Ogle, C. A. (2008). Neutral organocopper(III) complexes, *Chem. Commun.*, **10**, pp. 1176–1177.

56. Das, P., Sharma, D., Kumar, M., and Singh, B. (2010). Copper promoted CN and CO type cross-coupling reactions, *Curr. Org. Chem.*, **14**, pp. 754–783.

57. Kunz, K., Scholz, U., and Ganzer, D. (2003). Renaissance of Ullmann and Goldberg reactions-progress in copper catalyzed CN-, CO- and CS-coupling, *Synlett*, **15**, pp. 2428–2439.

58. Brasche, G., and Buchwald, S. L. (2008). C-H functionalization/C-N bond formation: Copper-catalyzed synthesis of benzimidazoles from amidines, *Angew. Chem. Int. Ed.*, **47**, pp. 1932–1934.

59. Huffman, L. M., and Stahl, S. S. (2008). Carbon-nitrogen bond formation involving well-defined aryl-copper (III) complexes, *J. Am. Chem. Soc.*, **130**, pp. 9196–9197.

60. King, A. E., Huffman, L. M., Casitas, A., Costas, M., Ribas, X., and Stahl, S. S. (2010). Copper-catalyzed aerobic oxidative functionalization of an arene C–H bond: Evidence for an aryl-copper (III) intermediate, *J. Am. Chem. Soc.*, **132**, pp. 12068–12073.

61. Ribas, X., Jackson, D. A., Donnadieu, B., Mahìa, J., Parella, T., Xifra, R., Hedman, B., Hodgson, K. O., Llobet, A., and Stack, T. D. P. (2002). Aryl C-H activation by Cu^{II} to form an organometallic Aryl-Cu^{III} species: A novel twist on copper disproportionation, *Angew. Chem. Int. Ed.*, **41**, pp. 2991–2994.

62. Casitas, A., King, A. E., Parella, T., Costas, M., Stahl, S. S., and Ribas, X. (2010). Direct observation of Cu^{I}/Cu^{III} redox steps relevant to Ullmann-type coupling reactions, *Chem. Sci.*, **1**, pp. 326–330.

63. Hickman, A. J., and Sanford, M. S. (2012). High-valent organometallic copper and palladium in catalysis, *Nature*, **484**, pp. 177–185.

64. Muñiz, K. (2009). High-oxidation-state palladium catalysis: New reactivity for organic synthesis, *Angew. Chem. Int. Ed.*, **48**, pp. 9412–9423.
65. Arnold, P. L., Sanford, M. S., and Pearson, S. M. (2009). Chelating N-heterocyclic carbene alkoxide as a supporting ligand for PdII/IV C-H bond functionalization catalysis, *J. Am. Chem. Soc.*, **131**, pp. 13912–13913.
66. Maleckis, A., and Sanford, M. S. (2011). Facial tridentate ligands for stabilizing Palladium (IV) complexes, *Organometallics*, **30**, pp. 6617–6627.
67. Iglesias, Á., Álvarez, R., de Lera, Á. R., and Muñiz, K. (2012). Palladium-catalyzed intermolecular C(sp3)-H amidation, *Angew. Chem. Int. Ed.*, **51**, pp. 2225–2228.
68. Pérez-Temprano, M. H., Racowski, J. M., Kampf, J. W., and Sanford, M. S. (2014). Competition between sp^3-C-N vs sp^3-C-F reductive elimination from PdIV complexes, *J. Am. Chem. Soc.*, **136**, pp. 4097–4100.
69. Ye, Y., Ball, N. D., Kampf, J. W., and Sanford, M. S. (2010). Oxidation of a cyclometalated Pd (II) dimer with "CF^{3+}": Formation and reactivity of a catalytically competent monomeric Pd(IV) aquo complex, *J. Am. Chem. Soc.*, **132**, pp. 14682–14687.
70. Cho, E. J., Senecal, T. D., Kinzel, T., Zhang, Y., Watson, D. A., and Buchwald, S. L. (2010). The palladium-catalyzed trifluoromethylation of aryl chlorides, *Science*, **328**, pp. 1679–1681.
71. Wang, X., Truesdale, L., and Yu, J.-Q. (2010). Pd(II)-catalyzed ortho-trifluoromethylation of arenes using TFA as a promoter, *J. Am. Chem. Soc.*, **132**, pp. 3648–3649.
72. Ye, B., and Cramer, N. (2012). Chiral cyclopentadienyl ligands as stereocontrolling element in asymmetric C–H functionalization, *Science*, **338**, pp. 504–506.
73. Song, G., Wang, F., and Li, X. (2012). C-C, C-O and C-N bond formation via rhodium (iii)-catalyzed oxidative C-H activation, *Chem. Soc. Rev.*, **41**, pp. 3651–3678.
74. Li, Y., Li, B. J., Wang, W. H., Huang, W. P., Zhang, X. S., Chen, K., and Shi, Z. J. (2011). Rhodium-catalyzed direct addition of aryl C-H bonds to N-sulfonyl aldimines, *Angew. Chem. Int. Ed.*, **50**, pp. 2115–2167.
75. Tauchert, M. E., Incarvito, C. D., Rheingold, A. L., Bergman, R. G., and Ellman, J. A. (2012). Mechanism of the rhodium (III)-catalyzed arylation of imines via C-H bond functionalization: Inhibition by substrate, *J. Am. Chem. Soc.*, **134**, pp. 1482–1485.

76. Lian, Y., Bergman, R. G., and Ellman, J. A. (2012). Rhodium (III)-catalyzed synthesis of phthalides by cascade addition and cyclization of benzimidates with aldehydes, *Chem. Sci.*, **3**, pp. 3088–3092.
77. Hyster, T. K., Knörr, L., Ward, T. R., and Rovis, T. (2012). Biotinylated Rh(III) complexes in engineered streptavidin for accelerated asymmetric C-H activation, *Science*, **338**, pp. 500–503.
78. List, B. (2007). Introduction: Organocatalysis, *Chem. Rev.*, **107**, pp. 5413–5415.
79. List, B., Lerner, R. A., and Barbas, C. F. (2000). Proline-catalyzed direct asymmetric aldol reactions, *J. Am. Chem. Soc.*, **122**, pp. 2395–2396.
80. Ahrendt, K. A., Borths, C. J., and MacMillan, D. W. C. (2000). New strategies for organic catalysis: The first highly enantioselective organocatalytic Diels-Alder reaction, *J. Am. Chem. Soc.*, **122**, pp. 4243–4244.
81. MacMillan, D. W. C. (2008). The advent and development of organocatalysis, *Nature*, **455**, pp. 304–308.
82. Bahmanyar, S., and Houk, K. N. (2003). Origins of opposite absolute stereoselectivities in proline-catalyzed direct Mannich and aldol reactions, *Org. Lett.*, **5**, pp. 1249–1251.
83. Bahmanyar, S., Houk, K. N., Martin, H. J., and List, B. (2003). Quantum mechanical predictions of the stereoselectivities of proline-catalyzed asymmetric intermolecular aldol reactions, *J. Am. Chem. Soc.*, **125**, pp. 2475–2479.
84. Allemann, C., Gordillo, R., Clemente, F. R., Cheong, P. H.-Y., and Houk, K. N. (2004). Theory of asymmetric organocatalysis of aldol and related reactions: Rationalizations and predictions, *Acc. Chem. Res.*, **37**, pp. 558–569.
85. Pihko, P. M., Laurikainen, K. M., Usano, A., Nyberg, A. I., and Kaavi, J. A. (2006). Effect of additives on the proline-catalyzed ketone–aldehyde aldol reactions, *Tetrahedron*, **62**, pp. 317–328.
86. Hayashi, Y., Tsuboi, W., Ashimine, I., Uroshima, T., Shoji, M., and Sakai, K. (2003). The direct and enantioselective, one-pot, three-component, cross-Mannich reaction of aldehydes, *Angew. Chem. Int. Ed.*, **42**, pp. 3677–3680.
87. Chowdari, N. S., Suri, J. T., and Barbas, C. F. (2004). Asymmetric synthesis of quaternary α- and β-amino acids and β-lactams via proline-catalyzed Mannich reactions with branched aldehyde donors, *Org. Lett.*, **6**, pp. 2507–2510.

88. Zhuang, W., Saaby, S., and Jørgensen, K. A. (2004). Direct organocatalytic enantioselective Mannich reactions of ketimines: An approach to optically active quaternary α-amino acid derivatives, *Angew. Chem. Int. Ed.*, **43**, pp. 4476–4478.
89. Mase, N., Nakai, Y., Ohara, N., Yoda, H., Takabe, K., Tanaka, F., and Barbas, C. F. (2006). Organocatalytic direct asymmetric aldol reactions in water, *J. Am. Chem. Soc.*, **128**, pp. 734–735.
90. Aratake, S., Itoh, T., Okano, T., Nagae, N., Sumija, T., Shoji, M., and Hayashi, Y. (2007). Highly diastereo- and enantioselective direct aldol reactions of aldehydes and ketones catalyzed by siloxyproline in the presence of water, *Chem. Eur. J.*, **13**, pp. 10246–10256.
91. Kanemitsu, T., Umehara, A., Haneji, R., Nagata, K., and Itoh, T. (2012). A simple proline-based organocatalyst for the enantioselective reduction of imines using trichlorosilane as a reductant, *Tetrahedron*, **68**, pp. 3893–3898.
92. Marigo, M., Fielenbach, D., Braunton, A., Kjærsgaard, A., and Jørgensen, K. A. (2005). Enantioselective formation of stereogenic carbon-fluorine centers by a simple catalytic method, *Angew. Chem. Int. Ed.*, **44**, pp. 3703–3706.
93. Erkkilä, A., and Pihko, P. M. (2007). Rapid organocatalytic aldehyde-aldehyde condensation reactions, *Eur. J. Org. Chem.*, **25**, pp. 4205–4216.
94. White, M. C. (2012). Adding aliphatic C–H bond oxidations to synthesis, *Science*, **335**, pp. 807–809.
95. Wang, D., Shuler, W. G., Pierce, C. J., and Hilinski, M. K. (2016). An iminium salt organocatalyst for selective aliphatic C–H hydroxylation, *Org. Lett.*, **18**, pp. 3826–3829.
96. Akiyama, T., Itoh, J., Yokota, K., and Fuchibe, K. (2004). Enantioselective Mannich-type reaction catalyzed by a chiral Brønsted acid, *Angew. Chem. Int. Ed.*, **43**, pp. 1566–1568.
97. Sickert, M., Abels, F., Lang, M., Sieler, J., Birkemeyer, C., and Schneider, C. (2010). The Brønsted acid catalyzed, enantioselective vinylogous Mannich reaction, *Chem. Eur. J.*, **16**, pp. 2806–2818.
98. MacMillan, D. W. C. (2008). The advent and development of organocatalysis, *Nature*, **455**, pp. 304–308.
99. Shapiro, N. D., Rauniyar, V., Hamilton, G. L., Wu, J., and Dean Toste, F. (2011). Asymmetric additions to dienes catalysed by a dithiophosphoric acid, *Nature*, **470**, pp. 245–249.

100. Palacio, C., and Connon, S. J. (2013). C-5'-substituted cinchona alkaloid derivatives catalyse the first highly enantioselective dynamic kinetic resolutions of azlactones by thiolysis, *Eur. J. Org. Chem.*, **24**, pp. 5398–5413.
101. Pereira, A. A., de Castro, P. P., de Mello, A. C., Ferreira, B. R. V., Eberlin, M. N., and Amarante, G. W. (2014). Brønsted acid catalyzed azlactone ring opening by nucleophiles, *Tetrahedron*, **70**, pp. 3271–3275.
102. Evano, G., Coste, A., and Jouvin, K. (2010). Ynamides: Versatile tools in organic synthesis, *Angew. Chem. Int. Ed.*, **49**, pp. 2840–2859.
103. Peng, B., Huang, X., Xie, L.-G., and Maulide, N. (2014). A Brønsted acid catalyzed redox arylation, *Angew. Chem. Int. Ed.*, **53**, pp. 8718–8721.
104. Vara, B. A., Struble, T. J., Wang, W., Dobish, M. C., and Johnston, J. N. (2015). Enantioselective small molecule synthesis by carbon dioxide fixation using a dual Brønsted acid/base organocatalyst, *J. Am. Chem. Soc.*, **137**, pp. 7302–7305.
105. Hopkinson, M. N., Richter, C., Schedler, M., and Glorius, F. (2014). An overview of N-heterocyclic carbenes, *Nature*, **510**, pp. 485–496.
106. Enders, D., Niemeier, O., and Henseler, A. (2007). Organocatalysis by N-heterocyclic carbenes, *Chem. Rev.*, **107**, pp. 5606–5655.
107. Fèvre, M., Pinaud, J., Gnanou, Y., Vignolle, J., and Taton, D. (2013). N-heterocyclic carbenes (NHCs) as organocatalysts and structural components in metal-free polymer synthesis, *Chem. Soc. Rev.*, **42**, pp. 2142–2172.
108. Bugaut, X., and Glorius, F. (2012). Organocatalytic umpolung: N-heterocyclic carbenes and beyond, *Chem. Soc. Rev.*, **41**, pp. 3511–3522.
109. Schedler, M., Wang, D.-S., and Glorius, F. (2013). NHC-catalyzed hydroacylation of styrenes, *Angew. Chem. Int. Ed.*, **52**, pp. 2585–2589.
110. Ryan, S. J., Candish, L., and Lupton, D. W. (2013). Acyl anion free N-heterocyclic carbene organocatalysis, *Chem. Soc. Rev.*, **42**, pp. 4906–4917.
111. Flanigan, D. M., Romanov-Michailidis, F., White, N. A., and Rovis, T. (2015). Enantioselective small molecule synthesis by carbon dioxide fixation using a dual Brønsted acid/base organocatalyst, *Chem. Rev.*, **115**, pp. 9307–9387.
112. Dong, X., Yang, W., Hu, W., and Sun, J. (2015). N-heterocyclic carbene catalyzed enantioselective α-fluorination of aliphatic aldehydes and α-chloro aldehydes: Synthesis of α-fluoro esters, amides, and thioesters, *Angew. Chem., Int. Ed.*, **54**, pp. 660–663.

113. Chen, J., and Huang, Y. (2014). Asymmetric catalysis with N-heterocyclic carbenes as non-covalent chiral templates, *Nat. Commun.*, **5**, pp. 3437.
114. Chen, J., Meng, S., Wang, L., Tang, H., and Huang, Y. (2015). Highly enantioselective Sulfa-Michael addition reactions using N-heterocyclic carbene as a non-covalent organocatalyst, *Chem. Sci.*, **6**, pp. 4184–4189.
115. Fu, Z., Xu, J., Zhu, T., Leong, W. W. Y., and Chi, Y. R. (2013). β-carbon activation of saturated carboxylic esters through N-heterocyclic carbene organocatalysis, *Nat. Chem.*, **5**, pp. 835–839.
116. Zhao, X. D., DiRocco, D. A., and Rovis, T. (2011). N-heterocyclic carbene and Brønsted acid cooperative catalysis: Asymmetric synthesis of trans-γ-lactams, *J. Am. Chem. Soc.*, **133**, pp. 12466–12469.
117. Harper, K. C., and Sigman, M. S. (2013). Asymmetric homoenolate additions to acyl phosphonates through rational design of a tailored N-heterocyclic carbene catalyst, *J. Org. Chem.*, **78**, pp. 2813–2818.
118. Po Jang, K., Hutson, G. E., Johnston, R. C., McCusker, E. O. Cheong, P. H.-Y., and Scheidt, K. A. (2014). Asymmetric homoenolate additions to acyl phosphonates through rational design of a tailored N-heterocyclic carbene catalyst, *J. Am. Chem. Soc.*, **136**, pp. 76–79.
119. Fèvre, M., Pinaud, J., Gnanou, Y., Vignolle, J., and Taton, D. (2013). N-heterocyclic carbenes (NHCs) as organocatalysts and structural components in metal-free polymer synthesis, *Chem. Soc. Rev.*, **42**, pp. 2142–2172.
120. Nyce, G. W., Glauser, T., Connor, E. F., Möck, A., Waymouth, R. M., and Hedrick, J. L. (2003). In situ generation of carbenes: A general and versatile platform for organocatalytic living polymerization, *J. Am. Chem. Soc.*, **125**, pp. 3046–3056.
121. Nederberg, F., Connor, E. F., Möller, M., Glauser, T., and Hedrick, J. L. (2001). New paradigms for organic catalysts: The first organocatalytic living polymerization, *Angew. Chem. Int. Ed.*, **40**, pp. 2712–2715.
122. Matsuoka, S.-I., Namera, S., and Suzuki, M. (2015). Oxa-michael addition polymerization of acrylates catalyzed by N-heterocyclic carbenes, *Polym. Chem.*, **6**, pp. 294–301.
123. Hong, M., Tang, X., Falivene, L., Caporaso, L., Cavallo, L., and Chen, E. Y.-X. (2016). Proton-transfer polymerization by N-heterocyclic carbenes: Monomer and catalyst scopes and mechanism for converting dimethacrylates into unsaturated polyesters, *J. Am. Chem. Soc.*, **138**, pp. 2021–2035.
124. Zhang, Q., Zhang, S., and Deng, Y. (2011). Recent advances in ionic liquid catalysis, *Green Chem.*, **13**, pp. 2619–2637.

125. Wasserscheid, P., and Keim, W. (2011). Ionic liquids—new "solutions" for transition metal catalysis, *Angew. Chem. Int. Ed.*, **39**, pp. 3772–3789.
126. Boon, J. A., Levisky, J. A., Pflug, J. L., and Wilkes, J. S. (1986). Friedel-Crafts reactions in ambient-temperature molten salts, *J. Org. Chem.*, **51**, pp. 480–483.
127. Fry, S. E., and Pienta, N. J. (1985). Effects of molten salts on reactions. Nucleophilic aromatic substitution by halide ions in molten dodecyltributylphosphonium salts, *J. Am. Chem. Soc.*, **107**, pp. 6399–6400.
128. Chauvin, Y., Gilbert, B., and Guibard, I. (1990). Catalytic dimerization of alkenes by nickel complexes in organochloroaluminate molten salts, *J. Chem. Soc., Chem. Commun.*, **23**, pp. 1715–1716.
129. Carlin, R. T. (1990). Complexation of Cp_2MCl_2 in a chloroaluminate molten salt: Relevance to homogeneous Ziegler-Natta catalysis, *J. Mol. Catal.*, **63**, pp. 125–129.
130. Wilkes, J. S., and Zaworotko, M. J. (1992). Air and water stable 1-ethyl-3-methylimidazolium based ionic liquids, *J. Chem. Soc., Chem. Commun.*, **13**, pp. 965–967.
131. Swartling, D., Ray, L., Compton, S., and Ensor, D. (2000). Preliminary investigation into modification of ionic liquids to improve extraction parameters, *Bull. Biochem. Biotechnol.*, **13**, pp. 1–6.
132. Dyson, P. J., Ellis, D. J., and Welton, T. (2001). A temperature-controlled reversible ionic liquid-water two phase-single phase protocol for hydrogenation catalysis, *Can. J. Chem.*, **79**, pp. 705–708.
133. Schleicher, J. C., and Scurto, A. M. (2009). Kinetics and solvent effects in the synthesis of ionic liquids: Imidazolium, *Green Chem.*, **11**, pp. 694–703.
134. Crowhurst, L., Lancaster, N. L., Perez-Arlandis, J. M., and Welton, T. (2004). Manipulating solute nucleophilicity with room temperature ionic liquids, *J. Am. Chem. Soc.*, **126**, pp. 11549–11555.
135. Nabavizadeh, S. M., Shahsavari, H. R., Sepehrpour, H., Hosseini, F. N., Jamali, S., and Rashidi, M. (2010). Oxidative addition reaction of diarylplatinum (II) complexes with MeI in ionic liquid media: A kinetic study, *Dalton Trans.*, **39**, pp. 7800–7805.
136. Cimpeanu, V., Parvulescu, V., Parvulescu, V. I., Thompson, J. M., and Hardacre, C. (2006). Thioethers oxidation on dispersed Ta-silica mesoporous catalysts in ionic liquids, *Catal. Today*, **117**, pp. 126–132.

137. Crowhurst, L., Mawdsley, P. R., Perez-Arlandis, J. M., Salter, P. A., and Welton, T. (2003). Solvent–solute interactions in ionic liquids, *Phys. Chem. Chem. Phys.*, **5**, pp. 2790–2794.
138. Jelicic, A., Garcia, N., Lohmannsroben, H.-G., and Beuermann, S. (2009). Prediction of the ionic liquid influence on propagation rate coefficients in methyl methacrylate radical polymerizations based on Kamlet–Taft solvatochromic parameters, *Macromolecules*, **42**, pp. 8801–8808.
139. Doherty, T. V., Mora-Pale, M., Foley, S. E., Linhardt, R. J., and Dordick, J. S. (2010). Ionic liquid solvent properties as predictors of lignocellulose pretreatment efficacy, *Green Chem.*, **12**, pp. 1967–1975.
140. Baker, S. N., Baker, G. A., and Bright, F. (2002). Temperature-dependent microscopic solvent properties of 'dry' and 'wet' 1-butyl-3-methylimidazolium hexafluorophosphate: Correlation with ET(30) and Kamlet–Taft polarity scales, *Green Chem.*, **4**, pp. 165–169.
141. Wu, Y., Sasaki, T., Kazushi, K., Seo, T., and Sakurai, K. (2008). Interactions between spiropyrans and room-temperature ionic liquids: Photochromism and solvatochromism, *J. Phys. Chem. B*, **112**, pp. 7530–7536.
142. Lee, J.-M., Ruckes, S., and Prausnitz, J. M. (2008). Solvent polarities and Kamlet–Taft parameters for ionic liquids containing a pyridinium cation, *J. Phys. Chem. B*, **112**, pp. 1473–1476.
143. Khupse, N. D., and Kumar, A. (2010). Contrasting thermosolvatochromic trends in pyridinium-, pyrrolidinium-, and phosphonium-based ionic liquids, *J. Phys. Chem. B*, **114**, pp. 376–381.
144. Coleman, S., Byrne, R., Minkovska, S., and Diamond, D. (2009). Thermal reversion of spirooxazine in ionic liquids containing the [NTf2]–anion, *Phys. Chem. Chem. Phys.*, **11**, pp. 5608–5614.
145. Zhang, S., Qi, X., Ma, X., Lu, L., and Deng, Y. (2010). Hydroxyl ionic liquids: The differentiating effect of hydroxyl on polarity due to ionic hydrogen bonds between hydroxyl and anions, *J. Phys. Chem. B*, **114**, pp. 3912–3920.
146. Sarkar, A., Trivedi, S., Baker, G. A., and Pandey, S. (2008). Multiprobe spectroscopic evidence for "hyperpolarity" within 1-butyl-3-methylimidazolium hexafluorophosphate mixtures with tetraethylene glycol, *J. Phys. Chem. B*, **112**, pp. 14927–14936.
147. Chauvin, Y., Mussmann, L., and Olivier, H. (1995). A novel class of versatile solvents for two-phase catalysis: Hydrogenation, isomerization, and hydroformylation of alkenes catalyzed by

rhodium complexes in liquid 1,3-dialkylimidazolium salts, *Angew. Chem. Int. Ed. Engl.*, **34**, pp. 2698–2700.

148. Suarez, P. A., Dullius, J. E., Einloft, S., De Souza, R. F., and Dupont, J. (1996). The use of new ionic liquids in two-phase catalytic hydrogenation reaction by rhodium complexes, *Polyhedron*, **15**, pp. 1217–1219.

149. Green, L., Hemeon, I., and Singer, R. D. (2000). 1-Ethyl-3-methylimidazolium halogenoaluminate ionic liquids as reaction media for the acylative cleavage of ethers, *Tetrahedron Lett.*, **41**, pp. 1343–1346.

150. Song, C. E., Oh, C. R., Roh, E. J., and Choo, D. J. (2000). Cr(salen) catalysed asymmetric ring opening reactions of epoxides in room temperature ionic liquids, *Chem. Commun.*, **18**, pp. 1743–1744.

151. Boon, J. A., Levisky, J. A., Pflug, J. L., and Wilkes, J. S. (1986). Friedel-Crafts reactions in ambient-temperature molten salts, *J. Org. Chem.*, **51**, pp. 480–483.

152. Song, C. E., Shim, W. H., Roh, E. J., and Choi, J. H. (2000). Scandium(III) triflate immobilised in ionic liquids: A novel and recyclable catalytic system for Friedel–Crafts alkylation of aromatic compounds with alkenes, *Chem. Commun.*, **17**, pp. 1695–1696.

153. Earle, M. J., McCormac, P. B., and Seddon, K. R. (1999). Diels–Alder reactions in ionic liquids. A safe recyclable alternative to lithium perchlorate–diethyl ether mixtures, *Green Chem.*, **1**, pp. 23–25.

154. Fischer, T., Sethi, A., Welton, T., and Woolf, J. (1999). Diels-Alder reactions in room-temperature ionic liquids, *Tetrahedron Lett.*, **40**, pp. 793–796.

155. Song, C. E., Shim, W. H., Roh, E. J., Lee, S. G., and Choi, J. H. (2001). Ionic liquids as powerful media in scandium triflate catalysed Diels–Alder reactions: Significant rate acceleration, selectivity improvement and easy recycling of catalyst, *Chem. Commun.*, **12**, pp. 1122–1123.

156. Rangits, G., and Kollár, L. (2005). Palladium-catalysed hydroalkoxycarbonylation of styrene in [BMIM][BF4] and [BMIM][PF6] ionic liquids, *J. Mol. Catal. A*, **242**, pp. 156–160.

157. Einloft, S., Dietrich, F. K., De Souza, R. F., and Dupont, J. (1996). Selective two-phase catalytic ethylene dimerization by NiII complexes/AlEtCl2 dissolved in organoaluminate ionic liquids, *Polyhedron*, **15**, pp. 3257–3259.

158. Dullius, J. E., Suarez, P. A., Einloft, S., de Souza, R. F., Dupont, J., Fischer, J., and De Cian, A. (1998). Selective catalytic hydrodimerization of

1,3-butadiene by palladium compounds dissolved in ionic liquids, *Organometallics*, **17**, pp. 815–819.

159. Earle, M. J., McCormac, P. B., and Seddon, K. R. (1998). Regioselective alkylation in ionic liquids, *Chem. Commun.*, **20**, pp. 2245–2246.

160. Chen, W., Xu, L., Chatterton, C., and Xiao, J. (1999). Palladium catalysed allylation reactions in ionic liquids, *Chem. Commun.*, **13**, pp. 1247–1248.

161. Gordon, C. M., and McCluskey, A. (1999). Ionic liquids: A convenient solvent for environmentally friendly allylation reactions with tetraallylstannane, *Chem. Commun.*, **15**, pp. 1431–1432.

162. Toma, Š., Gotov, B., Kmentová, I., and Solčániová, E. (2000). Enantioselective allylic substitution catalyzed by Pd0–ferrocenylphosphine complexes in [bmim][PF6] ionic liquid, *Green Chem.*, **2**, pp. 149–151.

163. Herrmann, W. A., and Böhm, V. P. (1999). Heck reaction catalyzed by phospha-palladacycles in non-aqueous ionic liquids, *J. Organometal. Chem.*, **572**, pp. 141–145.

164. Xu, L., Chen, W., and Xiao, J. (2000). Heck reaction in ionic liquids and the in situ identification of N-heterocyclic carbene complexes of palladium, *Organometallics*, **19**, pp. 1123–1127.

165. Mathews, C. J., Smith, P. J., and Welton, T. (2000). Palladium catalysed Suzuki cross-coupling reactions in ambient temperature ionic liquids, *Chem. Commun.*, **14**, pp. 1249–1250.

166. Keim, W., Vogt, D., Waffenschmidt, H., and Wasserscheid, P. (1999). New method to recycle homogeneous catalysts from monophasic reaction mixtures by using an ionic liquid exemplified for the Rh-catalysed hydroformylation of methyl-3-pentenoate, *J. Catal.*, **186**, pp. 481–484.

167. Wasserscheid, P., Waffenschmidt, H., Machnitzki, P., Kottsieper, K. W., and Stelzer, O. (2001). Cationic phosphine ligands with phenylguanidinium modified xanthene moieties—a successful concept for highly regioselective, biphasic hydroformylation of oct-1-ene in hexafluorophosphate ionic liquids, *Chem. Commun.*, **5**, pp. 451–452.

168. Favre, F., Olivier-Bourbigou, H., Commereuc, D., and Saussine, L. (2001). Hydroformylation of 1-hexene with rhodium in non-aqueous ionic liquids: How to design the solvent and the ligand to the reaction, *Chem. Commun.*, **15**, pp. 1360–1361.

169. Song, C. E., and Roh, E. J. (2000). Practical method to recycle a chiral (salen)Mn epoxidation catalyst by using an ionic liquid, *Chem. Commun.*, **10**, pp. 837–838.
170. Wheeler, C., West, K. N., Liotta, C. L., and Eckert, C. A. (2001). Ionic liquids as catalytic green solvents for nucleophilic displacement reactions, *Chem. Commun.*, **10**, pp. 887–888.
171. Laali, K. K., and Gettwert, V. J. (2001). Electrophilic nitration of aromatics in ionic liquid solvents, *J. Org. Chem.*, **66**, pp. 35–40.
172. Kabalka, G. W., and Malladi, R. R. (2000). Reduction of aldehydes using trialkylboranes in ionic liquids, *Chem. Commun.*, **22**, pp. 2191–2191.
173. Boulaire, V. L., and Grée, R. (2000). Wittig reactions in the ionic solvent [bmim][BF4], *Chem. Commun.*, **22**, pp. 2195–2196.
174. Peng, J., and Deng, Y. (2001). Cycloaddition of carbon dioxide to propylene oxide catalyzed by ionic liquids, *New J. Chem.*, **25**, pp. 639–641.
175. Handy, S. T., and Zhang, X. (2001). Organic synthesis in ionic liquids: The Stille coupling, *Org. Lett.*, **3**, pp. 233–236.
176. Kim, K. W., Song, B., Choi, M. Y., and Kim, M. J. (2001). Biocatalysis in ionic liquids: Markedly enhanced enantioselectivity of lipase, *Org. Lett.*, **3**, pp. 1507–1509.
177. Madeira Lau, R., Van Rantwijk, F., Seddon, K. R., and Sheldon, R. A. (2001). Lipase-catalyzed reactions in ionic liquids, *Org. Lett.*, **2**, pp. 4189–4191.
178. Narayana Reddy, P., Padmaja, P., Reddy, B. V. S., and Rambabu, G. (2015). Ionic liquid/water mixture promoted organic transformations, *RSC Adv.*, **5**, pp. 51035–51054.
179. Lee, J. C., Kim, J., Lee, S. B., Chang, S., and Jeong, Y. J. (2011). Efficient oxidation of benzylic alcohols with trichloroisocyanuric acid and ionic liquid in water, *Synth. Commun.*, 41, pp. 1947–1951.
180. Siyutkin, D. E., Kucherenko, A. S., Struchkova, M. I., and Zlotin, S. G. (2008). A novel (S)-proline-modified task-specific chiral ionic liquid: An amphiphilic recoverable catalyst for direct asymmetric aldol reactions in water, *Tetrahedron Lett.*, **49**, pp. 1212–1216.
181. Siyutkin, D. E., Kucherenko, A. S., and Zlotin, S. G. (2010). A new (S)-prolinamide modified by an ionic liquid moiety-a high performance recoverable catalyst for asymmetric aldol reactions in aqueous media, *Tetrahedron*, **66**, pp. 513–518.

182. Niehaus, D., Philips, M., Michael, A., and Wightman, R. M. (1989). Voltammetry of ferrocene in supercritical carbon dioxide containing water and tetrahexylammonium hexafluorophosphate, *J. Phys. Chem.*, **93**, pp. 6232–6236.
183. Fan, Y., and Qian, J. (2010). Lipase catalysis in ionic liquids/supercritical carbon dioxide and its applications, *J. Mol. Catal. B: Enzym.*, **66**, pp. 1–7.
184. Lozano, P., de Diego, T., Carrie, D., Vaultier, M., and Iborra, J. L. (2002). Continuous green biocatalytic processes using ionic liquids and supercritical carbon dioxide, *Chem. Commun.*, **7**, pp. 692–693.
185. Reetz, M. T., Wiesenhöfer, W., Franciò, G., and Leitner, W. (2002). Biocatalysis in ionic liquids: Batchwise and continuous flow processes using supercritical carbon dioxide as the mobile phase, *Chem. Commun.*, **9**, pp. 992–993.
186. Bogel-yukasik, R., Lourenco, N. M. T., Vidinha, P., Gomes da Silva, M. D. R., and Afonso, C. A. (2008). Lipase catalysed mono and di-acylation of secondary alcohols with succinic anhydride in organic media and ionic liquids, *Green Chem.*, **10**, pp. 243–248.
187. Kaar, J. L., Jesionowski, A. M., Berberich, J. A., Moulton, R., and Russell, A. J. (2003). Impact of ionic liquid physical properties on lipase activity and stability, *J. Am. Chem. Soc.*, **125**, pp. 4125–4131.
188. Lozano, P., de Diego, T., Guegan, J. P., Vaultier, M., and Iborra, J. L. (2001). Stabilization of α-chymotrypsin by ionic liquids in transesterification reactions, *Biotechnol. Bioeng.*, **75**, pp. 563–569.
189. Persson, M., and Bornscheuer, U. T. (2003). Increased stability of an esterase from Bacillus stearothermophilus in ionic liquids as compared to organic solvents, *J. Mol. Catal. B: Enzym.*, **22**, pp. 21–27.
190. Mesiano, A. J., Beckman, E. J., and Russell, A. J. (1999). Supercritical biocatalysis, *Chem. Rev.*, **99**, pp. 623–633.
191. Hintermair, U., Franicio, G., and Leitner, W. (2013). A fully integrated continuous-flow system for asymmetric catalysis: Enantioselective hydrogenation with supported ionic liquid phase catalysts using supercritical CO_2 as the mobile phase, *Chem. Eur. J.*, **19**, pp. 4538–4547.
192. Gallo, V., Mastrorilli, P., Nobile, C. F., Romanazzi, G., and Suranna, G. P. (2002). How does the presence of impurities change the performance of catalytic systems in ionic liquids? A case study: The Michael addition of acetylacetone to methyl vinyl ketone, *Dalton Trans.*, **23**, pp. 4339–4342.

193. Lee, S.-G. (2006). Functionalized imidazolium salts for task-specific ionic liquids and their applications, *Chem. Commun.*, **10**, pp. 1049–1063.
194. Fei, Z., Geldbach, T. J., Zhao, D., and Dyson, P. (2006). From dysfunction to bis-function: On the design and applications of functionalised ionic liquids, *Chem. Eur. J.*, **12**, pp. 2122–2130.
195. Giernoth, R. (2010). Task-specific ionic liquids, *Angew. Chem. Int. Ed.*, **49**, pp. 2834–2839.
196. Zhao, D., Fei, Z., Geldbach, T. J., Scopelliti, R., and Dyson, P. J. (2004). Nitrile-functionalized pyridinium ionic liquids: Synthesis, characterization, and their application in carbon-carbon coupling reactions, *J. Am. Chem. Soc.*, **126**, pp. 15876–15882.
197. Chiappe, C., Pieraccini, D., Zhao, D., Fei, Z., and Dyson, P. J. (2006). Remarkable anion and cation effects on Stille reactions in functionalised ionic liquids, *Adv. Synth. Catal.*, **348**, pp. 68–74.
198. Nockemann, P., Pellens, M., Van Hecke, K., Van Meervelt, L., Wouters, J., Thijs, B., Vanecht, E., Parac-Vogt, T. N., Mehdi, H., Schaltin, S., Fransaer, J., Zahn, S., Kirchner, B., and Binnemans, K. (2010). Cobalt(II) complexes of nitrile-functionalized ionic liquids, *Chem. Eur. J.*, **16**, pp. 1849–1858.
199. Yu, S., Lindeman, S., and Tran, C. D. (2008). Chiral ionic liquids: Synthesis, properties, and enantiomeric recognition, *J. Org. Chem.*, **73**, pp. 2576–2591.
200. Miao, W., and Chan, T.-H. (2005). Ionic-liquid-supported peptide synthesis demonstrated by the synthesis of Leu5-enkephalin, *J. Org. Chem.*, **70**, pp. 3251–3255.
201. de Kort, M., Tuin, A. W., Kuiper, S., Overkleeft, H. S., van der Marel, G. A., and Buijsman, R. C. (2004). Development of a novel ionic support and its application in the ionic liquid phase assisted synthesis of a potent antithrombotic, *Tetrahedron Lett.*, **45**, pp. 2171–2175.
202. Fraga-Dubreuil, J., and Bazureau, J. (2003). Efficient combination of task-specific ionic liquid and microwave dielectric heating applied to one-pot three component, *Tetrahedron*, **59**, pp. 6121–6130.
203. Kohlmann, C., Greiner, L., Leitner, W., Wandrey, C., and Lütz, S. (2009). Ionic liquids as performance additives for electroenzymatic syntheses, *Chem. Eur. J.*, **15**, pp. 11692–11700.
204. Hough, W. L., Smiglak, M., Rodríguez, H., Swatloski, R. P., Spear, S. K., Daly, D. T., Pernak, J., Grisel, J. E., Carliss, R. D., Soutullo, M. D., Davis, J. H., and Rogers, R. D. (2007). The third evolution of ionic

liquids: Active pharmaceutical ingredients, *New J. Chem.*, **31**, pp. 1429–1436.
205. Giacalone, F., and Gruttadauria, M. (2016). Covalently supported ionic liquid phases: An advanced class of recyclable catalytic systems, *ChemCatChem*, **8**, pp. 664–684.
206. Selvam, T., Machoke, A., and Schwieger, W. (2012). Supported ionic liquids on non-porous and porous inorganic materials—a topical review, *Appl. Catal. A Gen.*, **445**, pp. 92–101.
207. Mehnert, C. P. (2005). Supported ionic liquid catalysis, *Chem. Eur. J.*, **11**, pp. 50–56.
208. Jin, M. J., Taher, A., Kang, H. J., Choi, M., and Ryoo, R. (2009). Palladium acetate immobilized in a hierarchical MFI zeolite-supported ionic liquid: A highly active and recyclable catalyst for Suzuki reaction in water, *Green Chem.*, **11**, pp. 309–313.
209. Arya, K., Rawat, D. S., and Sasai, H. (2012). Zeolite supported Brønsted-acid ionic liquids: An eco approach for synthesis of spiro [indole-pyrido [3, 2-e] thiazine] in water under ultrasonication, *Green Chem.*, **14**, pp. 1956–1963.
210. Karimi, B., Elhamifar, D., Clark, J. H., and Hunt, A. J. (2010). Ordered mesoporous organosilica with ionic-liquid framework: An efficient and reusable support for the palladium-catalyzed Suzuki-Miyaura coupling reaction in water, *Chem. Eur. J.*, **16**, pp. 8047–8053.
211. Shin, J. Y., Lee, B. S., Jung, Y., Kim, S. J., and Lee, S. G. (2007). Palladium nanoparticles captured onto spherical silica particles using a urea cross-linked imidazolium molecular band, *Chem. Commun.*, **48**, pp. 5238–5240.
212. DeCastro, C., Sauvage, E., Valkenberg, M. H., and Hölderich, W. F. (2000). Immobilised ionic liquids as Lewis acid catalysts for the alkylation of aromatic compounds with dodecene, *J. Catal.*, **196**, pp. 86–94.
213. Hagiwara, H., Ko, K. H., Hoshi, T., and Suzuki, T. (2007). Supported ionic liquid catalyst (Pd-SILC) for highly efficient and recyclable Suzuki–Miyaura reaction, *Chem. Commun.*, **27**, pp. 2838–2840.
214. Kang, T., Feng, Q., and Luo, M. (2005). An active and recyclable polystyrene-supported N-heterocyclic carbene-palladium catalyst for the Suzuki reaction of arylbromides with arylboronic acids under mild conditions, *Synlett*, **15**, pp. 2305–2308.
215. Kim, J. W., Kim, J. H., Lee, D. H., and Lee, Y. S. (2006). Amphiphilic polymer supported N-heterocyclic carbene palladium complex for Suzuki cross-coupling reaction in water, *Tetrahedron Lett.*, **47**, pp. 4745–4748.

216. Park, M. J., and Lee, S. G. (2007). Palladium catalysts supported onto the ionic liquid-functionalized carbon nanotubes for carbon-carbon coupling, *Bull. Korean Chem. Soc.*, **28**, pp. 1925–1926.
217. Taher, A., Kim, J. B., Jung, J. Y., Ahn, W. S., and Jin, M. J. (2009). Highly active and magnetically recoverable Pd-NHC catalyst immobilized on Fe_3O_4 nanoparticle-ionic liquid matrix for Suzuki reaction in water, *Synlett*, **15**, pp. 2477–2482.
218. Huang, J., Jiang, T., Han, B., Wu, W., Liu, Z., Xie, Z., and Zhang, J. (2005). A novel method to immobilize Ru nanoparticles on SBA-15 firmly by ionic liquid and hydrogenation of arene, *Catal. Lett.*, **103**, pp. 59–62.
219. Ma, X., Zhou, Y., Zhang, J., Zhu, A., Jiang, T., and Han, B. (2008). Solvent-free Heck reaction catalyzed by a recyclable Pd catalyst supported on SBA-15 via an ionic liquid, *Green Chem.*, **10**, pp. 59–66.
220. Vangeli, O. C., Romanos, G. E., Beltsios, K. G., Fokas, D., Kouvelos, E. P., Stefanopoulos, K. L., and Kanellopoulos, N. K. (2010). Grafting of imidazolium based ionic liquid on the pore surface of nanoporous materials-study of physicochemical and thermodynamic properties, *J. Phys. Chem. B*, **114**, pp. 6480–6491.
221. Kim, D.-W., Lim, D.-O., Cho, D.-H., Koh, J.-C., and Park, D.-W. (2011). Production of dimethyl carbonate from ethylene carbonate and methanol using immobilized ionic liquids on MCM-41, *Catal. Today*, **164**, pp. 556–560.
222. Bivona, L. A., Fichera, O., Fusaro, L., Giacalone, F., Buaki-Sogo, M., Gruttadauria, M., and Aprile, C. (2015). A polyhedral oligomeric silsesquioxane-based catalyst for the efficient synthesis of cyclic carbonates, *Catal. Sci. Technol.*, **5**, pp. 5000–5007.
223. Montroni, E., Lombardo, M., Quintavalla, A., Trombini, C., Gruttadauria, M., and Giacalone, F. (2012). A liquid–liquid biphasic homogeneous organocatalytic aldol protocol based on the use of a silica gel bound multilayered ionic liquid phase, *ChemCatChem*, **4**, pp. 1000–1006.
224. Zhao, H., Yu, N., Ding, Y., Tan, R., Liu, C., Yin, D., Qiu, H., and Yin, D. (2010). Task-specific basic ionic liquid immobilized on mesoporous silicas: Efficient and reusable catalysts for Knoevenagel condensation in aqueous media, *Micropor. Mesopor. Mater.*, **136**, pp. 10–17.
225. Yang, H., Han, X., Li, G., and Wang, Y. (2009). N-Heterocyclic carbene palladium complex supported on ionic liquid-modified SBA-16: An efficient and highly recyclable catalyst for the Suzuki and Heck reactions, *Green Chem.*, **11**, pp. 1184–1193.

226. Ye, W., He, S., Ding, L., Yao, Y., Wan, L., Miao, S., and Xu, J. (2013). Supported ionic-liquid "semi-heterogeneous catalyst": An interfacial chemical study, *J. Phys. Chem. C*, **117**, pp. 7026–7038.
227. Liang, S., Zhou, Y., Liu, H., Jiang, T., and Han, B. (2010). Immobilized 1,1,3,3-tetramethylguanidine ionic liquids as the catalyst for synthesizing propylene glycol methyl ether, *Catal. Lett.*, **140**, pp. 49–54.
228. Liu, G., Hou, M., Wu, T., Jiang, T., Fan, H., Yang, G., and Han, B. (2011). Pd(II) immobilized on mesoporous silica by N-heterocyclic carbene ionic liquids and catalysis for hydrogenation, *Phys. Chem. Chem. Phys.*, **13**, pp. 2062–2068.
229. Gadenne, B., Hesemann, P., and Moreau, J. J. E. (2004). Supported ionic liquids: Ordered mesoporous silicas containing covalently linked ionic species, *Chem. Commun.*, **15**, pp. 1768–1769.
230. Miao, J., Wan, H., and Guan, G. (2011). Synthesis of immobilized Brønsted acidic ionic liquid on silica gel as heterogeneous catalyst for esterification, *Catal. Commun.*, **12**, pp. 353–356.
231. El Kadib, A., Hesemann, P., Molvinger, K., Brandner, J., Biolley, C., Gaveau, P., Moreau, J. J. E., and Brunel, D. (2009). Hybrid materials and periodic mesoporous organosilicas containing covalently bonded organic anion and cation featuring MCM-41 and SBA-15 structure, *J. Am. Chem. Soc.*, **131**, pp. 2882–2892.
232. Zhao, H., Yu, N., Wang, J., Zhuang, D., Ding, Y., Tan, R., and Yin, D. (2009). Preparation and catalytic activity of periodic mesoporous organosilica incorporating Lewis acidic chloroindate (III) ionic liquid moieties, *Micropor. Mesopor. Mater.*, **122**, pp. 240–246.
233. Chen, W., Zhang, Y., Zhu, L., Lan, J., Xie, R., and You, J. (2007). A concept of supported amino acid ionic liquids and their application in metal scavenging and heterogeneous catalysis, *J. Am. Chem. Soc.*, **129**, pp. 13879–13886.
234. Sun, J., Cheng, W., Fan, W., Wang, Y., Meng, Z., and Zhang, S. (2009). Reusable and efficient polymer-supported task-specific ionic liquid catalyst for cycloaddition of epoxide with CO_2, *Catal. Today*, **148**, pp. 361–367.
235. Zhi, H., Lü, C., Zhang, Q., and Luo, J. (2009). A new PEG-1000-based dicationic ionic liquid exhibiting temperature-dependent phase behavior with toluene and its application in one-pot synthesis of benzopyrans, *Chem. Commun.*, **20**, pp. 2878–2880.

236. Kim, S. Y., Kim, S., and Park, M. J. (2010). Enhanced proton transport in nanostructured polymer electrolyte/ionic liquid membranes under water-free conditions, *Nat. Commun.*, **1**, p. 88.
237. Baudoux, J., Perrigaud, K., Madec, P. J., Gaumont, A. C., and Dez, I. (2007). Development of new SILP catalysts using chitosan as support, *Green Chem.*, **9**, pp. 1346–1351.
238. Meng, Y., Gu, D., Zhang, F., Shi, Y., Yang, H., Li, Z., Yu, C., Tu, B., and Zhao, D. (2005). Ordered mesoporous polymers and homologous carbon frameworks: Amphiphilic surfactant templating and direct transformation, *Angew. Chem. Int. Ed.*, **117**, pp. 7215–7221.
239. Zhang, F., Meng, Y., Gu, D., Yan, Y., Yu, C., Tu, B., and Zhao, D. (2005). A facile aqueous route to synthesize highly ordered mesoporous polymers and carbon frameworks with Ia3-d bicontinuous cubic structure, *J. Am. Chem. Soc.*, **127**, pp. 13508–13509.
240. Zhang, W., Wang, Q., Wu, H., Wu, P., and He, M. (2014). A highly ordered mesoporous polymer supported imidazolium-based ionic liquid: An efficient catalyst for cycloaddition of CO_2 with epoxides to produce cyclic carbonates, *Green Chem.*, **16**, pp. 4767–4774.
241. Shin, J. Y., Kim, Y. S., Lee, Y., Shim, J. H., Lee, C., and Lee, S. G. (2011). Impact of anions on electrocatalytic activity in palladium nanoparticles supported on ionic liquid–carbon nanotube hybrids for the oxygen reduction reaction, *Chem. Asian J.*, **6**, pp. 2016–2021.
242. Campisciano, V., La Parola, V., Liotta, L. F., Giacalone, F., and Gruttadauria, M. (2015). Fullerene-ionic-liquid conjugates: A new class of hybrid materials with unprecedented properties, *Chem. Eur. J.*, **21**, pp. 3327–3334.
243. Zheng, X., Luo, S., Zhang, L., and Cheng, J. P. (2009). Magnetic nanoparticle supported ionic liquid catalysts for CO_2 cycloaddition reactions, *Green Chem.*, **11**, pp. 455–458.
244. Zhang, Q., Yang, F., Tang, F., Zeng, K., Wu, K., Cai, Q., and Yao, S. (2010). Ionic liquid-coated Fe_3O_4 magnetic nanoparticles as an adsorbent of mixed hemimicelles solid-phase extraction for preconcentration of polycyclic aromatic hydrocarbons in environmental samples, *Analyst*, **135**, pp. 2426–2433.
245. Marshall, S. T., O'Brien, M., Oetter, B., Corpuz, A., Richards, R. M., Schwartz, D. K., and Medlin, J. W. (2010). Controlled selectivity for palladium catalysts using self-assembled monolayers, *Nat. Mater.*, **9**, pp. 853–858.
246. Wang, Y., Ren, J. W., Deng, K., Gui, L. L., and Tang, Y. Q. (2000). Preparation of tractable platinum, rhodium, and ruthenium

nanoclusters with small particle size in organic media, *Chem. Mater.*, **12**, pp. 1622–1627.

247. Schrader, I., Warneke, J., Backenköhler, J., and Kunz, S. (2015). Functionalization of platinum nanoparticles with L-proline: Simultaneous enhancements of catalytic activity and selectivity, *J. Am. Chem. Soc.*, **137**, pp. 905–912.

248. Biswas, M., Dinda, E., Rashid, M. H., and Mandal, T. K. (2012). Correlation between catalytic activity and surface ligands of monolayer protected gold nanoparticles, *J. Colloid. Inter. Sci.*, **368**, pp. 77–85.

249. Hu, P., Chen, L., Kang, X., and Chen, S. (2016). Surface functionalization of metal nanoparticles by conjugated metal–ligand interfacial bonds: Impacts on intraparticle charge transfer, *Acc. Chem. Res.*, **49**, pp. 2251–2260.

250. Maity, P., Takano, S., Yamazoe, S., Wakabayashi, T., and Tsukuda, T. (2013). Binding motif of terminal alkynes on gold clusters, *J. Am. Chem. Soc.*, **135**, pp. 9450–9457.

251. Deming, C. P., Zhao, A., Song, Y., Liu, K., Khan, M. M., Yates, V. M., and Chen, S. W. (2015). Alkyne-protected AuPd alloy nanoparticles for electrocatalytic reduction of oxygen, *ChemElectroChem*, **2**, pp. 1719–1727.

252. Chen, W., and Chen, S. W. (2009). Oxygen electroreduction catalyzed by gold nanoclusters: Strong core size effects, *Angew. Chem. Int. Ed.*, **48**, pp. 4386–4389.

253. Zhu, Y., Qian, H. F., Drake, B. A., and Jin, R. C. (2010). Atomically precise $Au_{25}(SR)_{18}$ nanoparticles as catalysts for the selective hydrogenation of α,β-unsaturated ketones and aldehydes, *Angew. Chem. Int. Ed.*, **49**, pp. 1295–1298.

254. Tsukuda, T. (2012). Toward an atomic-level understanding of size-specific properties of protected and stabilized gold clusters, *Bull. Chem. Soc. Jpn.*, **85**, pp. 151–168.

255. Li, G., and Jin, R. C. (2013). Atomically precise gold nanoclusters as new model catalysts, *Acc. Chem. Res.*, **46**, pp. 1749–1758.

256. Negishi, Y. (2014). Award accounts: The chemical society of Japan award for young chemists for 2007: Toward the creation of functionalized metal nanoclusters and highly active photocatalytic materials using thiolate-protected magic gold clusters, *Bull. Chem. Soc. Jpn.*, **87**, pp. 375–389.

257. Wang, Y., Su, H. F., Xu, C. F., Li, G., Gell, L., Lin, S. C., Tang, Z. C., Häkkinen, H., and Zheng, N. F. (2015). An intermetallic $Au_{24}Ag_{20}$ superatom

nanocluster stabilized by labile ligands, *J. Am. Chem. Soc.*, **137**, pp. 4324–4327.

258. Wang, Y., Wan, X. K., Ren, L., Su, H., Li, G., Malola, S., Lin, S., Tang, Z., Häkkinen, H. K., Teo, B., Wang, Q.-M., and Zheng, N. (2016). Atomically precise alkynyl-protected metal nanoclusters as a model catalyst: Observation of promoting effect of surface ligands on catalysis by metal nanoparticles, *J. Am. Chem. Soc.*, **138**, pp. 3278–3281.

259. Kuschel, A., Luka, M., Wessig, M., Drescher, M., Fonin, M., Kiliani, G., and Polarz, S. (2010). Organic ligands made porous: Magnetic and catalytic properties of transition metals coordinated to the surfaces of mesoporous organosilica, *Adv. Funct. Mater.*, **20**, pp. 1133–1143.

260. Park, S. S., Moorthy, M. S., and Ha, C. S. (2014). Periodic mesoporous organosilicas for advanced applications, *NPG Asia Mater.*, **6**, p. e96.

261. Zhou, J., Zhou, R., Mo, L., Zhao, S., and Zheng, X. (2002). MCM-41 supported aminopropylsiloxane palladium (0) complex: A highly active and stereoselective catalyst for Heck reaction, *J. Mol. Catal. A: Chem.*, **178**, pp. 289–292.

262. Rat, M., Zahedi-Niaki, M. H., Kaliaguine, S., and Do, T. O. (2008). Sulfonic acid functionalized periodic mesoporous organosilicas as acetalization catalysts, *Micropor. Mesopor. Mater.*, **112**, pp. 26–31.

263. Li, C., Liu, J., Zhang, L., Yang, J., and Yang, Q. (2008). Mesoporous organosilicas containing disulfide moiety: Synthesis and generation of sulfonic acid functionality through chemical transformation in the pore wall, *Micropor. Mesopor. Mater.*, **113**, pp. 333–342.

264. Guo, X., Qin, Z., Fan, W., Wang, G., Zhao, R., Peng, S., and Wang, J. (2009). Zinc carboxylate functionalized mesoporous SBA-15 catalyst for selective synthesis of methyl-4,4′-di(phenylcarbamate), *Catal. Lett.*, **128**, pp. 405–412.

265. Farjadian, F., Hosseini, M., Ghasemi, S., and Tamami, B. (2015). Phosphinite-functionalized silica and hexagonal mesoporous silica containing palladium nanoparticles in Heck coupling reaction: Synthesis, characterization, and catalytic activity, *RSC Adv.*, **5**, pp. 79976–79987.

266. Gupta, G., Patel, M. N., Ferrer, D., Heitsch, A. T., Korgel, B. A., Jose-Yacaman, M., and Johnston, K. P. (2008). Stable ordered FePt mesoporous silica catalysts with high loadings, *Chem. Mater.*, **20**, pp. 5005–5015.

Chapter 3

Supermolecular Catalysts

Shangbin Jin,[a] Jiang He,[a] Qingmin Ji,[b] Jonathan P. Hill,[c] and Katsuhiko Ariga[c]

[a]*Huazhong University of Science and Technology, Luoyu Road 1037, Wuhan, China*
[b]*Herbert Gleiter Institute of Nanoscience, Nanjing University of Science and Technology, 200 Xiaolingwei, Nanjing 210094, China*
[c]*WPI Center for Materials Nanoarchitectonics, National Institute for Materials Science, 1-1 Namiki, Tsukuba, Ibaraki 305-0044, Japan*

jinsb@hust.edu.cn, jiqingmin@njust.edu.cn

3.1 Introduction

Supermolecular catalysis is not a well-defined field but generally refers to the application of supermolecular chemistry, especially molecular recognition and guest binding toward catalysis. For the term "supermolecular," as J. M. Lehn and F. Vogtle said, in contrast to molecular chemistry, which is predominantly based on the covalent bonding of atoms, supermolecular structures are based on the chemistry of the intermolecular bonds, covering the structures and functions of the entities formed by the association of two or more building blocks [1, 2]. Beyond the chemical molecules, supermolecular catalysts focus on the catalysis system which is made up of a discrete number of assembled molecular subunits and components. Supermolecular chemistry was originally inspired by nature systems, which utilize the reversible interactions, including hydrogen bonding, hydrophobic forces, van der Waals forces, π–π interaction, electrostatic forces, and metal coordination forces. However, at present, the driving forces for the organization of supermolecular catalysts may vary from traditional weak non-covalent bonding to strong covalent bonding. The important concepts of supermolecular chemistry (i.e., molecular self-assembly, molecular recognition, host–guest chemistry, molecular architectures) have also been applied for the construction of the supermolecular catalysis systems [3–6].

The development on supermolecular catalysts has great significance for not only the design of novel catalysis systems but also the understanding of catalysis mechanisms. The binding of reactants into conformations suitable for reactions and lower the transition state energy of reactions is extremely important to improve the efficiency of catalysis. On the other hand, supermolecular encapsulation systems such as capsules, hollow cages, and dendrimers may create microenvironments for cascade catalysis reactions or reactions not possible to progress on a macroscopic scale. In contrast to smart biocatalysis systems, which possess complex structures and certain conformations, supermolecular catalysts offer a simpler model for adjusting the catalytic performance (i.e., dramatically accelerate reaction and allow selective reactions to occur).

3.2 Biocatalysts: Enzymes

Life on Earth depends on chemical transformations, which can enable cellular function and the implementation of basic biological tasks. All these processes rely on enzymes to proceed. Enzymes operate as catalysts, kick-start the process, and occur at the necessary rate. In the beginning of the 20th century, enzymes were successfully isolated and used as biocatalysts outside living cells. As enzymes are capable of catalyzing chemical reactions rapidly, selectively, efficiently and under mild conditions, they are promising solutions to improve the overall sustainability of chemical manufacturing [7, 8].

All known enzymes are bio-macromolecules of proteins, which principally are made of chains of amino acids linked by peptide bonds. One of their most attractive and important properties is their specificity to the relative catalytic reactions. Some enzymes even exhibit absolute specificity for only one particular reaction [9, 10]. The recognition between enzymes and substrates or other ligands can be explained by two basic models: the "lock and key" model and "induced fit" model (Fig. 3.1) [11, 12]. For the "lock and key" model, the lock is the enzyme and the key is the substrate. Only the proper sized substrate can fit into the active site of the enzyme. The active site of the enzyme binds with the substrate by means of weak bonds or by hydrogen bonding and is responsible for the enzyme activity. The "induced fit" model assumes that the substrate plays a role in determining the final shape of the enzyme and the enzyme is partially flexible. The proper substrate is capable of changing the conformation of the active site of the enzyme and form enzyme-substrate complex. Therefore, the substrates too small or too large will not induce the proper alignment of the active site and cannot react. Enzymes enhance the rate of reaction by lowering the activation energy for a chemical reaction. The rate determination step is assumed to pass through a high-energy transition state. Enzymes will lower the energy difference between transition state and ground state, and thus promote the reaction.

Although enzymes exhibit several unique catalytic properties compared to other kinds of catalysts, their applications are still limited due to relatively easy denaturation to temperature, pH, solvents, and some inhibitor molecules. Some enzymes are still very expensive or require expensive additives. With the rapid increase of the knowledge on enzymes during the past decades, and fast development on the molecular synthesis and nanotechnology, it now allows the rational use of enzymes in many processes by biomimetic chemistry and new analytical techniques.

Beyond nature's enzymes, enzymes have evolved to "engineering" era. By the integration of newly developed technologies from rational design and directed evolution, new enzymes have been created with novel catalytic function, given dramatic enhancement in catalytic efficiency, stereoselectivity and stability of chemical synthesis [13, 14]. Here, we focus on the recent research from the three aspects on the design of enzymatic systems for catalysis reactions. We will describe the functionalization of artificial enzymes on the performance of biocatalysis processes, the design of enzymes based on the genetic engineering, and the construction of integrated enzyme systems for multifunctional applications.

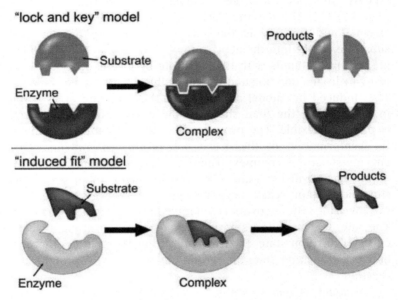

Figure 3.1 The reaction models for enzymes.

3.2.1 Engineering Artificial Enzymes for Selective Catalysis

Enzymes generally use the action of two or more well-placed functional groups to achieve catalysis on the binding substrates. The fitting between the enzyme and the substrate leads to substrate selectivity, reaction selectivity, and stereoselectivity. The bindings may be driven by metal coordinations, ion pairings, Lewis acid–base coordinations, hydrogen bonding or hydrophobic interactions. Natural enzymes are large protein structures of extreme molecular complexity but their mechanism of catalysis is frequently intriguingly simple with only a few amino acids involved in catalysis. Mimicking the structures of natural enzymes, artificial enzymes possess binding cavity and various functional groups are constructed [15, 16].

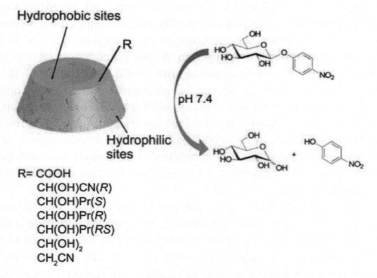

Figure 3.2 Modified β-CD as artificial enzyme for enhancing the hydrolysis reaction of nitrophenyl glycosides.

The most successful artificial enzymes with binding cavities are cyclodextrins (CDs). These water-soluble ring-shaped oligomers are composed of D-glucopyranoside units and form a truncated cone-shaped structure, which imitates the hydrophobic binding pocket typical of many enzymes (Fig. 3.2) [17, 18]. However, these mimics do not incorporate the catalytic groups; thus various functional groups have been introduced into CDs to affect the binding capability with substrates. The free –OH groups can be functionalized to give catalytic activity, while the rest of the CD molecule can act as a cavity for the specific bound of substrates inside. Easton et al. used β-CD itself as a chemzyme for 1,3-dipolar cycloaddition of nitrile oxides to alkynes [19]. They showed that the transient attachment of dipolarophiles to the primary rim of CD can dramatically affect the regioselectivity of products and accelerate the reaction rate up to near 500 times. Bols et al. synthesized CD dicyanohydrin as an artificial glycosidase [20]. The hydrolysis of aryl glycosides by this artificial enzyme was shown much enhanced (6–60 times) in contrast to CD with diacid. They pointed out that because the cyanohydrin 6-OH groups are fixed in the *gauche-trans* (*gt*) conformation pointing toward the binding site, the OH group becomes more acidic, which may facilitate the cleavage of the exocyclic oxygen and thus promote the catalysis reaction. As the cyanohydrin moiety is unstable, later the same group synthesized a substitution of the cyano function with an alternative electron-withdrawing groups [21]. The nitrile group was replaced with a trifluoromethyl group, since these two groups possess similar magnitude of electron-withdrawing effect (CF_3 group has a σ_I of 0.41; CN has a σ_I of 0.56). This modified CD compound was shown successfully to be an artificial enzyme and can increase the rate of hydrolysis of nitrophenyl glycosides with 14–90 folds.

Previous studies have indicated that CDs' host–guest interactions can increase the binding affinity with substrate and the cooperative binding between intramolecular CDs. Due to the chiral nature of the CD cavity, functionalized CDs can promote enantioselective molecular binding synthesis, and mediate chemical and photochemical enantio-differentiating reactions [22]. Cheng and Luo et al. developed chiral catalysts by combining β-CD with various chiral primary amine catalysts [23]. The resulting β-CD enamine catalysts could effectively promote asymmetric direct aldol reactions with excellent enantioselectivity in an aqueous buffer solution. They revealed that the initiation of the reaction is from the binding of substrates into the CD cavity via a synergistic action of hydrophobic interaction and noncovalent interaction with the side chain of the β-CD enamine catalyst. A rate-limiting enamine forming step is then involved, which is followed by the formation of C–C bond for product. The products then subsequently release from the cavity and thus the catalytic cycle is completed. Unlike enzymatic primary amino catalysis, the modified CD works much more favorably under slightly acidic conditions rather than enzymatic neutral conditions.

As a host cavity, CD can bind with guest molecules based on various binding strengths and by different kinetics. The groups on the rims may act as anchors to further built functional CDs with metal ions. The cooperative effects of CDs with coordinated metal ions can enhance the enantioselective catalytic capability. Yuan and Fujita et al. prepared the Ce(IV)–CD complexes by a EDTA-bridged CD dimers [24]. The dimers were proven to complex with Ce more efficiently by the corresponding monomer. They found that the cavity shape of cyclodextrin moieties and their cooperation displayed an important role in amplifying the chemiluminescence. The well-known chemiluminescence reaction of luminol with base and hydrogen peroxide was shown to be remarkably enhanced by this catalyst system. Further modification of either the cyclodextrin rims or the EDTA linker may alter significantly the catalytic abilities of the CD dimers. Wang and Mao et al. synthesized two pyridine-linked bis(β-CD) (bisCD) copper(II) complexes to mimic metallohydrolase and superoxide dismutase (Fig. 3.3) [25]. Hydrolytic kinetic resolution of three pairs of amino acid ester enantiomers at neutral pH indicated that the "back-to-back" configuration of bisCD copper complex contributes to the increase of the catalytic efficiency. The complex exhibited an enantiomer selectivity of 15.7 for N-Boc-phenylalanine 4-nitrophenyl esterenantiomers. They demonstrated that the enantioselective hydrolysis is directed by the cooperative roles of the intramolecular flanking chiral CD cavities and the coordinated copper ion. Shen et al. prepared a glutathione peroxidase mimic based on β-CD complexed with a metalloid telluride. The ditelluride β-CD dimer can catalyze the reduction of cumene peroxide by an aryl thiol 200,000 times more effectively than diphenyl diselenide, which is another similar enzyme mimic [26].

Enzyme engineering employs approaches of either rational design or directed evolution to modify enzyme properties, which is often a time-consuming process. A most exciting development for engineering enzyme is the application of structural genomics (Fig. 3.4) and various computational approaches to manipulate enzyme properties [27–29]. The better understanding on protein structures, dynamics, functions and their correlation reduces the time span for designing new natural enzyme analogues possessing functions or improved properties. Through the engineering, a number of properties of enzymes such as the yield and kinetics of enzyme, the ease of downstream processing and various safety aspects, etc. may be improved. Enzymes from dangerous microorganisms or from unhealthy plants or tissues may alter into safe high-production microorganisms. The acceleration on engineering enzymes may promote the generation of biocatalysts and is expected to broaden their applications more efficiently in complex catalytic processes or industrial processes.

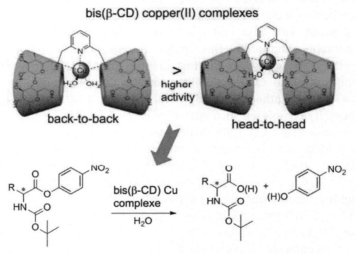

Figure 3.3 The scheme of the bis(β-CD) (bisCD) copper(II) complexes as artificial enzyme to mimic metallohydrolase and superoxide dismutase.

3.2.2 Genetic Engineering to Enzyme Technology

Nature provides a vast amount of enzyme resources. As the total number of microbial cells on Earth is estimated to be 1030, microbial hosts have been treated as a huge natural wealth for protein resources. Microorganisms such as *E. Coli* and *S. cerevisiae* have been used extensively as microbial hosts for the scale-up production of proteins. However, because the majority of microorganisms have not yet been isolated in pure culture, useful characterization of their enzymes remains quite difficult. Various tools have been expanded for new enzymes, which are (i) metagenome screening [30, 31], (ii) genome mining [32], and (iii) exploring the diversity of extremophiles [33]. Metagenomic screening can be based on either function or sequence approaches.

Function-based screening is a straightforward way to isolate genes that show the desired function by direct phenotypical detection, heterologous complementation, and induced gene expression [34]. The sequence-based screening is performed using either the polymerase chain reaction (PCR) or hybridization procedures [35]. The genome mining is based on the vast available information from genome sequence databases, which create large opportunity to find new natural products (including enzymes). Extremophiles are a very interesting source of enzymes with extreme stability under critical conditions. These tools facilitate the easy generation of chimeric enzymes with diverse substrate specificity.

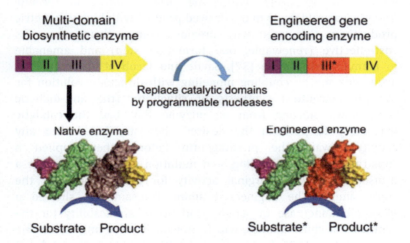

Figure 3.4 The scheme for engineering a multi-domain condensing enzyme by programmable nucleases [29].

Based on novel biosynthesis techniques, currently novel enzymes are generated with improved properties, such as activity and stability at different pH values and temperatures, increased or modified enantioselectivity, altered substrate specificity, stability in organic solvents, novel substrate specificity/activity and increased biological activity of protein pharmaceuticals or biological molecules. For example, proteases in the dairy industry are used for the manufacture of cheese. Calf rennin had been preferred in cheese making due to its high specificity, but microbial proteases produced by GRAS microorganisms such as

Mucor miehei, *Bacillus subtilis*, *Mucor pusillus* Lindt, and *Endothia parasitica* are gradually replacing it. The primary function of these enzymes in cheese making is to hydrolyze the specific peptide bond (Phe$_{105}$-Met$_{106}$) that generates *para*-k-casein and macropeptides [36]. Production of calf rennin (chymosin) in recombinant *Aspergillus niger var. awamori* amounted to about 1 gL^{-1} after nitrosoguanidine mutagenesis and selection for 2-deoxyglucose resistance. Further improvement was achieved by parasexual recombination, resulting in a strain producing 1.5 gL^{-1} from parents producing 1.2 gL^{-1}. 2-Methyl pentanol is an important intermediate for the manufacture of pharmaceuticals and liquid crystals. Gooding et al. developed an evolved ketoreductase enzyme to selectively reduce the (*R*)-enantiomer of racemic 2-methylvaleraldehyde to the desired product with high volumetric productivity. Compared with previous routes, their process is cost-effective (renewable, one formation step) and amenable to manufacturing scale [37]. Savile and Jacob et al. created an active enzyme by coupling modeling with directed evolution for the pharmaceutical synthesis of sitagliptin (the anti-diabetic compound). Starting from an enzyme that had the catalytic machinery to perform the desired chemistry but lacked any activity toward the prositagliptin ketone, they applied a substrate walking, modeling, and mutation approach to create a transaminase with marginal activity for the synthesis of the chiral amine. The engineered amine transaminase showed a 40,000-fold increase in activity and broad applicability for the synthesis of chiral amines, which previously were only accessible via resolution [38]. In comparison with the rhodium catalyzed processes, such biocatalytic process provides sitagliptin with a 10% to 13% increase in overall yield, a 53% increase in productivity and a 19% reduction in total waste. The process eliminated all heavy metals and reduced total manufacturing cost. The enzymatic reaction can be run in multi-purpose vessels, avoiding the need for specialized high-pressure hydrogenation equipment.

Enzyme cascades are also significant interesting systems for chemical biotechnology [39]. Cascade catalysis involves the coupling of more than two enzymes, which enable consecutive transformation of defined low-cost substrates to high-value products in one pot. A representative example of enzyme cascade is the process for the animation of alcohols to amines. Sattler et al. developed a three-enzyme cascade system of alcohol dehydrogenase, transaminase and L-alanine dehydrogenase for a direct conversion of primary alcohols to amines [40]. This system was redox-neutral without the need of external hydride source. This synthetic cascade is highly efficient for the animation of 1-hexanol and 3-phenyl-1-propanol (yield > 99%). By altering the substrate specificity of cascade enzyme, the system can be suitable for the conversion of different alcohols. Zhang et al. designed a non-natural synthetic enzymatic pathway with the maximum yield of 12 mol hydrogen per glucose [41]. This system contains 13 enzymes and works in three steps: (i) generation of phosphorylated monomer sugars such as glucose-6-phosphate; (ii) complete oxidation of glucose-6-phosphate to CO_2 and generate NADPH using enzymes from the pentose phosphate pathway, glycogenesis and glycolysis pathways; (iii) production of molecule hydrogen from NADPH by hydrogenase. This cell-free enzyme system can also be applied to various substrates, which proceeds effective cascade process, analogous to natural systems [42].

3.2.3 Integrated Multifunctional Enzyme Systems

In general, multifunctional enzyme means that the enzyme is functionalized in multiple ways to synthesize more than one kind of products. Many microorganisms possess multifunctional enzymes able to work efficiently for biological processes. These enzymes may be composed of a single polypeptide with several domains, which have multiple catalytic or binding functions. Besides utilizing protein or genome engineering to design multifunctional enzymes, an integrated multifunctional enzymatic system can be constructed by the coupling of diverse enzymes or other catalytic components. With substantial success has been achieved for the directed assembly of building blocks, the complex multifunctional systems can be organized from relatively more simplified and uncomplicated processes. Based on covalent or non-covalent interactions, enzymes and functional motifs can be well assembled into complex and functional architectures for not only biocatalytic reactions but also novel nanodevices.

Alzheimer's disease (AD) is characterized by a loss of brain function affecting memory, cognition, and behavior. Accumulation of misfolded amyloid-β (Aβ) peptide plaques, elevated levels of reactive oxygen species (ROS), and metal ion dyshomeostasis are the major features of AD. There is also a synergistic relationship between oxidative stress and Aβ accumulation. Therefore, effective therapeutic strategies need to simultaneously target Aβ, ROS, and metal ion accumulation [43, 44]. Proteases and superoxide dismutase (SOD) were found to have enzymatic functions for decreasing the total accumulation of Aβ peptides and Aβ-induced ROS, respectively. However, these enzymes have distinct subcellular localizations and are mediated by different cell signaling pathways, which makes it difficult to upregulate or transport intracellular native enzymes simultaneously in the disease state [45]. Qu et al. first reported the use of an integrated nanozyme with both protease and SOD-like activities for AD treatment [46]. Their design of the nanozyme was inspired from native serine protease. Polyoxometalate (POM) with Wells–Dawson structure (POMD) (for inhibition in vitro aggregation of Aβ) and hepta-peptides with the SOD-like activity were bound on the surface of gold nanoparticles (Au NPs) by steps. The AuNPs serve as a scaffold to create coupled POMD-peptide compound, which facilitate electron transfer between POMD and the hepta-peptides. In addition, the capability for blood–brain barrier (BBB) penetration is also improved, which is crucial for medical therapeutic applications of neurodegenerative disease. Their results indicated that this enzyme system possesses low toxicity and can exhibit excellent protease activities, SOD-like functionality (enhanced by chelated copper), and metal-ion chelation capability. It showed great potential to act as multifunctional therapeutic agents for the treatment of AD.

Over the past few decades, unprecedented progress had been made in the diagnosis and detection of many diseases, especially various forms of cancers. Wang and Xu prepared a dual-enzyme-loaded multifunctional hybrid nanogel system for dual-modality of pathological responsive ultrasound (US) imaging and enhanced T2-weighted magnetic resonance (MR) imaging [47]. Their system was composed of ROS-responsive enzymes (catalase and superoxide dismutase) and a polysaccharide cationic polymer (glycol chitosan) nanogel functionalized with superparamagnetic iron oxide nanoparticles (SPIO NPs) (Fig. 3.5). The hydrophilic/hydrophobic interactions and electrostatic attraction were driving forces for binding enzymes and SPIO NPs in the gel. This system can react with ROS in the biological system and generate bubbles for enhanced US imaging, while also accumulating in high amounts in acidic environments for enhancing T2-weighted MR imaging. The in vivo results indicated that signals obtained from US imaging from tumors 1 h after injection were enhanced approximately sevenfold. In the case of MR imaging, significantly darkened tumor signals with approximately 16% enhancement can be observed. As it is a fluid-like transport system, the hybrid nanogel can be a perfect candidate as efficient and safe probe for biocatalysis, biomedical diagnosis, and therapy.

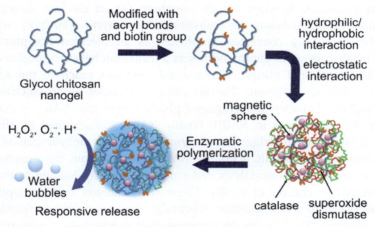

Figure 3.5 The scheme of dual-enzyme-loaded multifunctional hybrid nanogel system for biological imaging.

Quantum dots (QDs) have gained great interest in both fundamental research and technical applications owing to their unique size-dependent physical and electronic properties [48]. Eychmüller et al. developed enzyme-encapsulating quantum dot gels for both biocatalysis and biosensing [49]. The enzyme tyrosinase was mixed into the gel of CdTe QDs. The networks of the QDs can effectively serve as "cage" for the enzymes, preventing them from leaching. With both quantum confinement effect and high enzymatic activity well retained, the enzyme-encapsulating CdTe QD hydrogel serves simultaneously as signal-transforming and signal-recording unit in a biosensor. A more sensitive detection (four times better than CdTe QD sol) of dopamine was achieved. The sensing response can also be reproducible for months. This kind of system offers great potential in the development of various enzyme-based fluorescence biosensors and portable sensing devices.

The photocatalyst-enzyme coupled system is one of the most promising platforms for solar energy conversion by the photosynthesis of organic chemicals or fuel. Baeg et al. fabricated a photocatalyst-enzyme coupled artificial photosynthesis system in production of formic acid from CO_2 [50]. The graphene-based photocatalyst-enzyme coupled system was constructed by covalently bonding the chromophore (as an electron donor) and multianthraquinone substituted porphyrin (MAQSP) on the chemically converted graphene (as an electron acceptor). The absorption of photon occurs as a transition between localized orbitals around MAQSP. The created electrons reach to the Rh complex via graphene. The Rh complex shuttles as an electron mediator between the graphene photocatalyst and NAD^+, which is then converted to NADH. Finally, NADH is consumed by the CO_2 substrate to trigger its enzymatic (formate dehydrogenase) conversion to formic acid. The NAD^+ released from the enzyme can undergo photocatalysis cycle in the same way, leading to the photoregeneration of NADH. These two catalysis cycles couple integrally to work together effectively, ultimately yielding formic acid from CO_2. The results demonstrated successfully a new and promising graphene photocatalyst-based artificial photosynthetic system for the ultimate goal of solar energy utilization in tailor-made synthesis of fine chemicals and solar fuel from CO_2.

A remarkable progress has been envisaged in the field of enzyme systems and a great future for the upcoming artificial enzymes and complex multifunctional enzyme systems is prospected. Therefore, the study of tailored biocatalysts with reengineering enzymes and architectured enzyme systems remains an active and interesting research area. However, biocatalysis systems that are capable of performing pericyclic transformations with high levels of efficiency and selectivity are still challenging. Design of powerful catalysts that can achieve defined geometrically directed multiple functionalizations, good binding properties, and more environment-friendly processes is expected.

3.3 Metal-Organic-Framework (MOF)-Based Catalysts

Metal organic frameworks (MOFs) are crystalline porous solids composed of a coordinated network of metal ions with organic molecules. The spatial organization of these repeated coordination entities leads to a system of channels and cavities in the nanometer length scale. Correct selection of the structural units and the linkage pathways allows systematic modification of the pore structure of MOFs. Over the past decade, the elevated surface area and pore volume, and the flexibility of pore design, which are characteristics of MOFs, have sparked research aimed mainly at preparing new MOF structures and studying their applications in gas storage, separation, sensing, delivery and also including catalysis.

With containing uniform pores of molecular dimensions, MOFs are regarded as analogous to zeolites, which are crystallines made by aluminasilicates. Similar to zeolites, most syntheses of MOFs are based on hydrothermal or solvothermal techniques. In contrast to zeolites (purely inorganic materials), MOFs are constructed from bridging organic ligands. Since ligands in MOFs typically bind reversibly, the slow growth process often allows defects to be redissolved, resulting in millimeter-scale crystals with a near-equilibrium defect density. The coordination of additional metal ions to sites on the bridging ligands, and addition or removal of metal atoms to the metal site may also be allowed to execute, which facilitate the post-functionalization on the porous structures and might not be achieved easily by conventional synthesis.

It was known that zeolites play a major role in a range of catalytic applications, especially for the reactions in gas and petroleum refining. As a nanoporous material similar to zeolites, acting as catalysts is one of the earliest proposed applications for MOFs. Due to more flexibility on the manipulation of porous features and functionalities, MOFs have been considered to outperform zeolite materials in catalysis. Compared with standard zeolite and MOF material for a same sample reaction, the zeolites start to lose their efficiency after an hour, and are generally completely useless after 5 h, while the MOFs keep active after 6 h, and reach 99% yields after 2 h [51].

Besides possessing large surface areas and uniform pore sizes, which are relevant features of catalysts, MOFs may also exhibit distinguished and important characters on morphological functions in contrast to zeolites (Fig. 3.6). Since organic ligands are involved in the frameworks, MOFs can be synthesized in much greater chemical variety than zeolites. Due to the construction based on the interconnection between metal ions and organic linkers, MOFs may be weaker on structural stability than zeolites, but still competitive for specific reactions under milder conditions.

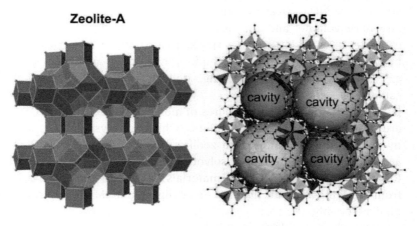

Figure 3.6 The comparison of zeolite (Zeolite-A) and MOF (MOF-5).

During the past two decades, a substantial foundation of MOFs on their synthesis, characterization, and applications has been developed. The chemical and structural versatility from a significant amount of MOFs gives them a great potential to act as candidates for new and unique catalysis [52]. To develop unique catalytic systems with these remarkable materials, various strategies have been prospected and are evaluated in rapid fashion. Here, we briefly summarize the strategies developed for the novel catalysis systems related to MOFs materials. The control of metal framework for MOF catalysts, post-functionalization of MOFs for selective catalysis, encapsulation of catalytic components in MOFs, and MOFs as platforms for the creation of inorganic catalysts are discussed.

3.3.1 Metal Framework for MOF Catalysts

MOFs are a class of compounds constituted by the binding of metal ions or clusters with organic molecules possessing coordinating moieties, such as amines, pyridines, carboxylates, sulfates, and phosphates. They form crystalline materials and may give one-, two-, or three-dimensional structures with a cavity size varying from microporous (smaller than 2 nm) to mesoporous (between 2 and 50 nm). Due to the inherent geometric regularity, the exact chemical composition and position of the atom in the frames can be estimated and a more homogeneous distribution of the active sites in MOFs can be easily achieved. With the use of different organic building blocks to combine with various inorganic ones, almost infinite possible structures may be generated. MOFs based on various compositions of the catalytic sites and organic building units are explored for catalysis applications (Fig. 3.7).

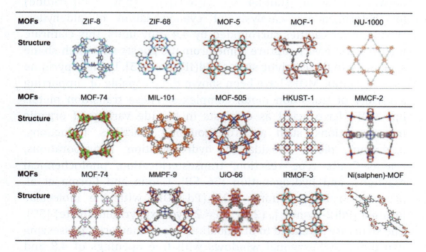

Figure 3.7 The MOFs of various morphologies for catalysis applications.

One of the most important properties of MOFs is their high porosity and high specific surface area, which are important features for the application as catalysts. The research showed that the theoretical upper limit for MOF surface areas may be

14600 m^2/g. Yaghi and Kim et al. reported the surface area of MOF based on Zn$_4$O(CO$_2$)$_6$ unit can reach 10400 m^2/g [53]. Unlike zeolites, whose pore size rarely exceeds 1 nm without significant loss in stability, the pore size of MOFs can be relatively large through the systematic expansion. It has been reported that the pore aperture of MOFs can be attained to the size regime of >32 Å [54], which make MOF possess higher diffusivity in contrast to zeolites. The larger pores may allow molecules to enter and exit MOFs more easily during catalysis process, prevent the reaction overheating, and thus protect MOF framework for long effective life. The high crystalline structure of MOF framework will make a homogeneous distribution of metal active sites. With the fine design on the organic linkers and metal atoms, the appropriate porous structures and the nature of the active sites can be regulated for liquid phase reactions, which is characteristic of fine chemistry.

The first reports of MOF-based catalysis might be reported by Fujita and Ogura et al. They synthesized a 2D square network material {[Cd(4,4'-bpy)$_2$](NO$_3$)$_2$)}$_\infty$ (bpy = bipyridine) for heterogeneous catalysis of cyanosilylation of aldehydes. Having inner cavities surrounded by 4,4'-bpy units, this clathrate material can easily capture some aromatic guests with high shape specificity to the catalytic sites of Cd(II) ions [55]. Using bipyridine as linker units, MOFs have been shown to be capable of supporting a variety of transition metal complexes. These transition metal-bpy-MOFs have acted as catalysts in a wide variety of organic transformations, such as cross-coupling reactions, reductions, oxidations, olefin epoxidation, hydrosilylation, hydroborations, Friedel–Crafts type reaction, transesterification, photochemical reactions, and hydrogenation [56–58]. Garcia and Corma et al. prepared a Pd-containing MOF (Pd-MOF) with the molecular formula [Pd(2-pymo)$_2$]$_n$ (2-pymo = 2-hydroxypyrimidinolate) [59]. The structure is related to 3D sodalite-type frameworks, possessing two different hexagonal windows with free openings of 4.8 and 8.8 Å and with a fraction of 42% of the crystal volume available to adsorbed guests. The Pd-MOF exhibits catalytic activity derived from tetracoordinated Pd ions and was found to be an active catalyst for alcohol oxidation, Suzuki coupling, and olefin hydrogenation. In sharp contrast with many other MOFs as catalysts, Pd-MOF has another remarkable advantage of inertness to moisture, which can be used in aqueous medium.

MOF-5 is an archetypal cubic MOF formed from Zn_4O nodes with 1,4-benzodicarboxylic acid struts. This compound apparently catalyzes the Friedel–Crafts tert-butylation of both toluene and biphenyl (butylating agent = tert-butyl chloride). Furthermore, *para*-alkylation is strongly favored over *ortho* alkylation, a behavior thought to reflect the encapsulation of reactants by the MOF. The catalytic activity is tentatively ascribed to the encapsulated zinc hydroxide clusters or to a hydrolytically degraded form of the parent material. Han et al. used MOF-5 in the presence of quaternary ammonium salts (Me_4NCl, Me_4NBr, Et_4NBr, $n-Pr_4NBr$, $n-Bu_4NBr$) to catalyze the coupling reaction of CO_2 with propylene oxide (PO) to produce propylene carbonate (PC) [60]. It was discovered that MOF-5 and quaternary ammonium salts had excellent synergetic effect in promoting the reaction. Cyclic carbonates can be produced under mild conditions in high yield. The coupling reaction is initiated by coordination of the Zn_4O clusters (as a Lewis acidic site) in MOF-5 with the oxygen atom of the epoxide, which can activate the epoxy ring. Then the Br^- generated from $n-BuN_4Br$ attacks the less-hindered carbon atom of the coordinated epoxides, followed by the ring-opening step. The oxygen anion of the opened epoxy ring can interact with the CO_2 and form an alkylcarbonate anion, which is converted into the corresponding cyclic carbonate through the ring-closing step. Meanwhile, the catalyst is regenerated. The catalytic reaction is not involved any organic solvent and the catalyst can be reused. It presented a greener, inexpensive, active, selective and stable catalytic system for the synthesis of cyclic carbonates from CO_2 and epoxides.

Férey et al. discovered mesoporous chromium(III) terephthalate CrMIL-101 and an iron(III) analogue, Fe-MIL-101 as catalysts [61]. This type of MOF has a rigid crystal structure, consisting of quasi-spherical cages of two modes (2.9 and 3.4 nm) with accessible windows of ca. 1.2 and 1.6 nm. They have fairly good resistance to common solvents and thermal stability (Cr-MIL-101 up to 300°C and Fe-MIL-101 up to 180°C). They were found to be active in the cyanosilylation, oxidations, and coupling reactions. Allylic oxidation of cyclic alkenes is of considerable value for the production of α,β-unsaturated alcohols and ketones, which are key intermediates in the manufacture of fine chemicals.

Kholdeeva et al. studied Cr-MIL-101 as a heterogeneous catalyst for the liquid-phase allylic oxidation of alkenes using tert-butyl hydroperoxide (TBHP) as oxidant [62]. The Cr-MIL-101 demonstrated high catalytic activity and product selectivities in the oxidation of a range of alkenes. The selectivity towards α,β-unsaturated ketones reaches 86–93%. The Cr-MIL-101 is stable for chromium leaching. It behaves as a true heterogeneous catalyst, and can be easily recovered without loss of catalytic performance.

Zeolite imidazolate frameworks (ZIFs) are also an important subclass of MOFs. They consist of tetrahedrally coordinated transition metal ions (e.g., Fe, Co, Cu, Zn) connected by imidazolate linkers. Since the metal–imidazole–metal angle is similar to the Si–O–Si angle in zeolites, ZIFs are thought to combine the advantages of both zeolites and conventional MOFs. ZIFs are chemically and thermally stable materials. ZIF-8 shows high thermal stability (up to 550°C) and remarkable chemical resistance to boiling alkaline water and organic solvents. Chizallet et al. studied ZIF-8 as a catalyst for the transesterification of vegetable oil [63]. They pointed out that the acid-basic sites at the external the surface (OH and NH groups, hydrogenocarbonates, low-coordinated Zn atoms, and free N-moieties belonging to linkers) play important roles in the catalytic activity. The prevailing role of Zn(II) species as acid sites, combined with N-moieties and OH groups as basic ones, determines the catalytic properties of ZIF-8. Phan et al. used ZIF-8 as an efficient heterogeneous catalyst for the Knoevenagel reaction [64]. By studying the condensation of benzaldehyde with malononitrile to form benzylidene malononitrile as the principal product, they found that the ZIF-8 catalyst was quite active in the Knoevenagel reaction. The reaction rate was higher than some previously reported Lewis acid catalysts, where longer reaction time or/and higher catalyst loading were required. The ZIF-8 catalyst can also easily isolated from the reaction mixture by simple filtration and reused without significant degradation in activity.

Zr-based MOF are notable porous crystalline materials, consisting of discrete $Zr_6O_4(OH)_4$ clusters linked by polycarboxylates. UiO-66/67/68 (UiO: University of Oslo) are Zr-based MOFs, which were first reported by Lillerud et al [65]. They have attracted great interest because of their remarkable stability at high temperatures and high pressures or in the presence of different solvents, acids, and bases. UiO-66 is obtained by connecting $Zr_6O_4(OH)_4$ inorganic cornerstones using 1,4-benzene-dicarboxylate (BDC) as linker, while UiO-67 using the longer 4,4′-biphenyl-dicarboxylate (BPDC) as linker. In the UiO-66 framework, terephthalate defects or missing linkers can create accessible Zr-sites for catalysis. De Vos et al. showed some UiO-66 MOFs acted as highly selective catalysts for the cross-aldol condensation between benzaldehyde and heptanal [66]. They prepared amino-functionalized UiO-66(NH_2) by simple replacement of terephthalic acid (linker) in UiO-66 with 2-aminoterephthalic acid (Fig. 3.8). UiO-66 and UiO-66(NH_2) have BET surface area of 891 m^2g^{-1} and 1206 m^2g^{-1}, respectively, indicating the highly accessible surface area of both materials. UiO-66 showed high and long-term (at least 3 month at 433 K) thermal stability. It collapses only at temperatures above 723 K, while UiO-66(NH_2) is stable up to 548 K. Both UiO-66 and UiO-66(NH_2) are also moisture stable. A maximum 85% and 93% reaction yield of selective synthesis of jasminaldehyde can be achieved for UiO-66 and UiO-66(NH_2). The transformation of the $Zr_6O_4(OH)_4$ unit to the Zr_6O_6 core opens potential access to one coordination site per Zr atom. This creation of Lewis acid sites may enhance catalytic activity of UiO-66 after dehydroxylation, activating either heptanal or benzaldehydeis and thus increasing the reaction rate.

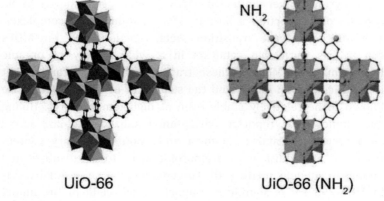

Figure 3.8 UiO-66 and UiO-66 (NH_2).

3.3.2 Post-Functionalization of MOFs for Selective Catalysis

The potential of MOFs in catalysis applications depends on the presence of chemical functionality either at the organic linkers and/or at the metal nodes. However, the preparation of highly functionalized MOFs has been largely limited by the solvothermal synthetic approaches. Fortunately, the reactive moieties of the linkers and/or nodes make functionality can be introduced into MOFs through post-synthetic strategies. The first report on chemically alternation of MOFs by the utility of postsynthetic approach was from the Cohen group in 2007. They proposed "postsynthetic modification" (PSM) for the reaction of basic isoreticular metal-organic framework (IRMOF-3) with acetic anhydride [67]. Subsequently, PSM is extensively developed for functional-group interference during MOF assembly [68, 69]. The integration of functional groups in MOFs by PSM can be via

(i) modification of linkers including covalent modification, deprotection of linker functionality, electron addition and concomitant incorporation of charge compensating ions;
(ii) modification of metal-containing nodes including incorporation of non-framework ligands, grafting to oxygen atoms in metal-oxide nodes, attachment of metal ions or complexes at node oxygen sites.

Bipyridyl linker units (bpy-MOFs) have been shown to be capable of supporting a variety of transition metal complexes. The utility of these transition metal–functionalized bpy-MOFs can act as recyclable catalysts in a wide variety of organic transformations. Most of these transformations are carried out in aprotic reaction media and the structures of the bpy-MOFs are known to be relatively stable even at high temperatures. Huang and Stanley et al. reported Pd(II)-functionalized MOF-253 acting as a recyclable catalyst to form all-carbon quaternary centers via conjugate additions of arylboronic acids to β,β-disubstituted enones in aqueous media [70]. The post metalation with Pd(OAc)$_2$ in MOF-253 has a significant impact on the yield of the model reaction of phenylboronic acid with 3-methylcyclohex-2-en-1-one.

Increasing the total Pd content in the reaction with the amount of phenylboronic acid may lead to a high production yield up to 99%. Van Der Voort et al. reported the catalytic performance of a Ga-based MOF (denoted as COMOC-4, same topology as MOF-253) with PSM by Cu ions. The catalytic performance of the resulting Cu^{2+}@COMOC-4 material was investigated for the aerobic oxidation of cyclo-hexene in the presence of the co-oxidant isobutyraldehyde. In comparison to Cu-BTC (BTC=benzene-1,3,5-tricarboxylate), Cu^{2+}@COMOC-4 shows the best catalytic performance in terms of selectivity towards cyclohexene oxide. Furthermore, no leaching of either Ga or Cu species was detected over four successive runs, which indicates the good stability and reusability of the catalyst [71].

Chirality has become increasingly significant in many fields such as pharmaceutics, chemical industry, agriculture, and clinical analysis. Even for the same substance, enantiomers may behave differently in the body or environment. Chiral MOFs are attractive candidates as heterogeneous asymmetric catalysts for the enantioselective process and producing optical pure fine chemicals [72, 73]. PSM may bring more choices on the design of MOFs bearing chiral functionality. It allows the fabrication of homochiral porous MOFs from either homochiral porous MOFs or achiral porous MOFs. Via the open channels or cavities, the chiral functions can be accessed and address the applications in chirotechnology.

Kim et al. fabricated chiral catalytic MOFs from an achiral framework of chromium based MIL-101 by post-synthesis and successfully used the chiral MOF for heterogeneous catalysis of the asymmetric aldol reactions (Fig. 3.9) [74]. By attaching L-proline-derived catalytic units to the open metal coordination sites, chiral units are incorporated into MIL-101 with keeping the parent framework intact. The new chiral MOFs, CMIL-1, $[Cr_3O(L1)_{1.8}(H_2O)_{0.2} F(bdc)_3]\cdot 0.15(H_2bdc)\cdot H_2O$ showed remarkable catalytic activity in asymmetric aldol reaction. For the reactions between various aromatic aldehydes and ketones, excellent yields of 60–90% and good enantioselectivity for R-isomers with 55–80% enantiomeric excess (*ee*) can be achieved. The enantioselectivity is even higher than the chiral ligands themselves.

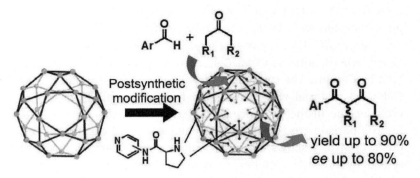

Figure 3.9 The formation of chiral MOF (MIL-101) by post synthesis [74].

Lin et al. created a strategy for the fabrication of catalytically active chiral porous MOFs by utilizing chiral bridging ligands containing orthogonal functional groups [75]. Axially chiral bridging ligand, which contains the bipyridyl primary functionality and orthogonal chiral dihydroxy secondary functionality, was used to construct homochiral porous MOFs. By PSM with Ti(O^{i}Pr)$_4$ to the chiral dihydroxy groups, the compound can afford active catalysis sites for the addition of ZnEt$_2$ to aromatic aldehydes and lead to the formation of chiral secondary alcohols. It showed a complete conversion of the addition of ZnEt$_2$ to 1-naphthaldehyde and the production of (R)-1-(1-naphthyl)propanol with 93% ee.

It has been demonstrated that the catalytic activity of the coordinatively saturated UiO-66 is due to the missing linkers in the structure. The activity of the structures can be significantly increased by using electron-withdrawing groups on the terephthalate linkers [76]. De vos et al. reported that a much more active UiO-66 catalyst can be created by the modification of the framework with modulators, like trifluoroacetic acid (TFA) [77]. The crystalline structure of UiO-66-X can be well maintained as small intergrown particles (~300–600 nm) after the addition of different equivalents (X) of TFA. The facile deprotonation of the strongly acidic TFA may make trifluoroacetate be incorporated both at the crystal boundaries and on the hexanuclear cluster inside the structure. The resulting charge imbalance on the opposing cluster may be balanced by replacing an OH$^-$ group in the [Zr$_6$(OH)$_n$O$_{8-n}$]$^{(8+n)+}$ cluster with an O^{2-} group. Because of the volatile and labile nature of TFA (b$_p$ = 72°C), thermal removal of the trifluoroacetate with a thermal treatment creates even more open sites (Fig. 3.10).

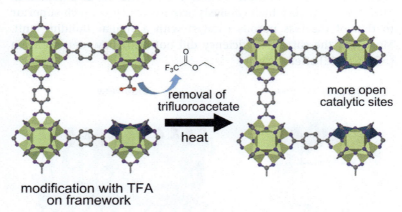

Figure 3.10 The thermal treatment on the modified MOF for more open catalytic sites [76].

N-heterocyclic carbenes (NHCs) have been used to combine with diverse metal atoms for a number of applications because they are both excellent s-donors and weak p-acceptors. NHC-based catalysts are highly active and very stable under various conditions and are much superior to the phosphine-based catalysts [78–80]. It has been reported that the MOF materials based on NHCs or their precursors of azoliums have interesting properties which are associated with the distinctive functions of MOFs and NHCs [81, 82]. Wu et al. presented the creation of a tubular MOF by functional NHCs for catalytic applications [83]. The metal-organic nanotube (MONT) was synthesized by linking up the bent organic ligands and the tetra-coordinated Zn cations under mild conditions. The MONT has a very large exterior wall diameter of 4.91 nm and an interior channel diameter of 3.32 nm (Fig. 3.11). Interlocking of the nanotubes gives rise to a 3D chiral framework containing 1D helical cylindered channels with diameter of 2.0 nm. After post-modification to deprotonate the imidazolium moieties, NHCs were formed in the 1D channels. The further treatment with Pd(OAc)$_2$ resulted the formation of Pd-carbene derivative in the MONT. Pd(II)-NHCs functionalized MONT have very interesting property by synergizing the functionality of nanotubes, MOFs, and NHCs. In the Suzuki–Miyaura cross-coupling, Heck coupling reaction and reduction of C–C multiple bonds, the Pd(II)-NHC-functionalized MONT showed very high catalysis activity, which is much superior to that of the less porous catalyst with a similar building unit. Moreover, the catalytic efficiency did not also show deterioration (>99% yield) after recovery.

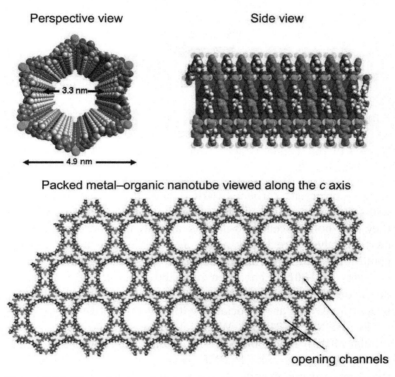

Figure 3.11 The scheme of chiral metal organic nanotube with large exterior wall diameter of 4.91 nm and an interior channel diameter of 3.32 nm [83].

Recently, a new post-synthesis route called building block replacement (BBR) has been also exploited. BBR involves replacement of key structural components of the MOF, including solvent-assisted linker exchange (SALE), non-bridging ligand replacement, and transmetalation at nodes or within linkers. BBR has opened up a very broad strategy for the synthesis of isostructural MOFs with functionalities. It also may provide access to MOFs featuring gradient compositions and chemical properties [84].

Cohen and Ott et al. described the incorporation of an organometallic Fe$_2$ complex that bears structural resemblance to the active site of [FeFe] H$_2$-ases into an MOF [85]. [FeFe](bdt)(CO)$_6$ (bdt = benzenedithiolate) has previously been shown to be an effective proton reduction catalyst in electro- and photochemical schemes. The decoration with two carboxylates at the bdt ligand makes the thermally unstable Fe$_2$ complex capable of being introduced into MOFs by solvent-assisted linker exchange with 1,4-benzenedicarboxylate (BDC) ligands in the parent MOF (Zr(IV)-based UiO-66 MOF). UiO-66-[FeFe](dcbdt)(CO)$_6$ was found to be a highly active hydrogen production catalyst in photochemical arrays with [Ru(bpy)$_3$]$^{2+}$ as a photosensitizer and ascorbate as an electron donor. The catalytic performance exceeds that of an analogous homogeneous reference system. Incorporation of the Fe$_2$ complex in UiO-66-[FeFe](dcbdt)-(CO)$_6$ can enhance the stability under the photocatalysis conditions, protecting reduced species from nonproductive charge recombination, and may promote disproportionation reactions to produce catalytically active dianion I^{2-}.

3.3.3 Encapsulation of Catalytic Components in MOFs

MOFs possess low framework density and high pore volume relative to other porous matrices. They thus can provide externally accessible nanosized cavities and channels for the further incorporation of substrates inside the crystal to facilitate the heterogeneous catalytic action of MOFs. Despite the functionalization of MOF structures through chemical synthesis or chemical interaction, the incorporation of catalytic active metal ions, guest molecules (via host–guest doping), or nanoparticles may also be an effective pathway to tailor functionality of MOFs for catalysis.

Catalytic oxidations based on ambient air are of considerable interest because they promise the removal of many toxic or odorous molecules using only air without needing energy sources, solvents, or other reactives. Oxidizable undesirable compounds include toxic industrial chemicals, chemical warfare agents, indoor air contaminants, and many environmental pollutants [86]. The incorporation of particular polyoxometalates (POMs) in large-pore MOFs has shown high catalytic activity for aerobic oxidations [87, 88]. To ensure optimal POM retention in MOFs, it requires proper MOF apertures for fitting the radius of the POM (with counterions). Hill et al. reported a MOF (MOF-199) with pore sizes closely matching a POM $[CuPW_{11}O_{39}]^{5-}$ guest. The MOFs self-assemble around the POMs and make the material have significantly different properties than either the POM or MOF component alone. The close matching of the MOF's pore size to POM's diameter makes this POM-MOF material, $[Cu_3(C_9H_3O_6)_2]_4$ $[\{(CH_3)_4N\}_4CuPW_{11}O_{39}H]$, show a substantial synergistic stabilization for both the MOF and the POM. Electrostatic interactions between the POMs and MOFs may also activate POM and increase the reduction potential. Therefore, a dramatic increase in the catalytic turnover rate of the POM for air-based oxidations was achieved. The detoxification of various sulfur compounds, including H_2S to S_8 under only ambient air, can be effectively performed by this POM-MOF catalyst.

Hatton et al. reported the combination of Cr(III) terephthalate MOF (MIL-100) with phosphotungstic acid (PTA) in water and its application for a variety of catalytic reactions [89]. Just by impregnating PTA in already synthesized MIL-101 in aqueous media, novel structures of the MIL-101/PTA composites with Keggin ions of PTA being aligned within the MIL-101 pores (small) and cavities (large) can be formed (Fig. 3.12). MIL-101/PTA composites were shown to promote the catalysis of Baeyer condensations and epoxidation reactions with high turnover frequency (TOF). For example, in the reaction for β-caryophyllene oxide synthesis, the TOF estimates approximately 22-fold higher than the TOF catalyzed by titanium-monosubstituted Keggin heteropolyanions loaded into MIL-101 cages. It was suggested that this improvement on the reaction selectivity is due to the immobilization of PTA anion within the MIL-101 cages. As the immobilization shields the acidic character of PTA anion, it thus may provide protection effect on the acid-sensitive epoxide from undergoing the ring-opening reaction.

Noble metal nanoparticles (NMNPs) usually show good catalytic performance. Preventing the sintering of NMNPs during reactions, minimizing their usage amounts, and prolonging the life-time have been a major research objective in catalyst discovery and optimization. Therefore, loading NMNPs in suitable supports is one of the most important pathways for optimizing their applications as the heterogeneous catalysts. MOF micro-crystals or nanocrystals can be good supports and are capable of stabilizing nanosized NMNPs or clusters. Efforts have been made for loading various metals (Pd, Ru, Cu, Pt, Ni, Ag, Au, etc.) in MOFs (ZIF-8, MOF-74, MIL-101, MOF-177, etc.).

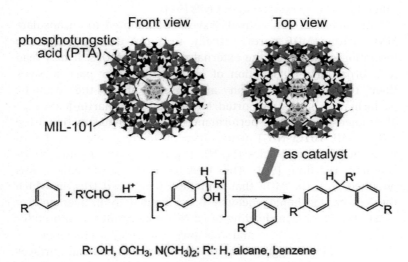

Figure 3.12 The scheme of MIL-101/PTA composite for the catalysis of solvent-free Baeyer reactions.

Choosing stable porous MOFs under experimental conditions is the most important step for utilizing MOFs as supports. In general, embedding metal nanoparticles (MNPs) within MOF pores is achieved by metal precursor infiltration followed by reduction or decomposition of the precursors to metal(0) atoms. It might require MOF supports to possess suitable pore structures and enough thermal/chemical stability. Fischer and co-workers prepared MOF-immobilized MNPs by using a solvent-free CVD technique to load organometallic precursors and subsequent molecular H_2 reduction [90]. They infiltrated a number of volatile organometallic precursors into different MOF pores by the sublimation process and developed a series of MNP@MOF nanocomposites after H_2 reduction. The CVD techniques have the merit of controlling the contents of metal precursors including high loading of metal precursors. It thus can avoid some of the deficiencies of liquid-phase (aqueous/organic solvents) impregnation of guest molecules in MOFs. De Vos et al. prepared Ag@MOF-5 by photocatalytic reduction of Ag(I). When an ethanolic solution of AgNO$_3$ is stirred in the presence of MOF-5 under UV irradiation, Ag NPs can be formed in MOF-5. This method is unique to fabricate MNPs in a specific part of the MOF crystal, leaving the other part of the crystal without NPs [91].

Although several methods have been adopted to encapsulate MNPs within MOF pores, actually, it is not easy to avoid the formation of MNPs on the external surfaces of MOFs. It may lead to a broad size distribution of MNPs with larger particle sizes than the pore dimension and might affect the catalytic performance of the supported MNPs. Zou and Martín-Matute et al. studied the catalytic performances of the amino-functionalized MOFs (MIL-101Cr-NH$_2$) with different loading amounts (4–16 wt%) of Pd nanoparticles (Pd NPs) in the Suzuki–Miyaura cross-coupling reaction [92]. The best catalytic performance was obtained with the MOF that contained 8 wt% Pd NPs, in which the NPs can be homogeneously distributed with an average size of 2.6 nm. The higher loading of Pd NPs may result into a broader size distribution in MOFs and thus lower the catalytic efficiency.

To avoid the aggregation of MNPs on the external surfaces of the MOF and make the encapsulation of MNPs more effectively, various strategies have been explored. Introduction of functional groups to the frameworks of MOFs or on the surface of MNPs can facilitate the confinement of the MNPs in the interior of MOFs by enhanced surface interactions. Hupp and Huo et al. developed an encapsulation strategy that allows several types of nanoparticles to be incorporated fully within a readily synthesized MOF, ZIF-8, in a well-dispersed fashion (Fig. 3.13) [93]. They modified nanoparticle surfaces with polyvinylpyrrolidone (PVP) and then carried out the crystallization of ZIF-8 in methanol at room temperature in the presence of PVP-modified nanoparticles under optimized experimental conditions. In contrast to previous reports on nanoparticles used as seeds to induce the nucleation of MOF crystals, their strategy relies on the successive adsorption of nanoparticles onto the continuously forming surfaces of the growing MOF crystals. This allows readily control over the spatial distribution of various nanoparticles or nanostructures. Pt (2.5, 3.3 and 4.1 nm), CdTe (2.8 nm), Fe$_3$O$_4$ (8 nm) and lanthanide-doped NaYF$_4$ (24 nm), Ag cubes (160 nm), polystyrene (PS) spheres (180 nm), β-FeOOH rods (22 nm × 160 nm), and lanthanide-doped NaYF$_4$ (50 nm × 310 nm) rods) can all be incorporated into ZIF-8 crystals by adjusting the time of nanoparticle addition during the MOF-formation reaction. The as-prepared hybrid materials exhibit active (catalytic, magnetic and optical) properties that derive from the incorporated nanoparticles as well as size- and alignment-selective behaviors from the well-defined microporous nature of the MOF component. Xu et al. also reported the encapsulation of noble-metal NPs in MOFs with carboxylic-acid-based ligands, such as UiO-66, NH$_3$-UiO-66, and NH3-MIL-53 [94]. The obtained NP/MOF composites exhibited excellent shape-selective catalytic properties in olefin hydrogenation, for aqueous reaction of 4-nitrophenol reduction, and enhanced molecular diffusion in CO oxidation. In the liquid-phase hydrogenation reaction of hexene, cyclooctene, trans-stilbene, cis-stilbene, triphenyl ethylene, and tetraphenyl ethylene, the Pt/UiO-66 composites showed high activity and selectivity on only hexene, while pure Pt NPs supported on carbon nanotubes displayed indiscriminating catalytic capability towards olefin hydrogenation of hexene, cyclooctene, cis-stilbene and trans-stilbene. The MOF matrix demonstrated convincing advantages as a catalytic carrier, not only preventing the aggregation of NPs but also imparting new functionalities to the catalyst composite.

Xu et al. reported unique strategy named a "double solvents" impregnation method (DSM) for the complete inclusion of Pt NPs within MIL-101 cavity [95]. The method involves a hydrophilic solvent (water) and a hydrophobic solvent (hexane). The former solvent contains the metal precursor with a volume set equal to or less than the pore volume of the MOF, which can be absorbed within the hydrophilic MOF pores, while the latter, in a large amount, plays an important role to suspend the adsorbent and facilitates the impregnation process. Because the inner surface areas of MOFs are much larger than the outer surface area, the small amount of aqueous metal precursor solution could enter the hydrophilic pores by capillary force, which greatly minimizes the deposition of precursors on the outer surfaces. By varying the concentration of the Pt precursor, various amounts of Pt NPs were loaded in MIL-101 with an average size of 1.8 ± 0.2 nm, much smaller than the cavity sizes. Using this method, other noble metals (Pd, Au, and Rh) NPs can also be fabricated within the MIL-101 cavity [96].

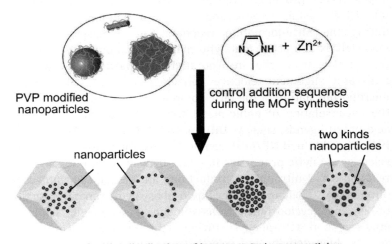

Figure 3.13 The scheme of the controlled encapsulation of nanoparticles in ZIF-8 crystals.

Based on a well distribution of metal nanoparticles (MNPs) inside, the MOF supports may also give synergistic effect on the catalytic performance. It was found that MOF supports can significantly affect the catalytic activity and product selectivity of MNPs. Haruta et al. introduced Au clusters to various MOF supports, which exhibited noticeably high catalytic activity toward liquid-phase aerobic alcohol oxidation even in the absence of any base [97]. They found that the catalytic activity of Au/MOFs was superior to that of Au/AC (activated carbon) and Au/SiO$_2$ under base-free conditions. For the aerobic oxidation of benzyl alcohol in methanol medium Au NPs in ZIF-8 and ZIF-90 showed 81% and 13% conversion with the selectivity toward methyl benzoate for 98% and 50%, respectively. It is proposed that the intercalated Au NPs in ZIF-90 catalyze the oxidation of nearby imidazolate-2-carboxyaldehyde groups to imidazolate-2-methylcarboxylate in methanol under oxygen pressure. Therefore, the interactions of substrates and products with Au NPs inside of the pores were hindered. The residual catalyst performance is considered only from the accessible Au NPs outside of the MOF pores.

It should be mentioned that high stability of MOF support is also critical for sustained catalytic activity. Kaskel et al. reported that the more stable Pd@MIL-101 showed higher activity in the hydrogenation of both styrene and cis-cyclooctene compared to unstable Pd@MOF-5 and was even higher than that of Pd/C with the same amount of Pd loading (1 wt%) [98]. Besides monometallic NPs, bimetallic NMPs can also be introduced into MOFs and exhibit higher catalysis activity than the monometallic counterparts. Kempe et al. synthesized PdNi@MIL-101 catalysts with different Pd/Ni compositions and investigated the catalytic performance for hydrogenation reactions [99]. By applying MOCVD and followed by reduction with H_2 at room temperature, MIL-101 cavity-conform PdNPs can be formed. When the reduction is carried out at 70°C, significantly smaller PdNPs to the size of the cavity are generated, which allows the loading of a second metal in the remaining space. The so-formed catalysts are active in the hydrogenation of dialkyl ketones. High catalytic efficiency is only observed if both metals operate synergistically and are nearly atomically dispersed. This bimetallic Ni/Pd NP catalyst can also be reused. The MIL-101 support combines ideally stability and also good access of the educts to the catalytically active sites.

Yolk-shell nanostructures have generated recent research interest due to their appealing structures and tunable physical and chemical properties. An integrated functions from the nanocrystal core, the nanostructured shell, and the cavity may bring novel synergistic effects during the applications in heterogeneous catalysis, photocatalysis, and biomedicine. For a yolk-shell catalyst, the metal core provides a catalytically active surface for the reaction and the porous shell serves as a barrier layer to prevent aggregation with neighboring metal cores during the reaction. Tsung et al. first introduced MOF into the yolk-shell nanostructures [100]. They developed a new synthetic strategy in which metal nanocrystals (NCs) were first coated with a layer of Cu_2O as the sacrificial template and then covered a layer with ~100 nm thickness of polycrystalline ZIF-8. The clean Cu_2O surface assists the formation of the ZIF-8 coating layer and can be etched off simultaneously by the protons generated during the process. By this strategy, nanocrystal@ ZIF-8 yolk-shell nanocomposites with different metal NC cores can be constructed. The nanocomposites have been applied as catalysts for the gas-phase hydrogenations of ethylene, cyclohexene, and cyclooctene. The ZIF-8 shell showed interesting size selectivity in ethylene hydrogenation versus cyclooctene hydrogenation, which is due to the cavity effect in the yolk-shell structure.

3.3.4 MOFs for the Construction of Inorganic Catalysts

The supermolecular structures of MOFs consist of organic ligands with metal ions linked together by coordination bonds. Thus, they can also be attractive carbon sources for the fabrication of carbon materials with novel morphologies. Very recently, MOF-derived carbon materials have been effectively used for energy conversion and storage applications due to their exceptionally high specific surface area and controllable pore textures. Multiple types of MOFs could be transformed into amorphous microporous carbons by direct pyrolysis under an inert atmosphere. However, during the carbonization process, the porous crystalline morphologies of MOF may suffer during the conversion into bulk carbon powder, which leads to the decrease in the effective special surface area.

To well maintain the original porous structures of MOF and control the dimensions of the final carbon structures for the application convenience, active template materials are utilized to direct the growth of MOFs and the formation of various complex carbon networks. Yu et al. used ultrathin tellurium nanowires (TeNWs) as templates for directed growth and assembly of ZIF-8 (Fig. 3.14) [101]. The formed ZIF-8 nanofibers were conveniently converted into highly porous doped carbon nanofibers after calcination. Compared with bulk porous carbon by direct carbonization of MOF crystals, these doped carbon nanofibers exhibit complex network structure, hierarchical pores, and high surface area. Such doped carbon materials with hierarchical pore structure and nanofibrous morphology can be used as an efficient electrocatalyst for oxygen reduction reaction (ORR) in alkaline media, exhibiting excellent electrocatalytic activity with long-term durability and good resistance to methanol crossover effects. The catalytic performance was even better than the benchmark Pt/C catalyst.

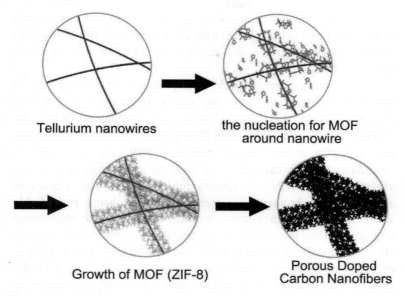

Figure 3.14 The scheme for nanowire-directed templating synthesis of ZIF-8 nanofibers and derived porous doped carbon nanofibers as catalyst.

Using MOFs as "precursors" has become an attractive pathway to obtain new nanostructured materials. MOFs were proved to be ideal sacrificial templates for fabricating porous carbon-based nanomaterials via thermal decomposition under controlled atmospheres [102–104]. In the thermolysis process, these metal ions of MOFs could be directly transformed into metal NPs, which are embedded in a ligand-derived porous carbon matrix. Li et al. presented a facile and efficient approach to fabricate non-noble metal@noble metal core–shell nanocatalysts by using MOFs-derived metal NPs as sacrificial templates [105]. ZIF-67 was first prepared from cobalt nitrate and 2-methylimidazole according to a facile room-temperature method. Co NPs embedded in N-doped carbon were then obtained by carbonizing ZIF-67 in an inert gas. They finally fabricated a nanocarbon with incorporation of Co and Pd (Co@Pd/NC) by the growth of Pd shell on the core of Co NPs through the replacement reaction. The highly exposed Pd atom on Co nanoparticles made it excellent as a highly active, extremely stable nanocatalyst toward the nitrobenzene hydrogenation reaction. The Co@Pd/NC showed

98% conversion of nitrobenzene after a 45 min reaction, which was far more active than the pristine ZIF-67 and MIL-101 supporting Pd NPs (with 40% and 3% conversion after 90 min reaction, respectively).

Easy aggregation of metal nanoparticles (MNPs) is one of the main problems for their application as catalysts, which may severely reduce the catalytic efficiency. Stabilizing MNPs by surface ligands and embedding MNPs inside protective shells are executed to solve the stability problems. MOFs are also excellent candidates to form carbonaceous protection layers on the surface of well-defined MNPs with specific functionalities. The conversion of MOFs to carbonaceous material may also bring remarkable advantages on much better chemical stability and mechanical strength during the applications. Zhou et al. synthesized Fe_3O_4@MC-Pd composite material derived from Fe_3O_4@MIL-100(Fe)/$PdCl_2$ at 450°C in nitrogen atmosphere [106]. The Fe_3O_4 NPs with the surface modification of carboxylic acid was used as a template to grow MIL-100(Fe) through a one-pot self-assembly strategy. The as-synthesized Fe_3O_4@MIL-100(Fe) was then encapsulated with Pd precursor of (H_2PdCl_4) and subsequently annealed in nitrogen atmosphere to form the magnetic porous carbon (MPC) containing Pd composite (Fe_3O_4@MC-Pd). The Fe_3O_4@MC-Pd exhibits excellent catalytic activity in reduction of methylene blue (MB) compared with most of the other reported catalysts. The as-synthesized nanocatalyst shows high efficiency in magnetic separation and recovery because of the high saturation magnetization and super-paramagnetic property. In addition, the chemical and thermal stability of Fe_3O_4@MC-Pd materials are better than those of Fe_3O_4@MIL-100(Fe).

Owing to the relatively versatile decomposition of MOFs by chemical or thermal treatments, MOFs have also been used as effective templates for some metal oxides with well-controlled size, shapes and porous structures. Lin et al. reported a simple, inexpensive, tunable, and scalable MOF-templated strategy for the synthesis of a mixed metal oxide nanocomposite with useful photophysical properties (Fig. 3.15) [107]. Fe-containing MIL-101 is coated with amorphous titania by acid-catalyzed hydrolysis and condensation of titanium(IV) bis(ammonium lactato)dihydroxide (TALH) in water, then calcined to produce crystalline Fe_2O_3@

TiO₂ composite nanoparticles. This material showed interesting photophysical properties as it enables photocatalytic hydrogen production from water using visible light, while neither component alone is able to do so. The versatile MOF-templated nanocomposite synthesis procedure can be readily exerted to prepare new nanocomposite materials with desirable synergistic properties by varying the type of MOF (both the metal ions and coordinating moieties) and the coating material.

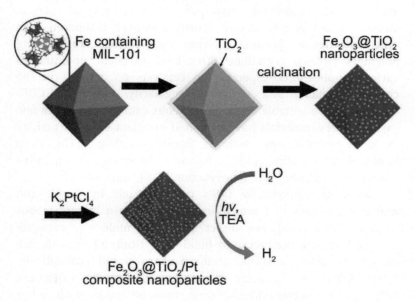

Figure 3.15 The MOF-templated strategy for the synthesis of a mixed metal oxide nanocomposites as photosynthesis catalyst.

3.4 Covalent Organic Framework-Based Catalysts

A new type of crystalline organic porous materials, covalent organic frameworks (COFs), are constructed with extending organic structures through strong covalent bonds by using reticular chemistry. Since Yaghi and co-workers first designed and synthesized COF materials in 2005 [108], extensive research has been focused on these new crystalline porous materials.

Until now, the typical COFs can be roughly classified into boron-containing COFs [109], imine-based COFs [110], triazine-based COFs [111]. In recent years, COFs have been used in a wide variety of applications, such as gas adsorption [112], catalysts and catalytic supports [113], semiconductive and photoconductive devices [114] because of their controllable organic structures, permanent porosities, low density, and crystallizability. Similar to other crystalline porous materials such as MOFs, COFs show tunable and predictable two- or three-dimensional porous structures and possess lower density. However, compared with other low density porous materials such as high cross-linking polymers (HCPs), crystalline COFs have excellent potential in various applications, especially in catalysis on account of the chemical tunable and structural regularity. Therefore, COFs would be more suitable for heterogeneous catalysts. In fact, a few cases of COFs materials have exhibited excellent catalytic activity in kinds of reactions, such as Suzuki coupling [115], Heck Sonogashira coupling [116], oxidation reaction [117], nitro reduction [118], photocatalytic reaction [119], and so on.

Generally, catalytically COFs can be built by direct and indirect strategies that are diffusely employed in heterogeneous catalysis [113, 120]. For direct strategies, molecular catalysts are used as building blocks to build catalytically COFs with rich catalytically active functional groups. So here we call it catalytically active COFs. For indirect strategies, catalytically COFs are prepared by post-modification or immobilization with other catalytically active species (metal ions or metal nanoparticles) and called as post-modification COFs. Here, we would like to summarize the recent research on the application of COFs as heterogeneous catalyst based on post-modification COFs or catalytically active COFs and the catalytic reaction types by using COFs as heterogeneous catalysis.

3.4.1 Classification of COFs as Heterogeneous Catalysts

COFs can serve as remarkable heterogeneous catalysts because of their versatile features. The uniform porosity, tunable structure, and rich active functionality possessed in COFs make them more suitable for reaching excellent catalytic activity or supporting for various catalytically active species. Furthermore, COFs might be more capable supports than other porous materials due to the strong covalent bond architecture, which acts as an efficient organic support for heterogeneous catalysis.

3.4.1.1 Post-modification COFs as catalysts

As discussed in the introduction part, two strategies have been used in preparing COFs during heterogeneous catalysis application. In recent years, most of the reported COFs for catalysts have been synthesized through post-modification methods.

Since the first discovery of COFs materials in 2005 [108], researchers have mostly focused on how to improve the surface area and tune pore sizes, accompany with some applications in gas adsorption and photoelectricity, until Wang and co-workers first reported the example of design, synthesis, and application of an imine-linked COF material for catalysis in 2011 [113], a new area for COFs as catalysts had been explored. In that report, it is pointed out that the high stability of COF-LZU1 to thermal treatments, water, and most of the organic solvents is an important issue for catalytic applications. Also, the imine-type ligands are available in incorporating Pd ions as shown in Fig. 3.16. When Pd/COF-LZU1 was applied in catalyzing the Suzuki–Miyaura coupling reaction, it showed remarkable catalytic activity and recyclability attributed to its uniform pore structure and strong interaction between imine groups and Pd ions.

After this finding, research on employing loaded COF materials for heterogeneous catalysis has increased. Jiang and co-workers [121] constructed a nitrogen-rich porphyrin-based COF (H_2P-Bph-COF) which acts as good support to incorporate Pd ions (Fig. 3.17). By integrating porphyrin units into predetermined long-rang ordered COFs, it is able to stabilize and uniformly disperse the Pd ions inside the COFs structure because of the abundant and periodically distributed N atoms from both the porphyrin units and the C–N bonds formed in H_2P-Bph-COF. What's more, the porphyrin units can be easily functionalized by various metal ions and substituents presenting one of the most attractive candidates for catalytic applications. The constructed Pd/H_2P-Bph-COF material shows a remarkable catalytic activity towards the Suzuki-coupling reaction between bromoarenes and arylboronic acids, achieving both good selectivity and yields. Introduction of nitrogen functionalities on COFs has been shown to effectively stabilize and disperse the metal ions.

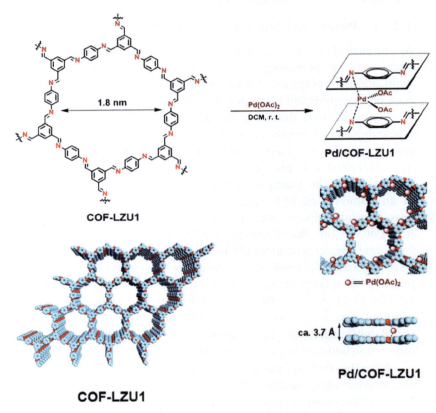

Figure 3.16 Construction of COF-LZU1 and Pd/COF-LZU1 [113].

Similar with the porphyrin group, Zhang and co-workers [122] introduced triazine functional group into the COF (COF-SDU1). The obtained COF-SDU1 with both imine and nitrogen-rich triazine functional groups could immobilize and disperse various metal ions well (Fig. 3.18). By a simple impregnation method, Pd(II) ions were stabilized and uniformly distributed into COF-SDU1 due to the two-dimensional eclipsed layer-sheet structure and nitrogen-rich content. The resulted Pd(II)/COF-SDU1 shows remarkable catalytic activity, selectivity and recyclability towards the silicon-based one-pot cross-coupling reaction of silanes and aryl iodide.

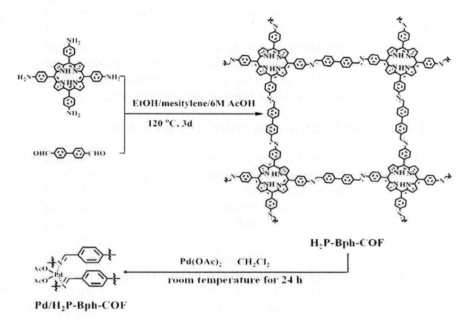

Figure 3.17 The preparation of H2P-Bph-COF and Pd/H2P-Bph-COF [121].

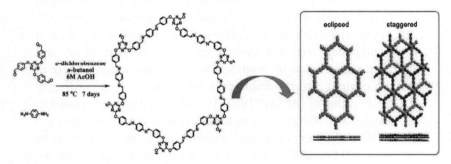

Figure 3.18 The preparation of COF-SDU1 by the condensation reaction between triazine and p-phenylenediamine [122].

Although COF materials are widely used as catalyst supports because of the well ordered structures and high surface area, the stability of COFs in the reaction media still remains a challenge. Banerjee and co-workers have immobilized gold nanoparticles into a highly stable, porous and crystalline COF (TpPa-1) [123]. The loading of uniformly distributed Au nanoparticles with 5 ± 3 nm particle sizes occupied the pore surface and interlayer spacings in the host TpPa-1. The as-synthesized Au(0)@TpPa-1 shows a superior catalytic activity towards nitrophenol reduction reaction than $HAuCl_4 \cdot 3H_2O$ and the excellent recyclability for more than six catalytic cycles with yields over 95% which demonstrated the advantages as heterogeneous catalysts.

3.4.1.2 Built-in catalytically active COFs as catalysts

COFs' structure can be easily turned by changing the building units. Many monomers with various function groups have been used to synthesize COFs with special functions which can be directly employed in heterogeneous catalysis without loading any metal ions or other catalytically active species. Although the direct strategy to prepare catalytically active COFs has limitations in the stability tolerance of the COF frameworks towards catalytically active functional groups [124], the research on the special catalytically active COFs has still received attention in recent years. Jiang and co-workers [125] reported the pore surface engineering strategy for the construction of covalently linked and highly active organocatalytic COFs and developed the first example of a covalently linked catalytically active COF. The [Pyr]X-H_2P-COFs were synthesized as shown in Fig. 3.19. A three-component reaction system consisting of H_2P, BPTA, and DHTA was developed to prepare the COFs. The ethynyl units of varying content into the framework were controlled according to the molar ratios of [BPTA]/([BPTA]+ [DHTA]) which are related to the density and composition of the functional groups on the pore walls. With further integrated ethynyl units with catalytically active pyrrolidine azide, which is a well-known organocatalyst for michael addition reactions, organocatalytic COFs is obtained. The [Pyr]$_X$-H_2P-COFs exhibited enhanced activity, good recyclability, and high capability to Michael addition reactions because of both crystalline-porous COF structure and catalytically active pyrrolidine derivatives.

Covalent Organic Framework-Based Catalysts | 135

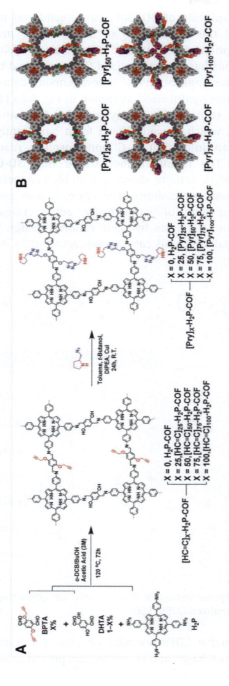

Figure 3.19 (A) The general strategy for the pore surface engineering of imine-linked COFs via a condensation reaction and click chemistries (the case for X = 50 was exemplified). (B) A graphical representation of [Pyr]$_X$-H$_2$P-COF with different densities of catalytic sites on the pore walls (gray: carbon, red: nitrogen, green: oxygen, purple: carbon atoms of the pyrrolidine unit; hydrogen is omitted for clarity) [125].

136 | *Supermolecular Catalysts*

The design and synthesis of 3D COFs are considered to be a great challenge and the applications of 3D-COFs are mostly focused on gas adsorption, until Yan and co-workers developed two new 3D microporous base-functionalized COFs and used them for the first time in the applications of heterogeneous catalysis. The related 3D COFs were synthesized as shown in Fig. 3.20 [126]. Two 3D COFs were studied for catalyzing the Knoevenagel condensation reaction, which is well known for a base-catalyzed model reaction and an important C–C bond-forming reaction in organic synthesis. The results demonstrated that 3D-COFs exhibited excellent catalytic activity, showing with high conversion, size selectivity, and good recyclability. Therefore, it also greatly facilitates the development of catalytically active COFs as promising heterogeneous catalysts by their rational design with porous structure, crystallinity, and catalytically active sites in the frameworks.

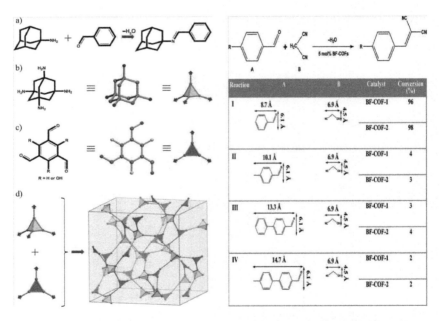

Figure 3.20 Schematic representation of the strategy for preparing 3D microporous base-functionalized COFs [126].

The catalytically active COFs are also effective catalysts for electrochemical reactions. Yaghi and co-workers reported modular optimization of COFs [127], in which the building units are cobalt porphyrin catalysts linked through imine bonds (Fig. 3.21), and their usage as a catalytic material for aqueous electrochemical reduction of CO_2 to CO. The ordered structure of COFs endows them with versatile synthesis, where the structural design and modification can be easily realized. Moreover, the electronic structure of the catalytic cobalt centers can be directly influenced by the COF framework; therefore, it gives great advantages in the selectivity and activity for catalysis. It thus exhibited that high efficiency and the activity was greatly enhanced by 26 times as compared with the cobalt porphyrin complex building block. Previous research also shows that the COF environment could directly modulate the electronic properties of molecular centers coupled into the extended lattice [128], which is the unique feature to the COFs.

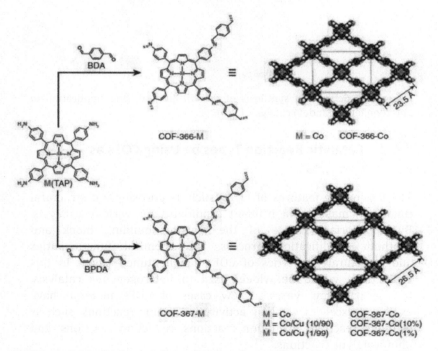

Figure 3.21 Design and synthesis of metalloporphyrin-derived 2D covalent organic frameworks [127].

Similar to this method, a novel synthetic strategy has been mentioned by Rahul Banerjee's group [129]. Comparing with the catalytically active monomers for the preparation of COFs, they chose a predesigned metal-anchored building block in which the metal nanoparticles could be generated in situ first and then form the COFs as shown in Fig. 3.22. According to their research, the formed Pd@TpBpy also showed superior catalytic performance and high recyclability.

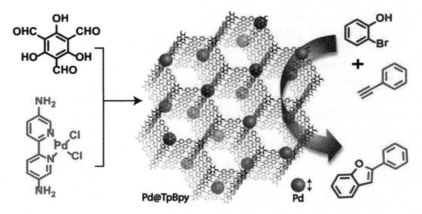

Figure 3.22 Schema of synthesis of Pd@TpBpy and their application in heterogeneous Tandem catalysis [129].

3.4.2 Catalytic Reaction Types by Using COFs as Catalysts

The fascinating features of COFs, such as porosity and structural tunability, make them brilliant candidates for various catalysts. By appropriate choice of the starting building block and synthetic modification process, the chemical functionalities and structural properties of COFs can be tuned easily. In this view, COFs would be widely used in heterogeneous catalysis. In the past few years, a few cases of COFs material have exhibited excellent catalytic activity in organic reactions, such as coupling reactions, oxidation reactions, reduction reactions, and photocatalytic reactions.

3.4.2.1 Cross-coupling reactions

Suzuki–Miyaura Coupling and Heck–Sonogashira coupling are the familiar coupling reactions to form the aromatic C–C bonds which can be widely used in producing many significant products such as drugs, materials, and optical devices. [130]. As mentioned above, both Wang [113] and Jiang [121] used postmodification COFs as catalysts for Suzuki–Miyaura coupling reaction. The modified COFs showed excellent catalytic activity and recyclability. Wang and co-workers also modified conjugated nanoporous polymer colloids (CNPCs), which allow for in situ incorporation of palladium nanocrystals to form the Pd@CNPC. The complex structure is validated to have outstanding catalytic activity and reusability with exceptionally high TOF (44,100 h^{-1}) for the Suzuki–Miyaura coupling reaction [131]. Rahul Banerjee found that TpPa-1 is the first COF matrix that can hold both Pd(0) nanoparticles and Pd ions complex without aggregation for catalytic purpose under both highly acidic and basic conditions [116] (Fig. 3.23). The highly dispersed Pd(0) nanoparticles exhibited excellent catalytic activity towards Cu-free Sonogashira and Heck coupling reactions under basic conditions and the Pd(0)@TpPa-1 catalyst was also efficient in one-pot sequential Heck–Sonogashira reactions (Fig. 3.23).

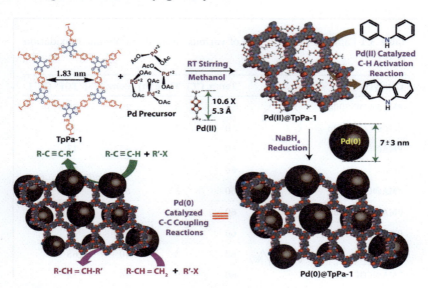

Figure 3.23 Synthesis of Pd(II) and Pd(0)-doped COFs (i.e., Pd(II)@TpPa-1 and Pd(0)@TpPa-1) and summary of their catalytic activity towards Sonogashira, Heck and oxidative biaryl couplings [116].

3.4.2.2 Oxidation reaction

COFs as heterogeneous catalysts are also widely used in oxidation reactions. A class of COFs that are formed by the trimerization of aromatic nitriles [132, 133], covalent triazine frameworks (CTF) exhibit very high amounts of nitrogen functionalities in the networks. Palkovits et al. applied a CTF loaded with a platinum salt as catalyst for methane oxidation in concentrated sulfuric acid at 215°C [134]. Laura Prati and co-workers used CTF as support for Pd nanoparticles [135]. The formed Pd/CTF showed better activity and in particular better stability compared to Pd supported on activated carbon during the liquid phase glycerol oxidation. Except for CTF, other COFs also have been applied in oxidation reactions. Jia et al. immobilizing 12-phosphomolybdic acid (PMA) onto COF-300 [117]. The existence of unique interpenetrated 3-D structure, suitable porous characteristics, and the abundant protonated imine groups in COF-300 are beneficial for dispersing and stabilizing the anions of PMA, which plays a critical role in highly active and stable heterogeneous PMA-functionalized COF-300 catalysts for olefin epoxidation (Table 3.1) [136].

Table 3.1 Catalytic performance of various catalysts in olefins epoxidation [136]

Catalysts	Substrates	Temp. (°C)	Time (h)	Conv. (%)	TOF[b] (h^{-1})
PMA@COF-300a		50	3	29	20
PMA@COF-300b		50	3	38	26
PMA@COF-300c		50	3	65	13
PMA		60	3	99	162
PMA@COF-300a		60	3	63	44
PMA@COF-300b		60	3	79	55
PMA@COF-300c		60	3	91	18
PMA		60	3	40	65
PMA@COF-300a		60	9	44	12
PMA@COF-300b		60	9	49	15
PMA@COF-300c		60	9	77	6

Catalysts	Substrates	Temp. (°C)	Time (h)	Conv. (%)	TOF[b] (h^{-1})
PMA		60	3	93	152
PMA@COF-300a		60	3	17	12
PMA@COF-300b		60	3	31	22
PMA@COF-300c		60	3	34	6

[a]Reaction conditions: catalyst PMA@COF-300a, PMA@COF-300b and PMA@COF-300c 10 mg; substrate cis-cyclooctene, 1-octene and cyclododecene 1.0 mmol; solvent CHCl$_3$ 2 mL; oxidant t-BuOOH 1.0 mmol.
[b]Turnover frequencies (TOF) are expressed in mol (olefin) $ mol(Mo)_1 $ h_1. The selectivity of the epoxide for all catalysts is greater than 99%.

3.4.2.3 Reduction reaction

As already mentioned, Rahul Banerjee et al. immobilized gold nanoparticles into a stable COF; the as-synthesized Au(0)@TpPa-1 shows high reactivity and recyclability for nitrophenol reduction reaction [123]. Omar M. Yaghi et al. reported the imine bond-linked COFs with cobalt porphyrin catalysts as building units, which exhibit high Faradaic efficiency (90%) and turnover numbers (up to 290,000, with initial turnover frequency of 9400 h^{-1}) during aqueous electrochemical reduction of CO$_2$ to CO [127]. In their research, the COFs covalent framework could directly influence the electronic structure of the catalytic cobalt centers.

3.4.2.4 Photocatalytic reaction

In the past decade, the research on photocatalytic systems has seen a rapid development [137], in which the carbon nitride polymers are promising polymer photocatalyst and are moderately active in hydrogen generation from water [138]. However, the major disadvantage of these polymers lies in their low crystallinity and generally low surface areas. Fortunately, the COFs with high crystalline structures and large surface areas are promising materials to mitigate these inherent deficiencies of carbon nitrides and therefore are potential photocatalysts to be explored [139]. Jiang and co-workers reported a squaraine-based COF applied in light-induced activation of oxygen and for the first time indicated the underlying potential of COFs as a photoactive material [140]. Bettina V. Lotsch et al. developed a new crystalline hydrazone-based TFPT-COF used as the proton reduction catalyst (PRS) in the presence of Pt, which is the first COF to show photocatalytic hydrogen evolution under visible light irradiation [119]. The triazine moieties in the TFPT-COF are considered to play an active role in the photocatalytic process.

In recent years, a new type of metal-free photocatalyst for photocatalytic carbon dioxide reduction has been developed by Baeg's group. They used a triazine-based 2D CTF to fabricate an efficient visible light active flexible film as photocatalyst [141]. The newly formed CTFs with robust polymer structures exhibited high light-harvesting capacity, appropriate band gap and energy levels, and highly conjugated π electron systems. These 2D layered structures are considered to be advantageous for charge transport of the charges, which is thought to be critical point for high photocatalytic activity. This work opens up a new strategy to prepare triazine-based CTFs for the application in the visible light active photocatalysis. In these conjugated polymers, the simple changes of the building blocks at the molecular level would lead to significant influence on electronic properties and morphologies of the materials, which greatly affect the photocatalytic performance [142]. Thus, it is highly important to be able to control the molecular structure of COFs for photocatalytic applications.

3.5 Supermolecular Nanoarchitectures for Catalysts

Mainly mimic enzymes, the field of supermolecular catalysis has grown significantly, providing new tools for selective transformations. In nature, much more complicated chemical reaction networks are proceeded effectively in complex systems, which involve the integration of the catalysts, the substrates, cofactors or sub-complex systems. Supramolecular organizations of catalysts or catalysis system and substrates are demonstrated to play a key role in a number of highly selective catalytic transformations. It may also provide new concepts for the construction of molecular devices, smart nanostructures and promote the step of organized catalysis systems to practical applications [143, 144]. The self-assembly of catalysis components into a programmed architecture is shown to exhibit interesting behavior and may affect certain chemical reactions. The confined effect of the organized architectures increases the local concentration of the reagents and thus accelerates reaction rates.

The confined architectures for catalysis can be derived from a variety of chemical systems and nanotechniques. They may be vesicles from amphiphiles, dendrimers, assembled hybrid layers, or hollow inorganic capsules. Here, we introduce the combination of catalysts in those self-assembled architectures, which can strongly affect the catalytic performance of chemical reactions. We will focus on the soft molecular architectures for the integration of catalysts and co-components, and their confined effects on the catalysis performance.

3.5.1 Assembled Vesicles for Catalysts

Lipid membranes are ubiquitous in all domains of life. Membranes are organizing structures needed to define physical boundaries, compartmentalize molecules within the cell, and provide sites for proteins to control transport and signaling. Numerous studies of artificial membranes have demonstrated the self-assembled bilayer vesicles from amphiphiles exhibiting properties reminiscent of cellular membranes [145–147]. Some studies even revealed that seeding artificial membranes with catalysts may trigger the generation of additional amphiphiles, drive the recruitment of lipids from surrounding environment, and cause vesicle budding and division [148–150]. Devaraj et al. reported the design of a simplified lipid membrane using a synthetic, membrane-embedded catalyst, which is capable of self-reproduction (Fig. 3.24) [151]. By substituting the complex network of biochemical pathways used in nature with a single autocatalyst that simultaneously drives membrane growth, their system continually transforms simpler, high-energy building blocks into new artificial membranes. To achieve simultaneous lipid and catalyst synthesis, a shared catalytic triazole coupling reaction is used to generate both triazole phospholipids and additional membrane-embedded copper-chelating oligotriazole catalysts. The oligotriazole TLTA (ligand) is capable of binding Cu^{1+} ions to form a catalytic complex (catalyst). The copper complex was shown to catalyze the synthesis of new ligand from 1-azidododecane (azide) and tripropargylamine (alkyne scaffold), and also the formation of triazole phospholipid from 1-azidododecane (azide) and an alkyne modified lysolipid. It indicated that this membrane-embedded catalyst is active on azide and alkyne reactive precursors, to synthesize additional phospholipid and oligotriazole ligands, and becomes new catalysts upon metallation. Although no organisms are thought to use the Cu-catalyzed cycloaddition reaction, integrating Cu-TLTA self-reproducing catalysts with model systems of fatty acid membranes is believed to enable future studies for exploring the role of phospholipid-synthesizing catalysts in the competitive growth and division of protocells [152].

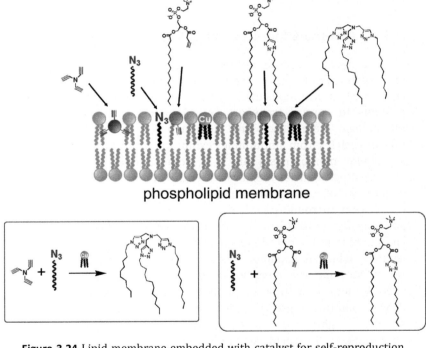

Figure 3.24 Lipid membrane embedded with catalyst for self-reproduction by click chemistry.

It has long been known that natural cell membranes can complex calcium ions via acidic lipids to initiate the formation of rafts or aggregation and fusion processes [153]. Through choosing proper amphiphiles or modified with a sufficiently hydrophobic tail, metals or their complexes can be incorporated into the membrane bilayer. The embedded metal catalysts in vesicle membranes was shown to be capable of changing their reactivity compared to in bulk solution. König's group prepared self-assembled catalytic systems based on amphiphilic derivatives of the Zn(II)-complex of 1,4,7,10-tetraazacyclododecane, which are well-known models for hydrolytic enzymes, promoting the hydrolysis of active phosphodiesters (Fig. 3.25) [154]. The complexes with C18-alkyl chain were incorporated in (1,2-distearoyl-sn-glycero-3-phosphocholine) DSPC vesicle membranes by noncovalent self-assembly. The vesicular catalysts containing tightly packed Zn^{2+}-domains within their bilayer membranes outperformed the identical system in homogenous aqueous solution by several orders of magnitude with respect to second-order-kinetics reaction rates and represented the most active artificial metal catalyst system reported so far. Self-assembled soft particles provided a more convenient preparation and higher activities compared to solid functional nanoparticles. They also show sufficient long-term stability in contrast to those readily cleaved otherwise stable substrates, like native plasmid DNA and non-activated oligonucleotide strands. The enhanced catalytic performance should be attributed by the high local concentration of catalytic centers in the membrane patches, which favors substrate attraction and conversion, decreases polarity at the vesicle-water interface, facilitates nucleophilic attack of phosphodiester substrates.

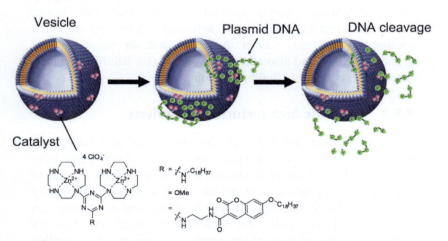

Figure 3.25 The vesicular catalyst containing Zn(II) complex as catalyst.

Recently, Harano and Nakamura et al. developed a vesicle with several-nanometer-thick bilayer by self-assembly of an organic fullerene amphiphile doped with a metathesis catalyst. The system was served as a nanosized chemical reactor in water. In the process, polymer was synthesized and assembled, depending on the affinity of the growing polymer and the organic groups on the amphiphile [155]. The anionic fullerene amphiphile with alkyl or perfluoro alkyl chain was shown to self-assemble in aqueous solution into vesicle with a uniform size (typically 30 nm in diameter). The fullerene bilayers keep the hydrophilic fullerene anion inside and exposes the hydrophobic groups to an aqueous environment to create three regions: surface, interior, and fullerene-rich core. With the doping of catalyst molecules in the fullerene bilayer, the ring-opening metathesis polymerization (ROMP) can control the assembly of polymer products from a norbornene diethyl ester in three different ways by the choosing the fluorous vesicle, the octyl vesicle, and the eicosanyl vesicle, depending on the miscibility of the growing polymer to the interior of the bilayer. For the fluorous vesicle, norbornene diethyl ester polymerized heterogeneously on the vesicle surface to afford 6 nm sized spherical particles, which were composed mostly of a single polymer chain. In contrast, the polymerization on the octyl vesicle occurred homogeneously near the fullerene core and resulted a hollow polymer capsule of 35 nm size composed of 20–30 tightly entangled polymer chains. However, for the eicosanyl vesicle, the polymerization produced a soft and sticky capsule. This catalysis system was shown to be tolerant for the functional groups of ROMP, and also allowed to polymerize halide-containing monomers, as well as a fluorescent copolymer.

3.5.2 Dendritic Architectures for Catalysts

Dendrimers were developed in the late 1970s and have been greatly exploded during the past two decades. Among the main potential applications of dendrimers, catalysis stands as one of the most promising ones. Dendrimers offer a unique opportunity to combine the advantages of homogeneous and heterogeneous catalysis and yet keep the well-defined molecular features required for the analysis of the catalytic processes. It is possible to tune the structure, size, shape, and solubility of dendrimers.

The location of catalytic sites can be at the core or at the periphery of dendrimers [156-159].

To across high activity energy of a reaction, it is preferred that a well-defined local domain and multi-electron redox system can be fabricated around the catalytic center. This kind of advanced multi-electron reservoir can be relatively easily constructed through the assembly of redox-active metallic centers into one molecule. The dendritic compound may provide a good platform for this kind of multi-electron catalysts. They are capable of fine assembling of metal complex, acting as a framework around various functional cores, which thus would accelerate concerted multi-electron transfer to the catalytic center and facilitate the activation of the substrate [160-163]. Yamamoto et al. presented the formation of multi-electron catalysts by a series of dendritic phenylazomethines with strong Lewis acids such as lanthanide trifluoromethanesulfonate [Ln(OTf)$_3$] (Fig. 3.26) [164]. The dendrimer has a cobalt porphyrin core, which bears four dendron units on the meso positions. The generation number was varied from 1 to 4. For the catalytic reduction of CO_2, the dendrimer catalysts in the presence of a strong Lewis acid showed an improved catalytic behavior in comparison with common macrocyclic cobalt complexes. These catalysts can activate the reduction of CO_2 through the mono-valent state of the cobalt complex, which require much lower overpotential in contrast to the common catalysts via the zero valent state. The improved performance is attributed by acceleration of multi-electron transfer from the coordinating metal ions around the catalytic center (cobalt porphyrin). The metal ions may also further activate the binding CO_2 molecule at the catalytic center.

Dendritic core–shell architectures have attracted increasing interest for their ability to act as unimolecular catalysts. Branched dendrimers may also provide capsular space for accommodating various materials like inorganic substances, metal ions, and organic compounds. By the addition of appropriate functional groups to the dendrimer, the metal can be coordinated within the dendrimer or form metal clusters by the reduction of metal ions dispersed in the dendrimer. Haag et al. reported the synthesis of novel dendritic architectures with perfluorinated shells derived from the covalent modification of perfect glyceroldendrimers, hyperbranched polyglycerols and hyperbranched polyethyleneimines [165]. They used this kind of system as a supramolecular support to immobilize a perfluoro-tagged catalyst for Suzuki coupling reactions [166]. The dendritic support showed a better solubility of the perfluoro-tagged palladium (Pd) complex in organic solvents and increased the yield of the reaction significantly. The high catalytic activity of this dendritic complex can be explained by its relative small nanoparticle size and the homogeneous dispersion the Pd complex in the reaction conditions. The recycling and multiple use of this catalyst is also straightforward by simple precipitation.

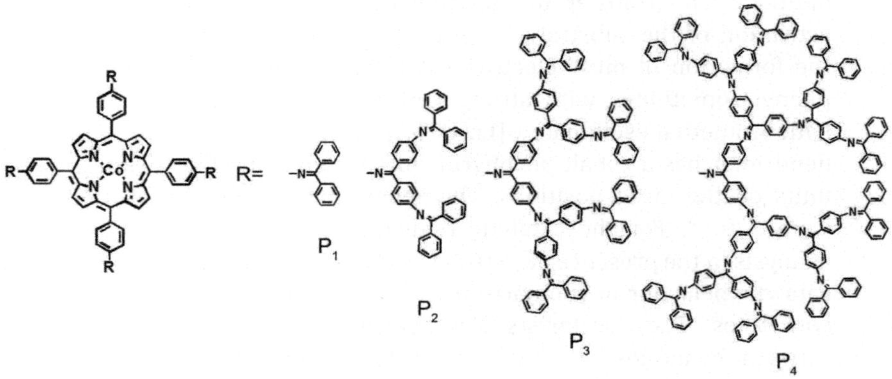

Figure 3.26 The structures of symmetric phenylazomethine dendrimers (P_n: n = 1–4) bearing a cobalt porphyrin core.

Yamamoto et al. loaded metal ion species in polyphenylazomethine dendrimers by a stoichiometrically stepwise fashion (Fig. 3.27) [167]. The fourth-generation phenylazomethine dendrimer has 30 imine and thus is capable of accommodating 30 molecules of $SnCl_2$. The controlled encapsulation is based on a defined gradient of basicity of the imine groups. It is also possible to assemble a set number of metal ions at predetermined locations. The research group also demonstrated that the metal encapsulated dendrimers used as an effective catalyst for polymerization [168]. The Co complexes and base units were packed in the dendritic phenylazomethines with a core–shell configuration. Due to the rigid π-conjugating structure of the phenylazomethine dendrimer, the metal center can retain its fine conformation. The additive-free aerobic oxidation of 2,6-difluorophenol using the dendrimer catalysts was achieved under very dilute catalytic conditions and with drastically decrease of the total waste.

Polyoxometalates (POMs) are a class of inorganic cage complexes with very interesting properties, which render them attractive for catalysis application. The catalytic properties of dendritic POM hybrids were based on electrostatic bonding between POMs (the most frequent one is Venturello ion $[PO_4\{WO(O_2)_2\}_4]^{3-}$) paired with dendritic cations. Dendritic POMs bearing Venturello ion exhibited good catalytic activity and recoverability in the oxidation of organic substrates such as alkenes, alcohols, and sulfides [169, 170]. Kortz and Nlate et al. used zirconium-peroxo-based dendritic POMs as efficient and recoverable sulfide oxidation catalysts [171]. The dendritic POMs are prepared by coupling zirconium-peroxotungstosilicate $[Zr_2(O_2)_2(SiW_{11}O_{39})_2]^{12-}$ with ammonium dendrons by electrostatic bonding. In contrast to the potassium salt of $[Zr_2(O_2)_2(SiW_{11}O_{39})_2]_{12-}$, the dendritic counterparts are soluble in common organic solvents, which is an important feature for the use of dendritic hybrids in homogeneous catalysis. The oxidation of sulfides to the corresponding sulfoxides and sulfones by these dendritic catalysts showed good to excellent conversion. They are also stable and recoverable without any appreciable loss of activity.

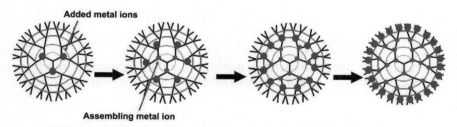

Figure 3.27 The radial stepwise complexation of metal ions in dendrimers.

Various efficient, water-soluble, ruthenium–benzylidene olefin, metathesis catalysts for metathesis of olefins in water have been reported. In some cases for olefin metathesis of hydrophobic substrates in pure water, ultrasonification of surfactants is required to get a homogeneously dispersion state. Astruc et al. reported the catalysis of ring-closing metathesis

(RCM), cross metathesis (CM) and enyne metathesis (EYM) of hydrophobic substrates in water and air under ambient or mild conditions using very low catalytic amounts (0.083 mol%) of a dendritic catalysis system [172]. The system is based on a suitably designed dendrimer terminated by 27 tetraethyleneglycol tethers and Grubbs' second-generation olefin-metathesis catalyst (Fig. 3.28). The water-soluble dendrimer allows the decrease in the amount of commercial ruthenium (Ru) catalyst for olefin RCM reactions in water. The aqueous solution of dendrimer can also be recovered and reused at least 10 times without significant decrease of metathesis product yield. The dendrimer plays the protection role as a nanoreactor towards the catalytically active species, in particular the sensitive Ru-methylene intermediate in the metathesis catalytic cycle, and also prevents catalyst decomposition in the presence of an olefin substrate.

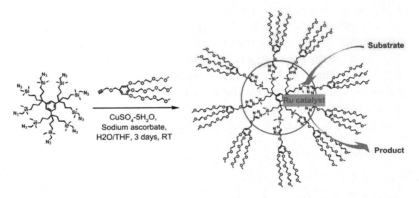

Figure 3.28 The water-soluble dendrimer dendritic catalysis system constructed by 1→3 connectivity and terminated by 27 tetraethyleneglycol tethers.

Dendritic catalysis is a rapidly growing field. In more sophisticated designs, catalysis proceeds in the dendritic interior. The dendrimer works like a nanoreactor, and thus in a biomimetic enzyme-type fashion [173]. The dendritic ligand design plays a crucial role for the catalytic activity or the combination with catalytic components, which may remarkably affect the catalytic performance. Recent enthusiasm toward organocatalysis has also largely involved dendrimer chemistry for those "green chemistry" applications.

3.5.3 Layered Architectures for Enzymes

Forming materials precisely from atoms and molecules into complex three-dimensional systems remains a relatively difficult matter. Of several possible strategies, layer-by-layer (LbL) assembly is a facile and easily achievable approach to integrate nanomaterials into controlled layer structures. Various materials, including polymers, inorganic substances, and supramolecular assemblies, as well as a wide range of catalysts or enzymes, may be assembled in LbL structures. The precise control on layer sequences and spacings leads to novel properties depart from those of corresponding single component or bulk materials [144, 174].

In the case of electrostatic LbL assembly, relatively high concentrations of the substances in solution lead to excess adsorption of the substances where charge neutralization and resaturation leads to charge reversal. Alternation of the surface charge results in a continuous assembly between positively and negatively charged materials. Other interactions such as metal coordination, hydrogen bonding, covalent bonding, supramolecular inclusion, biospecific recognition, charge transfer complex formation, and stereo-complex formation are also available in LbL assembly processes. Many kinds of deposition techniques, including spin coating and spraying, can also be used. Since the LbL technique does not require special conditions, biomaterials can be assembled appropriately to maintain their specific activities. It also proved that biomaterials embedded within LbL films show enhanced stabilities and small substance molecules can freely diffuse within the films.

The integration of enzymes into layer structures may facilitate the construction of catalytic systems with multi-activities or enzyme-based functional devices. Kunitake et al. prepared a multi-enzyme reactor by LbL technique (Fig. 3.29) [175, 176]. Two enzymes, glucoamylase (GA) and glucose oxidase (GOD), were assembled in the same film on an ultrafilter substrate, and an aqueous solution of starch was passed from the top of the reactor. Starch was hydrolyzed into glucose through hydrolysis of the glycoside bond by GA, and the glucose produced was converted to gluconolactone by GOD with hydrogen peroxide as a co-product. Unreacted starch was filtered out by the ultrafilter because of its high molecular weight. The great freedom in the film construction allowed us to realize the importance of film construction in enzymatic performance. The highest yield of products was obtained using films with a specific order of enzyme layers where the order of the enzymes corresponded with the order of the enzymatic reaction sequence. In addition, the separation between two layers of enzymes avoids any detrimental interference between them. This system well presented the superiority of controlled layering over random immobilization.

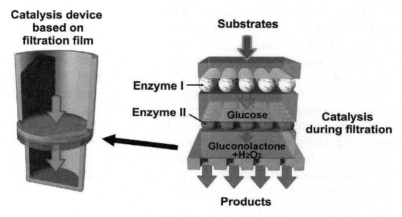

Figure 3.29 A filter catalysis system containing two enzymes.

Various types of enzyme-based systems based on the LbL technique have been fabricated. The design of the layer sequences in the system may affect the performance of enzymes. Shutave et al. prepared layered hemoglobin/polystyrene sulfonate $(Hb/PSS)_n$ films with a functional protective barrier to achieve long-time catalytic efficiency [177]. A catalase (Cat) layer was assembled on the top of the $(Hb/PSS)_n$ multilayers film. The protective barrier coating may prevent the contact of the outermost layers of Hb with hydrogen peroxide during the reaction, and thus play a crucial role for the performance of enzymes. High activity in hydrogen peroxide decomposition can be achieved in this system and Hb activity can be preserved for a longer period. Leblanc et al. prepared LbL films with an organophosphorus hydrolase that is capable of hydrolyzing a large variety of organophosphorus (OP) compounds [178]. OP derivatives are found in insecticides and pesticides, which are poisonous if ingested, inhaled, or absorbed through the skin. Organophosphorus hydrolase (OPH) is an enzyme that exhibits the ability to hydrolyze a large variety of organophosphorus compounds by producing less toxic products. Several bilayers of chitosan (CS) and poly(thiophene-3-acetic acid) (PTAA) were prepared to produce a stable supramolecular ultrathin film for OPH. It offers a cushion support to favor the adsorption and the stability of OPH. The bioactivity of OPH is shown enhanced even under harsh conditions, capable of a fast detection of paraoxon with quick recovery times. These examples again emphasize the importance of layer design in obtaining high performances of enzyme thin films.

Self-healing materials, which enable an autonomous repair response to damage, are highly desirable for the long-term reliability of woven or nonwoven textiles. Polyelectrolyte LbL films are of considerable interest as self-healing coatings due to the mobility of the components comprising the film. Dressick and Demirel et al. fabricated a stable self-healing film by alternative dip coating of polystyrenesulfonate (PSS), Squid ring teeth (SRT) proteins and urease enzyme (UE) [179]. The incorporation of SRTs brings the LbL film with unique self-healing property and high elastic modulus in both dry and wet conditions. The LbL architecture also provides a platform to stabilize the immobilized enzyme. The UE in the self-healing film showed to effectively promote the hydrolysis of urea into two ammonium species and can retain enzyme activity even after repairing large defects. Such complex films may enable a coating for multifunctional applications in programs such as Second Skin, by which advanced fabrics protect wearers from chemical and biological warfare agents by restricting the pores of the clothing to block out harmful agents [180].

Besides integrated enzymes into layered films, using a colloidal template, multi-functional capsular structures with enzymes loaded inside or in shell can be fabricated after final removal of the template. Li et al. immobilized ATP synthase into composite capsules by LbL assembly [181, 182]. They prepared polyelectrolyte LbL films on melamine-formaldehyde particles and destroyed the core giving hollow capsules covered with lipid bilayer and ATP synthase. Production of ATP from ADP was realized when a proton gradient through the membrane was applied. Caruso and coworkers invented capsosome as a microreactor with thousands of subcompartments [183]. Phospholipid liposomes (DMPC/DPPC, 80:20 wt%), including model enzyme β-lactamase, were first stabilized on a template particle by sandwiching the liposomes between a cholesterol-modified poly(L-lysine) (PLLc) precursor layer and a poly(methacrylic acid)-co-(cholesteryl methacrylate) (PMAc) capping layer (Fig. 3.30). Then five bilayers of PVP and thiol-modified PMA (PMASH) were sequentially adsorbed. They were then cross-linked and the template core was finally removed. Destruction of liposome structures by certain surfactants such as Triton X induced release of entrapped enzyme within polyelectrolyte capsule. The β-lactamase intercalated in the capsosomes well maintained its activity during the formation process. Thus the release of the catalytic function can be controlled by breaking the intact liposome in the shells.

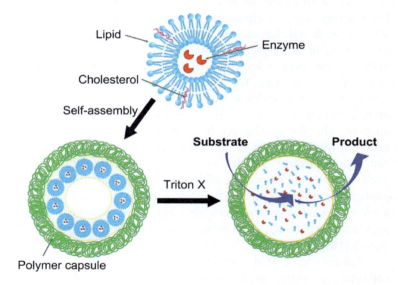

Figure 3.30 The capsosome composed of the enzyme preloaded into liposomes and embedded inside a polymer capsules.

Zhou et al. reported LbL assembled enzyme multilayers on magnetic microparticles (NNPs) exhibited improved total catalytic activity over those immobilized on flat surfaces (Fig. 3.31) [184]. Based on the strong, well-established avidin–biotin interactions, horseradish peroxidase (HRP) and glucose oxidase (GOX) was assembled into enzyme multilayers on Fe$_3$O$_4$ nanoparticles with bovine serum albumin (BSA) modified on surface, which acted as a biocompatible linker with enzyme. The enzymatic activity was shown to be increased with the number of enzyme layers on the nanoparticles. The coupled bienzyme–MNP system is shown to be suitable for rapid colorimetric quantification of glucose with 10 µM sensitivity by direct visual inspection. The sensitivity and the bienzyme coupling efficiency may also be facilely adjusted by the number of enzyme layers, or the rigid linkers in the LbL architectures.

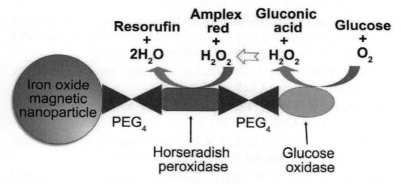

Figure 3.31 The scheme for the multilayer bienzyme–magnetic nanoparticle system based on layered adsorption.

3.6 Summary and Perspectives

The development of supermolecular catalysis systems has been under intensive investigation in recent years. Noticeable and impressive results have been achieved for supermolecular-based catalysts. The precise control over the catalytic sites in the supermolecular structures or architectures expands the new thoughts for the design of molecular catalysts and catalysis devices. It makes a great potential for further construction of unique and multi-functional catalysis systems.

Although supramolecular catalysis systems such as artificial enzymes and MOFs have been highly promising for chemical production, there are still quite a lot of challenges, including costs, stability, and scalability. Therefore, more efforts are needed to remarkably improve their overall performance for satisfying the requirements of practical applications. With more understanding on the formation of supramolecular architectures, and various synthesis techniques, undoubtedly a wide variety of multi-dimensional architectures can be available and foster the design of novel supramolecular catalysts with ever higher efficiency.

References

1. Huang, F., and Anslyn, E. V. (2015). Introduction: supramolecular chemistry, *Chem. Rev.*, **115**, pp. 6999–7000.
2. Lehn, J.-M. (1988). Supramolecular chemistry-scope and perspectives molecules, supermolecules, and molecular devices (Nobel Lecture), *Angew. Chem. Int. Ed. Engl.*, **27**, pp. 89–112.
3. Lehn, J.-M. (1985). Supramolecular chemistry: Receptors, catalysts, and carriers, *Science*, **227**, pp. 849–856.
4. Sanders, J. K. M. (1998). Supramolecular catalysis in transition, *Chem. Eur. J.*, **4**, pp. 1378–1383.
5. Meeuwissen, J., and Reek, J. N. H. (2010). Supramolecular catalysis beyond enzyme mimics, *Nat. Chem.*, **2**, pp. 615–621.
6. Leeuwen, P. W. N. M. (2008). *Supramolecular Catalysis*, Wiley-VCH Verlag GmbH & Co. KGaA (Germany) pp. 1–24.
7. Breslow, R. (1995). Biomimetic chemistry and artificial enzymes: Catalysis by design, *Acc. Chem. Res.*, **28**, pp. 146–153.
8. Wong, C.-H., and Whitesides, G. M. (1994). *Enzymes in Synthetic Organic Chemistry* (Academic Press, UK).
9. Kohen, A., and Klinman, J. P. (1998). Enzyme catalysis: Beyond classical paradigms, *Acc. Chem. Res.*, **31**, pp. 397–404.
10. Koeller, K. M., and Wong, C.-H. (2001). Enzymes for chemical synthesis, *Nature*, **409**, pp. 232–240.
11. Koshland, D. E. (1958). Application of a theory of enzyme specificity to protein synthesis, *Proc. Natl. Acad. Sci.*, **44**, pp. 98–104.
12. Eyring, H., and Stearn, A. E. (1939). The application of the theory of absolute reaction rates to proteins, *Chem. Rev.*, **24**, pp. 253–270.
13. Taylor, S. V., Kast, P., and Hilvert, D. (2001). Investigating and engineering enzymes by genetic selection, *Angew. Chem. Int. Ed.*, **40**, pp. 3310–3335.

14. Davids, T., Schmidt, M., Böttcher, D., and Bornscheuer, U. T. (2013). Strategies for the discovery and engineering of enzymes for biocatalysis, *Curr. Opin. Chem. Biol.*, **17**, pp. 215–220.
15. Motherwell, W. B., Bingham, M. J., and Six, Y. (2001). Recent progress in the design and synthesis of artificial enzymes, *Tetrahedron*, **57**, pp. 4663–4686.
16. Wei, H., and Wang, E. (2013). Nanomaterials with enzyme-like characteristics (nanozymes): Next-generation artificial enzymes, *Chem. Soc. Rev.*, **42**, pp. 6060–6093.
17. Breslow, R., and Dong, S. D. (1998). Biomimetic reactions catalyzed by cyclodextrins and their derivatives, *Chem. Rev.*, **98**, pp. 1997–2012.
18. Del Valle, E. M. M. (2004). Cyclodextrins and their uses: A review, *Proc. Biochem.*, **39**, pp. 1033–1046.
19. Barr, L., Lincoln, S. F., and Easton, C. J. (2006). Reversal of regioselectivity and enhancement of rates of nitrile oxide cycloadditions through transient attachment of dipolarophiles to cyclodextrins, *Chem. Eur. J.*, **12**, pp. 8571–8580.
20. Ortega-Caballero, F., Rousseau, C., Christensen, B., Petersen, T. E., and Bols, M. (2005). Remarkable supramolecular catalysis of glycoside hydrolysis by a cyclodextrin cyanohydrin, *J. Am. Chem. Soc.*, **127**, pp. 3238–3239.
21. Bjerre, J., Hauch Fenger, T., Marinescu, L. G., and Bols, M. (2007). Synthesis of some trifluoromethylated cyclodextrin derivatives and analysis of their properties as artificial glycosidases and oxidases, *Eur. J. Org. Chem.*, **4**, pp. 704–710.
22. Easton, C. J., and Lincoln, S. F. (1996). Chiral discrimination by modified cyclodextrins, *Chem. Soc. Rev.*, **25**, pp. 163–170.
23. Hu, S., Li, J., Xiang, J., Pan, J., Luo, S., and Cheng, J. P. (2010). Asymmetric supramolecular primary amine catalysis in aqueous buffer: Connections of selective recognition and asymmetric catalysis, *J. Am. Chem. Soc.*, **132**, pp. 7216–7228.
24. Yuan, D. Q., Lu, J., Atsumi, M., Yan, J. M., Kai, M., and Fujita, K. (2007). Cerium complexes of cyclodextrin dimers as efficient catalysts for luminol chemiluminescence reactions, *Org. Biomol. Chem.*, **5**, pp. 2932–2939.
25. Xue, S. S., Zhao, M., Ke, Z. F., Cheng, B. C., Su, H., Cao, Q., Cao, Z.-K., Wang, J., Ji, L.-N., and Mao, Z.-W. (2016). Enantioselective hydrolysis of amino acid esters promoted by bis (β-cyclodextrin) copper complexes, *Sci. Rep.*, **6**, p. 22080.

26. Dong, Z., Liu, J., Mao, S., Huang, X., Yang, B., Ren, X., Luo, G., and Shen, J. (2004). Aryl thiol substrate 3-carboxy-4-nitrobenzenethiol strongly stimulating thiol peroxidase activity of glutathione peroxidase mimic, 2,2′-ditellurobis(2-deoxy-β-cyclodextrin), *J. Am. Chem. Soc.*, **126**, pp. 16395–16404.

27. Romero-Rivera, A., Garcia-Borràsb, M., and Osuna, S. (2017). Computational tools for the evaluation of laboratory-engineered biocatalysts, *Chem. Commun.*, **53**, pp. 284–297.

28. Chen, Z., and Zeng, A. P. (2016). Protein engineering approaches to chemical biotechnology, *Curr. Opin. Biotechnol.*, **42**, pp. 198–205.

29. Leitão, A. L., Costa, M. C., and Enguita, F. J. (2017). Applications of genome editing by programmable nucleases to the metabolic engineering of secondary metabolites, *J. Biotech.*, **241**, pp. 50–60.

30. Uchiyama, T., and Miyazaki, K. (2009). Functional metagenomics for enzyme discovery: Challenges to efficient screening, *Curr. Opin. Biotechnol.*, **20**, pp. 616–622.

31. Ferrer, M., Beloqui, A., Timmis, K. N., and Golyshin, P. N. (2009). Metagenomics for mining new genetic resources of microbial communities, *J. Mol. Microbiol. Biotechnol.*, **16**, pp. 109–123.

32. Kaul, P., and Asano, Y. (2012). Strategies for discovery and improvement of enzyme function: State of the art and opportunities, *Microb. Biotechnol.*, **5**, pp. 18–33.

33. Schiraldi, C., and De Rosa, M. (2002). The production of biocatalysts and biomolecules from extremophiles, *Trends Biotechnol.*, **20**, pp. 515–521.

34. Li, S., Yang, X., Yang, S., Zhu, M., and Wang, X. (2012). Technology prospecting on enzymes: Application, marketing and engineering, *Comput. Struct. Biotechnol. J.*, **2**, pp. 1–11.

35. Uchiyama, T., and Miyazaki, K. (2009). Functional metagenomics for enzyme discovery: Challenges to efficient screening, *Curr. Opin. Biotechnol.*, **20**, pp. 616–622.

36. Rao, M. B., Tanksale, A. M., Ghatge, M. S., and Deshpande, V. V. (1998). Molecular and biotechnological aspects of microbial proteases, *Microbiol. Mol. Biol. Rev.*, **62**, pp. 597–635.

37. Gooding, O. W., Voladri, R., Bautista, A., Hopkins, T., Huisman, G., Jenne, S., Ma, S., Mundorff, E. C., Savile, M. M., and Truesdell, S. J. (2010). Development of a practical biocatalytic process for (R)-2-methylpentanol, *Org. Process Res. Dev.*, **14**, pp. 119–126.

38. Savile, C. K., Janey, J. M., Mundorff, E. C., Moore, J. C., Tam, S., Jarvis, W. R., Colbeck, J. C., Krebber, A., Fleits, F. J., Brands, J., Devine, P. N., Huisman, G. W., and Hughes, G. J. (2010). Biocatalytic asymmetric synthesis of chiral amines from ketones applied to sitagliptin manufacture, *Science*, **329**, pp. 305–309.

39. Muschiol, J., Peters, C., Oberleitner, N., Mihovilovic, M. D., Bornscheuer, U. T., and Rudroff, F. (2015). Cascade catalysis-strategies and challenges en route to preparative synthetic biology, *Chem. Commun.*, **51**, pp. 5798–5811.

40. Sattler, J. H., Fuchs, M., Tauber, K., Mutti, F. G., Faber, K., Pfeffer, J., Haas, T., and Kroutil, W. (2012). Redox self-sufficient biocatalyst network for the amination of primary alcohols, *Angew. Chem. Int. Ed.*, **51**, pp. 9156–9159.

41. Zhang, Y. H. P., Evans, B. R., Mielenz, J. R., Hopkins, R. C., and Adams, M. W. (2007). High-yield hydrogen production from starch and water by a synthetic enzymatic pathway, *Plos One*, **2**, p. e456.

42. Martín del Campo, J. S., Rollin, J., Myung, S., Chun, Y., Chandrayan, S., Patiño, R., Adams, M. W., and Zhang, Y. H. P. (2013). High-yield production of dihydrogen from xylose by using a synthetic enzyme cascade in a cell-free system, *Angew. Chem. Int. Ed.*, **52**, pp. 4587–4590.

43. Zhang, M., Mao, X., Yu, Y., Wang, C. X., Yang, Y. L., and Wang, C. (2013). Nanomaterials for reducing amyloid cytotoxicity, *Adv. Mater.*, **25**, pp. 3780–3801.

44. Qing, G., Zhao, S., Xiong, Y., Lv, Z., Jiang, F., Liu, Y., Chen, H., Zhang, M. X., and Sun, T. L. (2014). Chiral effect at protein/graphene interface: A bioinspired perspective to understand amyloid formation, *J. Am. Chem. Soc.*, **136**, pp. 10736–10742.

45. Grasso, G., Giuffrida, M. L., and Rizzarelli, E. (2012). Metallostasis and amyloid β-degrading enzymes, *Metallomics*, **4**, pp. 937–949.

46. Gao, N., Dong, K., Zhao, A., Sun, H., Wang, Y., Ren, J., and Qu, X. (2016). Polyoxometalate-based nanozyme: Design of a multifunctional enzyme for multi-faceted treatment of Alzheimer's disease, *Nano Res.*, **9**, pp. 1079–1090.

47. Wang, X., Niu, D., Li, P., Wu, Q., Bo, X., Liu, B., Bao, S., Su, T., Xu, H., and Wang, Q. (2015). Dual-enzyme-loaded multifunctional hybrid nanogel system for pathological responsive ultrasound imaging and T2-weighted magnetic resonance imaging, *ACS Nano*, **9**, pp. 5646–5656.

48. Green, M. (2004). Semiconductor quantum dots as biological imaging agents, *Angew. Chem. Int. Ed.*, **43**, pp. 4129–4131.
49. Yuan, J., Wen, D., Gaponik, N., and Eychmüller, A. (2013). Enzyme-encapsulating quantum dot hydrogels and xerogels as biosensors: Multifunctional platforms for both biocatalysis and fluorescent probing, *Angew. Chem. Int. Ed.*, **52**, pp. 976–979.
50. Yadav, R. K., Baeg, J. O., Oh, G. H., Park, N. J., Kong, K. J., Kim, J., Hwang, D. W., and Biswas, S. K. (2012). A photocatalyst-enzyme coupled artificial photosynthesis system for solar energy in production of formic acid from CO_2, *J. Am. Chem. Soc.*, **134**, pp. 11455–11461.
51. Li, B., Leng, K., Zhang, Y., Dynes, J. J., Wang, J., Hu, Y., Ma, D., Shi, Z., Zhu, L., Zhang, D., Sun, Y., Chrzanowski, M., and Ma, S. (2015). Metal-organic framework based upon the synergy of a Brønsted acid framework and Lewis acid centers as a highly efficient heterogeneous catalyst for fixed-bed reactions, *J. Am. Chem. Soc.*, **137**, pp. 4243–4248.
52. Lee, J., Farha, O. K., Roberts, J., Scheidt, K. A., Nguyen, S. T., and Hupp, J. T. (2009). Metal-organic framework materials as catalysts, *Chem. Soc. Rev.*, **38**, pp. 1450–1459.
53. Furukawa, H., Ko, N., Go, Y. B., Aratani, N., Choi, S. B., Choi, E., Yazaydin, A. O., Snurr, R. Q., O'Keeffe, M., Kim, J., and Yaghi, O. M. (2010). Ultrahigh porosity in metal-organic frameworks, *Science*, **329**, pp. 424–428.
54. Deng, H., Grunder, S., Cordova, K. E., Valente, C., Furukawa, H., Hmadeh, M., Gándara, F., Whalley, A. C., Liu, Z., Asahina, S., Kazumori, H., O'Keeffe, M., Terasaki, O., Stoddart, J. F., and Kazumori, H. (2012). Large-pore apertures in a series of metal-organic frameworks, *Science*, **336**, pp. 1018–1023.
55. Fujita, M., Kwon, Y. J., Washizu, S., and Ogura, K. (1994). Preparation, clathration ability, and catalysis of a two-dimensional square network material composed of cadmium(II) and 4,4′-bipyridine, *J. Am. Chem. Soc.*, **116**, pp. 1151–1152.
56. Corma, A., García, H., and Xamena, F. X. L. (2010). Engineering metal organic frameworks for heterogeneous catalysis, *Chem. Rev.*, **110**, pp. 4606–4655.
57. Farrusseng, D., Aguado, S., and Pinel, C. (2009). Metal–organic frameworks: Opportunities for catalysis, *Angew. Chem. Int. Ed.*, **48**, pp. 7502–7513.
58. Dhakshinamoorthy, A., and Garcia, H. (2012). Catalysis by metal nanoparticles embedded on metal-organic frameworks, *Chem. Soc. Rev.*, **41**, pp. 5262–5284.

59. Xamena, F. X. L., Abad, A., Corma, A., and Garcia, H. (2007). MOFs as catalysts: Activity, reusability and shape-selectivity of a Pd-containing MOF, *J. Catal.*, **250**, pp. 294–298.
60. Song, J., Zhang, Z., Hu, S., Wu, T., Jiang, T., and Han, B. (2009). MOF-5/n-Bu^4 NBr: An efficient catalyst system for the synthesis of cyclic carbonates from epoxides and CO_2 under mild conditions, *Green Chem.*, **11**, pp. 1031–1036.
61. Férey, G., Mellot-Draznieks, C., Serre, C., Millange, F., Dutour, J., Surblé, S., and Margiolaki, I. (2005). A chromium terephthalate-based solid with unusually large pore volumes and surface area, *Science*, **309**, pp. 2040–2042.
62. Taylor-Pashow, K. M., Rocca, J. D., Xie, Z., Tran, S., and Lin, W. (2009). Postsynthetic modifications of iron-carboxylate nanoscale metal-organic frameworks for imaging and drug delivery, *J. Am. Chem. Soc.*, **131**, pp. 14261–14263.
63. Chizallet, C., Lazare, S., Bazer-Bachi, D., Bonnier, F., Lecocq, V., Soyer, E., Quoineaud, A.-A., and Bats, N. (2010). Catalysis of transesterification by a nonfunctionalized metal–organic framework: Acido-basicity at the external surface of ZIF-8 probed by FTIR and ab initio calculations, *J. Am. Chem. Soc.*, **132**, pp. 12365–12377.
64. Tran, U. P., Le, K. K., and Phan, N. T. (2011). Expanding applications of metal-organic frameworks: Zeolite imidazolate framework ZIF-8 as an efficient heterogeneous catalyst for the Knoevenagel reaction, *ACS Catal.*, **1**, pp. 120–127.
65. Cavka, J. H., Jakobsen, S., Olsbye, U., Guillou, N., Lamberti, C., Bordiga, S., and Lillerud, K. P. (2008). A new zirconium inorganic building brick forming metal organic frameworks with exceptional stability, *J. Am. Chem. Soc.*, **130**, p. 13850.
66. Vermoortele, F., Ameloot, R., Vimont, A., Serre, C., and De Vos, D. (2011). An amino-modified Zr-terephthalate metal–organic framework as an acid-base catalyst for cross-aldol condensation, *Chem. Commun.*, **47**, pp. 1521–1523.
67. Wang, Z., and Cohen, S. M. (2007). Postsynthetic covalent modification of a neutral metal-organic framework, *J. Am. Chem. Soc.*, **129**, pp. 12368–12369.
68. Deng, H., Doonan, C. J., Furukawa, H., Ferreira, R. B., Towne, J., Knobler, C. B., Wang, B., and Yaghi, O. M. (2010). Multiple functional groups of varying ratios in metal-organic frameworks, *Science*, **327**, pp. 846–850.

69. Cohen, S. M. (2012). Postsynthetic methods for the functionalization of metal–organic frameworks, *Chem. Rev.* 112, pp. 970–1000.
70. Van Zeeland, R., Li, X., Huang, W., and Stanley, L. M. (2016). MOF-253-Pd(OAc)2: A recyclable MOF for transition-metal catalysis in water, *RSC Adv.*, **6**, pp. 56330–56334.
71. Liu, Y. Y., Leus, K., Bogaerts, T., Hemelsoet, K., Bruneel, E., Van Speybroeck, V., and Van Der Voort, P. (2013). Bimetallic-organic framework as a zero-leaching catalyst in the aerobic oxidation of cyclohexene, *ChemCatChem*, **5**, pp. 3657–3664.
72. Ma, L., Abney, C., and Lin, W. (2009). Enantioselective catalysis with homochiral metal–organic frameworks, *Chem. Sov. Rev.*, **38**, pp. 1248–1256.
73. Liu, Y., Xuan, W., and Cui, Y. (2010). Engineering homochiral metal-organic frameworks for heterogeneous asymmetric catalysis and enantioselective separation, *Adv. Mater.*, **22**, pp. 4112–4135.
74. Banerjee, M., Das, S., Yoon, M., Choi, H. J., Hyun, M. H., Park, S. M., Seo, G., and Kim, K. (2009). Postsynthetic modification switches an achiral framework to catalytically active homochiral metal-organic porous materials, *J. Am. Chem. Soc.*, **131**, pp. 7524–7525.
75. Wu, C. D., Hu, A., Zhang, L., and Lin, W. (2005). A homochiral porous metal-organic framework for highly enantioselective heterogeneous asymmetric catalysis, *J. Am. Chem. Soc.*, **127**, pp. 8940–8941.
76. Valenzano, L., Civalleri, B., Chavan, S., Bordiga, S., Nilsen, M. H., Jakobsen, S., Lillerud, K. P., and Lamberti, C. (2011). Disclosing the complex structure of UiO-66 metal organic framework: A synergic combination of experiment and theory, *Chem. Mater.*, **23**, pp. 1700–1718.
77. Vermoortele, F., Bueken, B., Le Bars, G., Van de Voorde, B., Vandichel, M., Houthoofd, K., Vimont, A., Daturi, M., Waroquier, M., Speybroeck, V. V., Kirschhock, C., and De Vos, D. E. (2013). Synthesis modulation as a tool to increase the catalytic activity of metal–organic frameworks: The unique case of UiO-66 (Zr), *J. Am. Chem. Soc.*, **135**, pp. 11465–11468.
78. Nair, V., Menon, R. S., Biju, A. T., Sinu, C. R., Paul, R. R., Jose, A., and Sreekumar, V. (2011). Employing homoenolates generated by NHC catalysis in carbon–carbon bond-forming reactions: State of the art, *Chem. Soc. Rev.*, **40**, p. 5336.
79. Fortman, G. C., and Nolan, S. P. (2011). N-Heterocyclic carbene (NHC) ligands and palladium in homogeneous cross-coupling catalysis: A perfect union, *Chem. Soc. Rev.*, **40**, pp. 5151–5169.

80. Nair, V., Vellalath, S., and Babu, B. P. (2008). Recent advances in carbon–carbon bond-forming reactions involving homoenolates generated by NHC catalysis, *Chem. Soc. Rev.*, **37**, pp. 2691–2698.
81. Oisaki, K., Li, Q., Furukawa, H., Czaja, A. U., and Yaghi, O. M. (2010). A metal-organic framework with covalently bound organometallic complexes, *J. Am. Chem. Soc.*, **132**, pp. 9262–9264.
82. Kong, G. Q., Xu, X., Zou, C., and Wu, C. D. (2011). Two metal–organic frameworks based on a double azolium derivative: Post-modification and catalytic activity, *Chem. Commun.*, **47**, pp. 11005–11007.
83. Kong, G. Q., Ou, S., Zou, C., and Wu, C. D. (2012). Assembly and post-modification of a metal-organic nanotube for highly efficient catalysis, *J. Am. Chem. Soc.*, **134**, pp. 19851–19857.
84. Deria, P., Mondloch, J. E., Karagiaridi, O., Bury, W., Hupp, J. T., and Farha, O. K. (2014). Beyond post-synthesis modification: Evolution of metal-organic frameworks via building block replacement, *Chem. Soc. Rev.*, **43**, pp. 5896–5912.
85. Pullen, S., Fei, H., Orthaber, A., Cohen, S. M., and Ott, S. (2013). Enhanced photochemical hydrogen production by a molecular diiron catalyst incorporated into a metal–organic framework, *J. Am. Chem. Soc.*, **135**, pp. 16997–17003.
86. Britt, D., Tranchemontagne, D., and Yaghi, O. M. (2008). Enhanced photochemical hydrogen production by a molecular diiron catalyst incorporated into a metal-organic framework, *Proc. Natl. Acad. Sci., U. S. A.*, **105**, pp. 11623–11627.
87. Murakami, M., Hong, D., Suenobu, T., Yamaguchi, S., Ogura, T., and Fukuzumi, S. (2011). Catalytic mechanism of water oxidation with single-site ruthenium-heteropolytungstate complexes, *J. Am. Chem. Soc.*, **133**, pp. 11605–11613.
88. Han, J. W., and Hill, C. L. (2007). A coordination network that catalyzes O_2-based oxidations, *J. Am. Chem. Soc.*, **129**, pp. 15094–15095.
89. Bromberg, L., Diao, Y., Wu, H., Speakman, S. A., and Hatton, T. A. (2012). Chromium (III) terephthalate metal organic framework (MIL-101): HF-free synthesis, structure, polyoxometalate composites, and catalytic properties, *Chem. Mater.*, **24**, pp. 1664–1675.
90. Meilikhov, M., Yusenko, K., Esken, D., Turner, S., Van Tendeloo, G., and Fischer, R. A. (2010). Chromium (III) terephthalate metal organic framework (MIL-101): HF-free synthesis, structure, polyoxometalate composites, and catalytic properties, *Eur. J. Inorg. Chem.*, **24**, pp. 3701–3714.

91. Ameloot, R., Roeffaers, M. B., De Cremer, G., Vermoortele, F., Hofkens, J., Sels, B. F., and De Vos, D. E. (2011). Metal–organic framework single crystals as photoactive matrices for the generation of metallic microstructures, *Adv. Mater.*, **23**, pp. 1788–1791.

92. Pascanu, V., Yao, Q., Bermejo Gómez, A., Gustafsson, M., Yun, Y., Wan, W., Samain, L., Zou, X., and Martín-Matute, B. (2013). Sustainable catalysis: Rational Pd loading on MIL-101Cr-NH$_2$ for more efficient and recyclable Suzuki-Miyaura reactions, *Chem. Eur. J.*, **19**, pp. 17483–17493.

93. Lu, G., Li, S., Guo, Z., Farha, O. K., Hauser, B. G., Qi, X., Wang, Y., Wang, X., Han, S., Liu, X., DuChene, J. S., Zhang, H., Zhang, Q., Chen, X., Ma, J., Loo, S. C. J., Wei, W. D., Yang, Y., Hupp, J. T., and Huo, F. (2012). Imparting functionality to a metal-organic framework material by controlled nanoparticle encapsulation, *Nat. Chem.*, **4**, pp. 310–316.

94. Zhang, W., Lu, G., Cui, C., Liu, Y., Li, S., Yan, W., Xing, C., Chi, Y. R., Yang, Y., and Huo, F. (2014). A family of metal-organic frameworks exhibiting size-selective catalysis with encapsulated noble-metal nanoparticles, *Adv. Mater.*, **26**, pp. 4056–4060.

95. Aijaz, A., Karkamkar, A., Joon Choi, Y., Tsumori, N., Rönnebro, E., Autrey, T., Shioyama, H., and Xu, Q. (2012). Immobilizing highly catalytically active Pt nanoparticles inside the pores of metal-organic framework: A double solvents approach, *J. Am. Chem. Soc.*, **134**, pp. 13926–13929.

96. Yadav, M., and Xu, Q. (2013). Catalytic chromium reduction using formic acid and metal nanoparticles immobilized in a metal-organic framework, *Chem. Commun.*, **49**, pp. 3327–3329.

97. Ishida, T., Nagaoka, M., Akita, T., and Haruta, M. (2008). Deposition of gold clusters on porous coordination polymers by solid grinding and their catalytic activity in aerobic oxidation of alcohols, *Chem. Eur. J.*, **14**, pp. 8456–8460.

98. Henschel, A., Gedrich, K., Kraehnert, R., and Kaskel, S. (2008). Catalytic properties of MIL-101, *Chem. Commun.*, **35**, pp. 4192–4194.

99. Hermannsdörfer, J., Friedrich, M., Miyajima, N., Albuquerque, R. Q., Kümmel, S., and Kempe, R. (2012). Ni/Pd@ MIL-101: Synergistic catalysis with cavity-conform Ni/Pd nanoparticles, *Angew. Chem. Int. Ed.*, **51**, pp. 11473–11477.

100. Kuo, C. H., Tang, Y., Chou, L. Y., Sneed, B. T., Brodsky, C. N., Zhao, Z., and Tsung, C. K. (2012). Yolk–shell nanocrystal@ ZIF-8 nanostructures for gas-phase heterogeneous catalysis with selectivity control, *J. Am. Chem. Soc.*, **134**, pp. 14345–14348.

101. Zhang, W., Wu, Z. Y., Jiang, H. L., and Yu, S. H. (2014). Nanowire-directed templating synthesis of metal-organic framework nanofibers and their derived porous doped carbon nanofibers for enhanced electrocatalysis, *J. Am. Chem. Soc.*, **136**, pp. 14385–14388.
102. Kim, T. K., Lee, K. J., Cheon, J. Y., Lee, J. H., Joo, S. H., and Moon, H. R. (2013). Nanoporous metal oxides with tunable and nanocrystalline frameworks via conversion of metal-organic frameworks, *J. Am. Chem. Soc.*, **135**, pp. 8940–8946.
103. Das, R., Pachfule, P., Banerjee, R., and Poddar, P. (2012). Metal and metal oxide nanoparticle synthesis from metal organic frameworks (MOFs): Finding the border of metal and metal oxides, *Nanoscale*, **4**, pp. 591–599.
104. Zhong, W., Liu, H., Bai, C., Liao, S., and Li, Y. (2015). Base-free oxidation of alcohols to esters at room temperature and atmospheric conditions using nanoscale Co-based catalysts, *ACS Catal.*, **5**, pp. 1850–1856.
105. Shen, K., Chen, L., Long, J., Zhong, W., and Li, Y. (2015). MOFs-templated Co@Pd core-shell NPs embedded in N-doped carbon matrix with superior hydrogenation activities, *ACS Catal.*, **5**, pp. 5264–5271.
106. Bao, C., Zhou, L., Shao, Y., Wu, Q., Ma, J., and Zhang, H. (2015). Palladium-loaded magnetic core-shell porous carbon nanospheres derived from a metal-organic framework as a recyclable catalyst, *RSC Adv.*, **5**, pp. 82666–82675.
107. deKrafft, K. E., Wang, C., and Lin, W. (2012). Metal-organic framework templated synthesis of Fe_2O_3/TiO_2 nanocomposite for hydrogen production, *Adv. Mater.*, **24**, pp. 2014–2018.
108. Cote, A. P., Benin, A. I., Ockwig, N. W., O'keeffe, M., Matzger, A. J., and Yaghi, O. M. (2005). Porous, crystalline, covalent organic frameworks, *Science*, **310**, pp. 1166–1170.
109. Kalidindi, S. B., Wiktor, C., Ramakrishnan, A., Weßing, J., Schneemann, A., Van Tendeloo, G., and Fischer, R. A. (2013). Lewis base mediated efficient synthesis and solvation-like host–guest chemistry of covalent organic framework-1, *Chem. Commun.*, **49**, pp. 463–465.
110. Uribe-Romo, F. J., Hunt, J. R., Furukawa, H., Klöck, C., O'Keeffe, M., and Yaghi, O. M. (2009). A crystalline imine-linked 3-D porous covalent organic framework, *J. Am. Chem. Soc.*, **131**, pp. 4570–4571.
111. Kuhn, P., Antonietti, M., and Thomas, A. (2008). Porous, covalent triazine-based frameworks prepared by ionothermal synthesis, *Angew. Chem. Int. Ed. Engl.*, **47**, pp. 3450–3453.

112. Furukawa, H., and Yaghi, O. M. (2009). Storage of hydrogen, methane, and carbon dioxide in highly porous covalent organic frameworks for clean energy applications, *J. Am. Chem. Soc.*, **131**, pp. 8875–8883.

113. Ding, S. Y., Gao, J., Wang, Q., Zhang, Y., Song, W. G., Su, C. Y., and Wang, W. (2011). Construction of covalent organic framework for catalysis: Pd/COF-LZU1 in Suzuki–Miyaura coupling reaction, *J. Am. Chem. Soc.*, **133**, pp. 19816–19822.

114. Cai, S. L., Zhang, Y. B., Pun, A. B., He, B., Yang, J., Toma, F. M., Sharp, I. D., Yaghi, O. M., Fan, J., Zheng, S.-R., Zhang, W.-G., and Liu, Y. (2014). Tunable electrical conductivity in oriented thin films of tetrathiafulvalene-based covalent organic framework, *Chem. Sci.*, **5**, pp. 4693–4700.

115. Hausoul, P. J., Eggenhuisen, T. M., Nand, D., Baldus, M., Weckhuysen, B. M., Gebbink, R. J. K., and Bruijnincx, P. C. (2013). Development of a 4,4'-biphenyl/phosphine-based COF for the heterogeneous Pd-catalysed telomerisation of 1,3-butadiene, *Catal. Sci. Technol.*, **3**, pp. 2571–2579.

116. Pachfule, P., Panda, M. K., Kandambeth, S., Shivaprasad, S. M., Díaz, D. D., and Banerjee, R. (2014). Multifunctional and robust covalent organic framework–nanoparticle hybrids, *J. Mater. Chem. A*, **2**, pp. 7944–7952.

117. Gao, W., Sun, X., Niu, H., Song, X., Li, K., Gao, H., Zhang, W., Yu, J., and Jia, M. (2015). Phosphomolybdic acid functionalized covalent organic frameworks: Structure characterization and catalytic properties in olefin epoxidation, *Micropor. Mesopor. Mater.*, **213**, pp. 59–67.

118. Jiang, H. L., Akita, T., Ishida, T., Haruta, M., and Xu, Q. (2011). Synergistic catalysis of Au@Ag core-shell nanoparticles stabilized on metal-organic framework, *J. Am. Chem. Soc.*, **133**, pp. 1304–1306.

119. Stegbauer, L., Schwinghammer, K., and Lotsch, B. V. (2014). A hydrazone-based covalent organic framework for photocatalytic hydrogen production, *Chem. Sci.*, **5**, pp. 2789–2793.

120. Zhang, Y., and Riduan, S. N. (2012). Functional porous organic polymers for heterogeneous catalysis, *Chem. Soc. Rev.*, **41**, pp. 2083–2094.

121. Hou, Y., Zhang, X., Sun, J., Lin, S., Qi, D., Hong, R., Li, D., Xiao, X., and Jiang, J. (2015). Good Suzuki-coupling reaction performance of Pd immobilized at the metal-free porphyrin-based covalent organic framework, *Micropor. Mesopor. Mater.*, **214**, pp. 108–114.

122. Lin, S., Hou, Y., Deng, X., Wang, H., Sun, S., and Zhang, X. (2015). A triazine-based covalent organic framework/palladium hybrid for

one-pot silicon-based cross-coupling of silanes and aryl iodides, *RSC Adv.*, **5**, pp. 41017–41024.

123. Pachfule, P., Kandambeth, S., Díaz, D. D., and Banerjee, R. (2014). Highly stable covalent organic framework-Au nanoparticles hybrids for enhanced activity for nitrophenol reduction, *Chem. Commun.*, **50**, pp. 3169–3172.

124. Shinde, D. B., Kandambeth, S., Pachfule, P., Kumar, R. R., and Banerjee, R. (2015). Bifunctional covalent organic frameworks with two dimensional organocatalytic micropores, *Chem. Commun.*, **51**, pp. 310–313.

125. Xu, H., Chen, X., Gao, J., Lin, J., Addicoat, M., Irle, S., and Jiang, D. (2014). Catalytic covalent organic frameworks via pore surface engineering, *Chem. Commun.*, **50**, pp. 1292–1294.

126. Fang, Q., Gu, S., Zheng, J., Zhuang, Z., Qiu, S., and Yan, Y. (2014). 3D microporous base-functionalized covalent organic frameworks for size-selective catalysis, *Angew. Chem. Int. Ed. Engl.*, **53**, pp. 2878–2882.

127. Lin, S., Diercks, C. S., Zhang, Y. B., Kornienko, N., Nichols, E. M., Zhao, Y., Kornienko, N., Nichol, E. M., Zhao, Y., Paris, A. R., Kim, D., Yang, P., Yaghi, O. M., and Chang, C. J. (2015). Covalent organic frameworks comprising cobalt porphyrins for catalytic CO_2 reduction in water, *Science*, **349**, pp. 1208–1213.

128. De Groot, F., Vankó, G., and Glatzel, P. (2009). The 1s x-ray absorption pre-edge structures in transition metal oxides, *J. Phys. Condens. Matter.*, **21**, pp. 104207.

129. Bhadra, M., Sasmal, H. S., Basu, A., Midya, S. P., Kandambeth, S., Pachfule, P., Balaraman, E., and Banerjee, R. (2017). Predesigned metal-anchored building block for in situ generation of Pd nanoparticles in porous covalent organic framework: Application in heterogeneous tandem catalysis, *ACS Appl. Mater. Interfaces*, **9**, pp. 13785–13792.

130. Corbet, J. P., and Mignani, G. (2006). Selected patented cross-coupling reaction technologies, *Chem. Rev.*, **106**, pp. 2651–2710.

131. Zhang, P., Weng, Z., Guo, J., and Wang, C. (2011). Solution-dispersible, colloidal, conjugated porous polymer networks with entrapped palladium nanocrystals for heterogeneous catalysis of the Suzuki–Miyaura coupling Reaction, *Chem. Mater.*, **23**, pp. 5243–5249.

132. Kuhn, P., Forget, A., Su, D., Thomas, A., and Antonietti, M. (2008). From microporous regular frameworks to mesoporous materials with

ultrahigh surface area: Dynamic reorganization of porous polymer networks, *J. Am. Chem. Soc.*, **130**, pp. 13333–13337.

133. Kuhn, P., Forget, A., Hartmann, J., Thomas, A., and Antonietti, M. (2009). Template-free tuning of nanopores in carbonaceous polymers through ionothermal synthesis, *Adv. Mater.*, **21**, pp. 897–901.

134. Palkovits, R., Antonietti, M., Kuhn, P., Thomas, A., and Schüth, F. (2009). Solid catalysts for the selective low-temperature oxidation of methane to methanol, *Angew. Chem. Int. Ed. Engl.*, **48**, pp. 6909–6912.

135. Chan-Thaw, C. E., Villa, A., Katekomol, P., Su, D., Thomas, A., and Prati, L. (2010). Covalent triazine framework as catalytic support for liquid phase reaction, *Nano Lett.*, **10**, pp. 537–541.

136. Uribe-Romo, F. J., Hunt, J. R., Furukawa, H., Klöck, C., O'Keeffe, M., and Yaghi, O. M. (2009). A crystalline imine-linked 3-D porous covalent organic framework, *J. Am. Chem. Soc.*, **131**, pp. 4570–4571.

137. Chen, X., Shen, S., Guo, L., and Mao, S. S. (2010). Semiconductor-based photocatalytic hydrogen generation, *Chem. Rev.*, **110**, pp. 6503–6570.

138. Maeda, K., Wang, X., Nishihara, Y., Lu, D., Antonietti, M., and Domen, K. (2009). Photocatalytic activities of graphitic carbon nitride powder for water reduction and oxidation under visible light, *J. Phys. Chem. C*, **113**, pp. 4940–4947.

139. Ding, S. Y., and Wang, W. (2013). Covalent organic frameworks (COFs): From design to applications, *Chem. Soc. Rev.*, **42**, pp. 548–568.

140. Nagai, A., Chen, X., Feng, X., Ding, X., Guo, Z., and Jiang, D. (2013). A squaraine-linked mesoporous covalent organic framework, *Angew. Chem. Int. Ed. Engl.*, **52**, pp. 3770–3774.

141. Yadav, R. K., Kumar, A., Park, N. J., Kong, K. J., and Baeg, J. O. (2016). A highly efficient covalent organic framework film photocatalyst for selective solar fuel production from CO_2, *J. Mater. Chem. A*, **4**, pp. 9413–9418.

142. Haase, F., Banerjee, T., Savasci, G., Ochsenfeld, C., and Lotsch, B. V. (2017). Structure-property-activity relationships in a pyridine containing azine-linked covalent organic framework for photocatalytic hydrogen evolution, *Faraday Discuss*, pp. 1–18.

143. Ariga, K., Ji, Q., Nakanishi, W., and Hill, J. P. (2015). Thin film nanoarchitectonics, *J. Inorg. Organomet. Polym.*, **25**, pp. 466–479.

144. Ariga, K., Ji, Q., Mori, T., Naito, M., Yamauchi, Y., Abe, H., and Hill, J. P. (2013). Enzyme nanoarchitectonics: Organization and device application, *Chem. Soc. Rev.*, **42**, pp. 6322–6345.

145. Chen, I. A., Roberts, R. W., and Szostak, J. W. (2004). The emergence of competition between model protocells, *Science*, **305**, pp. 1474–1476.

146. Wick, R., Walde, P., and Luisi, P. L. (1995). Light microscopic investigations of the autocatalytic self-reproduction of giant vesicles, *J. Am. Chem. Soc.*, **117**, pp. 1435–1436.
147. Brea, R. J., Cole, C. M., and Devaraj, N. K. (2014). In situ vesicle formation by native chemical ligation, *Angew. Chem. Int. Ed. Engl.*, **53**, pp. 14102–14105.
148. Adamala, K., and Szostak, J. W. (2013). Competition between model protocells driven by an encapsulated catalyst, *Nat. Chem.*, **5**, pp. 495–501.
149. Wick, R., and Luisi, P. L. (1996). Enzyme-containing liposomes can endogenously produce membrane-constituting lipids, *Chem. Biol.*, **3**, pp. 277–285.
150. Suzuki, K., Toyota, T., Takakura, K., and Sugawara, T. (2009). Sparkling morphological changes and spontaneous movements of self-assemblies in water induced by chemical reactions, *Chem. Lett.*, **38**, pp. 1010–1015.
151. Hardy, M. D., Yang, J., Selimkhanov, J., Cole, C. M., Tsimring, L. S., and Devaraj, N. K. (2015). Self-reproducing catalyst drives repeated phospholipid synthesis and membrane growth, *Proc. Natl. Acad. Sci. U. S. A.*, **112**, pp. 8187–8192.
152. Budin, I., and Szostak, J. W. (2011). Physical effects underlying the transition from primitive to modern cell membranes, *Proc. Natl. Acad. Sci. U. S. A.*, **108**, pp. 5249–5254.
153. Papahadjopoulos, D., Nir, S., and Duzgunes, N. (1990). Molecular mechanisms of calcium-induced membrane fusion, *J. Bioenerg. Biomembr.*, **22**, pp. 157–179.
154. Gruber, B., Kataev, E., Aschenbrenner, J., Stadlbauer, S., and Kçnig, B. (2011). Vesicles and micelles from amphiphilic zinc (II)-cyclen complexes as highly potent promoters of hydrolytic DNA cleavage, *J. Am. Chem. Soc.*, **133**, pp. 20704–20707.
155. Gorgoll, R. M., Harano, K., and Nakamura, E. (2016). Nanoscale control of polymer assembly on a synthetic catalyst-bilayer system, *J. Am. Chem. Soc.*, **138**, pp. 9675–9681.
156. Oosterom, G. E., Reek, J. N. H., Kamer, P. C. J., and Van Leeuwen, P. W. N. M. (2001). Transition metal catalysis using functionalized dendrimers, *Angew. Chem. Int. Ed.*, **40**, pp. 1828–1849.
157. Smith, D. K., and Diederich, F. (1998). Functional dendrimers: Unique biological mimics, *Chem. Eur. J.*, **4**, pp. 1353–1361.
158. Astruc, D., and Chardac, F. (2001). Dendritic catalysts and dendrimers in catalysis, *Chem. Rev.*, **101**, pp. 2991–3024.

159. Twyman, L. J., King, A. S. H., and Martin, I. K. (2002). Catalysis inside dendrimers, *Chem. Soc. Rev.*, **31**, pp. 69–82.
160. Astruc, D., Boisselier, E., and Ornelas, C. (2010). Dendrimers designed for functions: From physical, photophysical, and supramolecular properties to applications in sensing, catalysis, molecular electronics, photonics, and nanomedicine, *Chem. Rev.*, **110**, pp. 1857–1959.
161. Helms, B. A., and Frechet, J. M. J. (2006). The dendrimer effect in homogeneous catalysis, *Adv. Syn. Catal.*, **348**, pp. 1125–1148.
162. Bagul, R. S., and Jayaraman, N. (2014). Multivalent dendritic catalysts in organometallic catalysis, *Inorg. Chim. Acta*, **409**, pp. 34–52.
163. Wang, D., and Astruc, D. (2013). Dendritic catalysis—Basic concepts and recent trends, *Coord. Chem. Rev.*, **257**, pp. 2317–2334.
164. Imaoka, T., Tanaka, R., and Yamamoto, K. (2006). Synergetic activation of carbon dioxide molecule using phenylazomethine dendrimers as a catalyst, *J. Poly. Sci. Part A: Polym. Chem.*, **44**, pp. 5229–5236.
165. Garcia-Bernabé, A., Kràmer, M., Olah, B., and Haag, R. (2004). Syntheses and phase-transfer properties of dendritic nanocarriers that contain perfluorinated shell structures, *Chem. Eur. J.*, **10**, pp. 2822–2830.
166. Garcia-Bernabè, A., Tzschucke, C. C., Bannwarth, W., and Haag, R. (2005). Supramolecular immobilization of a perfluoro-tagged Pd-catalyst with dendritic architectures and application in Suzuki reactions, *Adv. Synth. Catal.*, **347**, pp. 1389–1394.
167. Yamamoto, K., Higuchi, M., Shiki, S., Tsuruta, M., and Chiba, H. (2002). Stepwise radial complexation of imine groups in phenylazomethine dendrimers, *Nature*, **415**, pp. 509–511.
168. Yamamoto, K., Kawana, Y., Tsuji, M., Hayashi, M., and Imaoka, T. (2007). Additive-free synthesis of poly (phenylene oxide): Aerobic oxidative polymerization in a base-condensed dendrimer capsule, *J. Am. Chem. Soc.*, **129**, pp. 9256–9257.
169. Plault, L., Hauseler, A., Nlate, S., Astruc, D., Ruiz, J., Gatard, S., and Neumann, R. (2004). Synthesis of dendritic polyoxometalate complexes assembled by ionic bonding and their function as recoverable and reusable oxidation catalysts, *Angew. Chem. Int. Ed.*, **43**, pp. 2924–2948.
170. Jahier, C., Cantuel, M., McClenaghan, N. D., Buffeteau, T., Cavagnat, D., Agbossou, F., Carraro, M., Bonchio, M., and Nlate, S. (2009). Enantiopure dendritic polyoxometalates: Chirality transfer from dendritic wedges

to a POM cluster for asymmetric sulfide oxidation, *Chem. Eur. J.*, **15**, pp. 8703–8708.

171. Jahier, C., Mal, S. S., Kortz, U., and Nlate, S. (2010). Dendritic zirconium-peroxotungstosilicate hybrids: Synthesis, characterization, and use as recoverable and reusable sulfide oxidation catalysts, *Eur. J. Inorg. Chem.*, **10**, pp. 1559–1566.

172. Diallo, A. K., Boisselier, E., Liang, L., Ruiz, J., and Astruc, D. (2010). Dendrimer-induced molecular catalysis in water: The example of olefin metathesis, *Chem. Eur. J.*, **16**, pp. 11832–11835.

173. Wang, D., and Astruc, D. (2013). Dendritic catalysis—Basic concepts and recent trends, *Coord. Chem. Rev.*, **257**, pp. 2317–2334.

174. Ariga, K., Ji, Q., Hill, J. P., Bando, Y., and Aono, M. (2012). Forming nanomaterials as layered functional structures toward materials nanoarchitectonics, *NPG Asia Mater.*, **4**, p. e17.

175. Onda, M., Lvov, Y., Ariga, K., and Kunitake, T. (1996). Sequential reaction and product separation on molecular films of glucoamylase and glucose oxidase assembled on an ultrafilter, *J. Ferment. Bioeng.*, **82**, pp. 502–506.

176. Onda, M., Lvov, Y., Ariga, K., and Kunitake, T. (1996). Sequential actions of glucose oxidase and peroxidase in molecular films assembled by layer-by-layer alternate adsorption, *Biotechnol. Bioeng.*, **51**, pp. 163–167.

177. Shutava, T. G., Kommireddy, D. S., and Lvov, Y. M. (2006). Layer-by-layer enzyme/polyelectrolyte films as a functional protective barrier in oxidizing media, *J. Am. Chem. Soc.*, **128**, pp. 9926–9934.

178. Constantine, C. A., Mello, S. V., Dupont, A., Cao, X., Santos, D., Oliveira, O. N., Strixino, F. T., Pereira, E. C., Cheng, T.-C., Defrank, J. J., and Leblanc, R. M. (2003). Layer-by-layer self-assembled chitosan/poly(thiophene-3-acetic acid) and organophosphorus hydrolase multilayers, *J. Am. Chem. Soc.*, **125**, pp. 1805–1809.

179. Gaddes, D., Jung, H., Pena-Francesch, A., Dion, G., Tadigadapa, S., Dressick, W. J., and Demirel, M. C. (2016). Self-healing textile: Enzyme encapsulated layer-by-layer structural proteins, *ACS Appl. Mater. Interfaces*, **8**, pp. 20371–20378.

180. Daniels, G. A., and Petrovich, P. A. (2012). Conformable self-healing ballistic armor. US Patent 7,966,923.

181. Duan, L., He, Q., Wang, K., Yan, X., Cu,i Y., Möhwald, H., and Li, J. (2007). Adenosine triphosphate biosynthesis catalyzed by F0F1

ATP synthase assembled in polymer microcapsules, *Angew. Chem. Int. Ed.*, **46**, pp. 6996–7000.

182. Qi, W., Duan, L., Wang, K., Yan, X., Cui, Y., He, Q., and Li, J. (2008). Motor protein CF0F1 reconstituted in lipid-coated hemoglobin microcapsules for ATP synthesis, *Adv. Mater.*, **20**, pp. 601–605.

183. Städler, B., Chandrawati, R., Price, A. D., Chong, S.-F., Breheney, K., Postma, A., Connal, L. A., Zelikin, A. N., and Caruso, F. (2009). A microreactor with thousands of subcompartments: Enzyme-loaded liposomes within polymer capsules, *Angew. Chem. Int. Ed.*, **48**, pp. 4359–4362.

184. Garcia, J., Zhang, Y., Taylor, H., Cespedes, O., Webb, M. E., and Zhou, D. (2011). Multilayer enzyme-coupled magnetic nanoparticles as efficient, reusable biocatalysts and biosensors, *Nanoscale*, **3**, pp. 3721–3730.

Chapter 4

Organic Polymers, Oligomers, and Catalysis

Gaulthier Rydzek and Amir Pakdel

WPI Center for Materials Nanoarchitectonics,
National Institute for Materials Science,
1-1 Namiki, Tsukuba, Ibaraki 305-0044, Japan

RYDZEK.Gaulthier@nims.go.jp

4.1 Introduction

In recent years, most traditional catalysts such as organocatalysts, organo-metallic complexes, and nanoparticles have been combined with polymeric supports. In spite of the fact that porous inorganic supports still hold the leads in supported catalysis field, polymer supports have also attracted significant attention due to their capability for providing original and adaptive design strategies. Catalyst supports based on organic oligomers, dendrimers and polymers became extremely popular in the last decade because of their versatility:

Soft Matters for Catalysts
Edited by Qingmin Ji and Harald Fuchs
Copyright © 2020 Jenny Stanford Publishing Pte. Ltd.
ISBN 978-981-4774-66-6 (Hardcover), 978-1-351-27284-1 (eBook)
www.jennystanford.com

- Polymers offer a broad variety of functionalization points whose nature and density can be tuned according to the catalyst needs.
- The structure of polymeric supports is adjustable both at the molecular and supramolecular levels. The activity and durability of resulting catalyst can be dramatically improved by a proper design.
- Polymeric materials can be soluble, dispersible, and insoluble, giving access to homogeneous, heterogeneous and "mixed system" catalysis.

Despite these advantages, supports based on polymers were traditionally considered only as relatively inert materials providing some convenient porosity. However, past decades have witnessed a change of perspectives with the idea, and polymeric materials are now believed to be able to improve a catalyst activity, durability, and recyclability. If this trend continues, one can expect that several polymer-supported catalysts will reach cost-efficiency level and have an impact in real world. This chapter offers an overview of current and emerging strategies used for performing homogeneous and heterogeneous catalysis with the help of polymers, oligomers, and dendrimers.

4.2 Definition and Characteristics of Polymer-Supported Catalysts

In the early 20th century, polymers were thought to be composed of aggregated colloids from small molecules. Nowadays, polymers are accepted as a mixture of large molecules formed by a repetition of one or several basic units, the monomers, linked together by covalent, coordination, or supramolecular bonds. The number of monomeric units is both large (superior to 20,000 for polymers, inferior for oligomers) and variable. Therefore, a polymer batch is often characterized by a polydisperse distribution of molecular weights. This concept of monomeric units linked together to form a polymer with a specific chemical structure was gradually accepted from the 1920s with the work of H. Staudinger, who received the Nobel prize in 1953. Depending on the monomer arrangement, in a linear, branched, or reticulated manner, polymers with very different physical and chemical properties can be achieved even if their chemical composition is the same (Fig. 4.1).

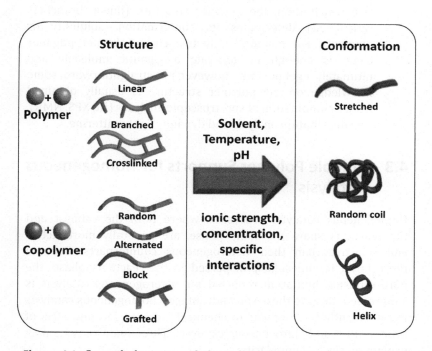

Figure 4.1 General depiction of the relationship between monomers, polymers, copolymers and their chain conformation. Adjusting each of these parameters allows tuning the chemical properties of the polymer matrix as well as its nano- and micro-structure arising from inter-chain interactions.

Flory's investigations in this field (Nobel prize in 1974) allowed enouncing the fundamental physical and chemical principles of modern polymer science. Consequently, an efficient and reproducible use of polymers necessitates knowing the following parameters:

- Chemical composition: the exact composition of a polymer can be challenging to analyze, for instance in the case of a copolymer or when pendant chains have been functionalized.

- Average size and distribution: usually determined by gel permeation chromatography. However, solubility issues may limit the measurement of these parameters.
- Polymer structure: for a given chemical composition and size distribution, the polymer structure (linear, branched, reticulated) determines its conformation, solubility or reactivity. For instance, only two cross-links per polymer chain is enough to generate a gigantic molecule and ultimately a gel polymer. However, not straightforward, some knowledge on the polymer structure is usually obtained from a combination of spectroscopic analyses (IR, XPS, NMR), size distribution analysis and dynamic light scattering.

4.3 Soluble Polymer Supports for Homogeneous Catalysis

Homogeneous catalytic processes, where both the catalyst and the reactants share the same phase, are generally more active and selective than their heterogeneous counterparts. Because the catalyst is molecularly dispersed in the reactants phase, the pore diffusion limitation reported for heterogeneous catalysis is suppressed. Despite these solid advantages, homogeneous catalysis is significantly less popular in chemical industry (around 20% of processes) than heterogeneous catalysis. Three main reasons may explain this lack of popularity:

- **Economic reasons**. Most homogeneous catalysts are using either well-designed organocatalysts or transition metals, including relatively expensive palladium, ruthenium, rhodium, and iridium. To complex these metals, it is often needed to design specific ligands, for instance when enantioselective catalysis is required. Pharmaceutical applications use ligands whose price reaches between 5000 and 20,000 euros/kg. Fortunately, some inexpensive ligands such as triphenylphosphine also provide good results for simple reactions.
- **Stability issues**. Since pricy materials are used, it becomes crucial to decrease the amount of catalyst needed to perform the process, for instance by reaching high turnover numbers.

This depends mainly on two parameters: the rate of the reaction and the stability of the catalyst. Some homogeneous catalysts could reach turnover numbers as high as 106, but this is usually possible only in mild conditions, limiting their applicability [1].
- **Separation issues**. Since the catalyst, reactant, and products are in the same phase, separation at end of the process, when it is possible, is expensive and challenging.

Employment of homogenous catalysis is, however, growing rapidly, predominantly within the polymer and pharmaceuticals industries. The use of soluble polymers and oligomers is generally intended to introduce a stimulus-responsive separation of catalysts from the reaction phase by solvent change. This ultimately allows reusing the catalyst and decreasing both the process cost and the product contaminations.

4.3.1   PEG-Bound Catalysts

Poly(ethylene glycol) (PEG) is a linear polymer composed of ethylene oxide as a monomer. It is gaining attention in several fields, due to its solubility in water, biocompatibility, and limited cost. In most cases, practical applications use a "short" oligomeric PEG chain (typically around 5000 to 10,000 M_w), allowing working at higher concentrations without forming a gel. These oligomers are soluble in several solvents, including water, toluene, dichloromethane, and dimethylformamide. However, PEG is insoluble in hexane, diethyl ether, isopropyl alcohol and cold ethanol. PEG represents thus a very attractive substrate for binding homogeneous catalysts that can be separated easily by changing the solvent. Consequently, a large number of strategies were developed for grafting organocatalysts on terminal position of PEG chains. However, before choosing PEG as a catalyst support, one should keep in mind the following properties of this oligomer:
- **Stability**. PEG is a remarkably stable oligomer. Its backbone structure is barely unaffected by pH changes in water. However, oxidation can occur and PEG chain-terminal hydroxyl groups begin to degrade into formic acid at 70°C, at least for short oligomers [2].

- **Interferences during the catalysis reaction.** The accessibility of grafted catalysts can be compromised by the PEG chain conformation. To counteract this effect, several groups used linkers to "separate" the catalyst from PEG chains [3]. The PEG chain size and the solvent are also two key parameters that can influence the coiling process.
- **Loading rates.** Most studies on PEG-bound catalysts are using linear PEG chains. This often restricts the number of catalysts to only one per chain, a very low number in comparison with grafting numbers on pendant-chains polymers for instance.
- **Solubility.** PEG is a very soluble oligomer, typically reaching solubility parameters above 60% (in weight) in water. However, the viscosity of obtained solutions increases with the molecular weight of chains. In the case of highly viscous PEG-supported catalyst mixtures, it becomes uncertain whether the catalyst will still behave as a homogeneous catalyst. This effect tends also to limit the catalyst-loading capability in PEG solutions.

The most popular homogeneous catalysts have been coupled to PEG chains, owing to the relatively easy modification of hydroxyl terminal functions. Thus a large range or organocatalysts and organometallic catalysts based on Salen, porphyrin, or phosphine ligands have been used. In most cases, the turnover efficiency of PEG-coupled catalysts could reach values comparable to non-coupled catalysts. The following pages give an overview of the most widely studied PEG-bound catalysts.

Organocatalysts have been emerging rapidly during the last decades as a greener alternative to organometallic catalysts. It allows working in aerobic and wet conditions while of course eliminating the problem of metal leaching. However, the ligand stability upon the catalyst recycling is an issue that is not systematically addressed, and most investigations have been restricted to 5 catalysis cycles.

2,2,6,6-tetramethyl-1-piperidinyloxyl (TEMPO) is a very popular organocatalyst for selective oxidation of primary alcohols to aldehydes in presence of stoichiometric oxidants such as 1,3,5-trichloro-2,4,6-triazinetrione, (bis(acetoxy)iodo)benzene and NaOCl. Several groups developed various grafting approaches of one TEMPO group per PEG chain, ranging from direct S_N2 reaction [4] to the use of linkers [5].

Reactions were conducted in dichloromethane, acetic acid or water/dichloromethane mixtures. TEMPO grafted on PEG retains a catalytic activity comparable to unsupported TEMPO, reaching a turnover frequency as high as 0.333 s^{-1}. By itself, this result is enthusing since it means that grafted TEMPO still behaves as a homogeneous catalyst. Recycling is ensured by precipitation of PEG in diethyl ethers. In most studies, 3 to 6 catalysis cycles could be performed without significant activity loss. However, calculating a precise turnover number for TEMPO–grafted PEG catalysts may necessitate performing more catalysis cycles. Both the grafting strategy and the length of the PEG chain influence the catalytic performances. On the one hand, using a linker between the TEMPO group and the PEG chain seems to protect it from hindrance by PEG's chain conformation [4]. On the other hand, longer PEG chains (higher than 5000 M_w) precipitate more easily, facilitating the catalyst recycling. Interestingly, when 2 or 4 TEMPO per PEG chain were grafted, turnover frequencies increased up to fivefold, illustrating the superiority of branched architectures when neighboring active centers can act cooperatively (Fig. 4.2) [6].

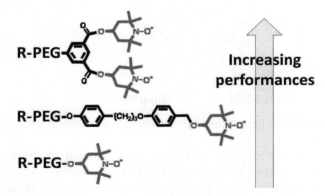

Figure 4.2 Immobilization strategies of TEMPO (depicted in red) on PEG chains. Catalysis performances were improved by using either a linker or reticulated structures.

The group of amino acids is very popular for designing oganocatalysts since the chirality of amino acids chirality gives access to enantioselective metal-free catalysis. This ability makes amino acid-based catalysts promising candidates for processes in the pharmaceutical industry. For instance, proline and tyrosine derivatives have been extensively studied for catalyzing of the aldol reaction between ketones and aldehydes. Early studies in 2002 reported the coupling of a proline derivative to a PEG_{5000} chain [7]. The corresponding PEG-supported catalyst was tested for hydroxyacetone aldolization with cyclohexanal, reaching a modest 45% yield after 60 h, but an enantiomeric excess of 96%. The enantioselectivity was maintained after recycling the catalyst three times. Such PEG-supported proline catalysts were later tested for a range of couplings including Michael, and iminoaldol reactions [8, 9].

Among the most popular organometallic ligand for catalysis, phosphine, diphenylethylenediamine, porphyrins and 1,1'-Bi-2-naphthol (BINOL) have been coupled to PEG chains and tested for epoxidation, hydrogenation, and coupling reactions (Fig. 4.3).

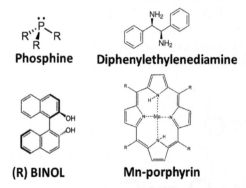

Figure 4.3 Popular ligands used for performing organometallic catalysis.

BINOL was coupled to PEG_{5000} and allowed the formation of a calcium-based catalyst [10]. It was tested for the asymmetric Michael coupling of cyclopentenone with dimethylmalonate in toluene, reaching high yields but poor enantiomeric excess (around 40%). The PEG-bound catalyst precipitated upon addition of diethyl ether in the reaction mixture. It was then reused three times without significant decrease of the yield.

Porphyrins are common ligands for transition metals, allowing performing catalytic reactions in organic solvents. Manganese porphyrin complexes were coupled to PEG$_{5000}$ chains (one porphyrin per PEG) and tested for alkenes oxidation reactions in presence of PhIO [11]. Cyclooctene was typically oxidized with a 96% yield in the presence of this PEG-bound catalyst. Recycling was performed seven times and it still exhibited a 63% yield during the last cycle.

Diphenylethylenediamines are popular chiral ligands that give access to asymmetric reactions. For instance, diphenylethylenediamine complexes with Ru(II) ions have been studied for enantioselective hydrogenation reaction of ketones [12]. This type of complex was coupled to a PEG$_{2000}$ chain for achieving recycling properties. For instance, aromatic ketones were converted in the corresponding alcohols with this catalyst, reaching yields and regioselectivities typically higher than 90%, even after three recycling cycles. However, significant Ru leaching up to 5–10% of the loaded metal seemed to occur during the catalyst recovery.

Phosphine ligands are extremely popular for designing palladium catalysts. They have been coupled on PEG chains since the early 1960s. Consequently, abundant literature is available on that system. Recent examples include Suzuki coupling, amination reactions, and asymmetric allylic alkylation reactions. Upon recycling of the PEG-supported phosphine catalysts, the sensitivity of phosphine to oxidation may limit the catalysis efficiency [13]. This sensitivity was decreased by attaching the phosphine moieties to dicyclohexyl(2′,6′-dimethoxy-2-biphenylyl) groups. The resulting PEG-bound catalyst could complex palladium ions and catalyze the aryl amination of 2,5-dimethylchlorobenzene by morpholine [14]. The turnover frequency obtained was 1000-fold greater than that of the same catalyst bound to SiO$_2$. However, the yield still decreased when the catalyst was recycled.

4.3.2 Non-Cross-Linked Polystyrene-Bound Catalysts

Cross-linked polystyrene is probably the most popular polymeric platform for catalyst immobilization, and played a key role in the development of peptide synthesis and combinatorial chemistry. Consequently, a substantial number of catalyst coupling methods have been transferred to linear non-cross-linked polystyrene.

This polymer is soluble in THF, benzene, dichloromethane, diethyl ether and chloroform, while insoluble in water, methanol, and hexane [15]. It appears to be very complementary to PEG by giving access to a larger number of organic solvents and higher operating temperatures. However, contrary to PEG "direct" functionalization methods, incorporation of catalysts in linear polystyrene necessitates the prior introduction of functional monomers during the polymerization (yielding to a copolymer), or a post-polymerization functionalization procedure. Non-cross-linked polystyrene-bound catalysts are extensively used for coupling reactions, such as Suzuki, Stille, and Heck couplings. Earliest attempts consisted of incorporating triphenylphosphine derivative in polystyrene. This was achieved by copolymerization of styrene with either 4-styryl-diphenylphosphine or 4-diphenylarsinostyrene (Fig. 4.4) [16, 17]. The obtained copolymers could then be used as ligands for palladium in coupling reactions. For instance, Suzuki reaction performed with this copolymer was achieved with similar yields (around 85%) as with non-grafted triphenylarsine [16, 17] while allowing recycling five times by precipitation in methanol. Similar copolymer derivatives were proven efficient for the Pd-catalyzed Stille coupling of 4-bromoanisole with phenyltributyltin [18]. More complicated monomers, such as palladacycle-containing styrene could also be used for copolymerization with styrene, giving access to the Heck coupling reaction in the presence of palladium acetate [19, 20].

Another approach for grafting catalysts on polystyrene consists of modifying it after its (co)-polymerization (Fig. 4.4). For instance, the copolymer composed of chloromethylstyrene and styrene was used to anchor SPhos phosphine and ammonium catalysts. The corresponding polymer-bound catalysts were used for Suzuki coupling, nucleophilic substitution, halogen exchange, and alkylation of phenols [14, 21]. This post-polymerization anchoring approach was used later to bind proline moieties to linear polystyrene. The resulting functional polymer was used for catalyzing an aldol coupling and recycled five times without decreasing the yield (70%) nor the diastereoselectivity and the enantioselectivity (more than 90%) of the reaction [22].

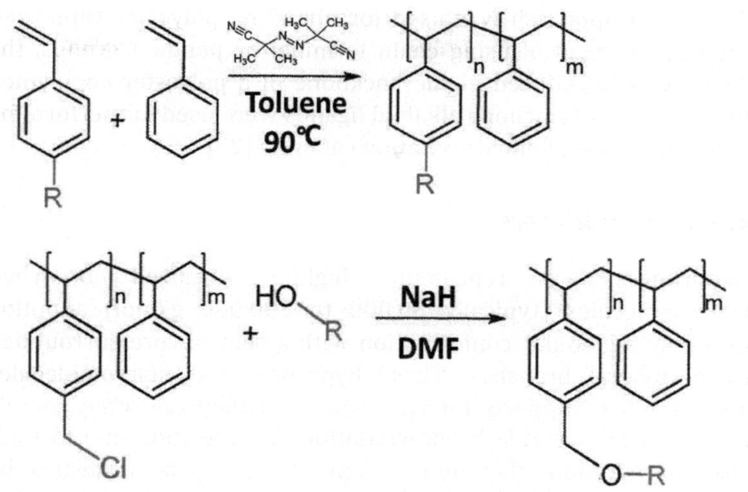

Figure 4.4 Strategies of ligand immobilization on a non-cross-linked polystyrene by copolymerization or post-functionalization. R is typically a diphenylphosphine or a diphenylarsine ligand.

4.3.3 Other Soluble Polymer-Based Catalysts

Several polymers, including poly(methacrylate), polyester and poly(norbornene), could be dissolved in water or organic solvents and were used for homogeneous catalysis. Hydrogenation reactions catalyzed by ruthenium complexes were performed in an isopropanol/toluene mixture by using both pyridine and bipyridine bearing poly(norbornene) polymers as a support [23]. Catalytic reduction of cyclohexanone reached a turnover number of 300, decreasing by less than 5% upon recycling. Poly(norbornenes) were also modified with TEMPO and Salen ligands as pendant groups. This library of modified poly(norbornene) opened the route for catalyzing various reactions including epoxidation of aromatic olefins with Mn-salen [24], addition of CN to unsaturated imides with Al-salen [25], and oxidation of primary alcohols to aldehyde with TEMPO [26]. Also used as pendant groups, norephedrine ligands were incorporated into a poly(methacrylate) derivative. The corresponding polymer was used for designing Ru(II) hydrogenation catalysts allowing the conversion of aromatic ketones to the corresponding alcohols [27].

A different approach was used for obtaining polyester-supported catalysts. Instead of using chain terminal or pendant groups, the ligand was introduced in the backbone of a polyester copolymer. In that case, bis-cinchona alkaloid ligands were used for performing Os-complex based dihydroxylation catalysis [28].

4.3.4 Dendrimers

Dendrimers are repeatedly highly organized branched macromolecules (typically 50,000 to 200,000 g/mol) adopting generally a globular conformation with a central core surrounded by peripheral branches. These hyperbranched macromolecules are attractive supports for catalysts since they are often soluble but easily recoverable by precipitation and filtration due to their size. In addition, they offer numerous catalysis strategies by either confining reactant in their inner cavities, or immobilizing catalysts in the core moiety and on the peripheral groups (Fig. 4.5). Because of the close proximity between peripheral groups, cooperative effect between catalysts on the surface of dendrimers frequently occurs [29]. Several families of dendrimers have been developed during the past two decades, including poly(amidoamines) (PAMAM), poly(propyleneimine), and poly(benzylethers). Controlling the following parameters is the key point to design a dendrimer-based catalyst:

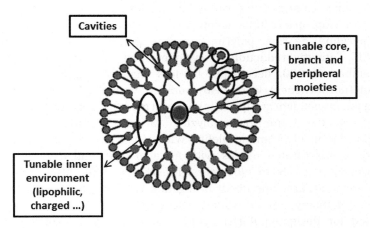

Figure 4.5 Typical structure of a dendrimer and its main functionalization points.

- The recoverability, conformation, and catalysis cooperativity largely depend on the size of the dendrimer, i.e., its polymerization degree, called specifically generation number. Contrary to polymers, dendrimers tend to be characterized by a monodisperse weight distribution.
- The internal architecture of the dendrimer, induced by the initial core, the type of dendrons and the synthesis approach used. Intentionally designing an architecture-controlled dendrimer with channels and cavities is a catalysis strategy discussed in the "heterogeneous catalyst" section.
- The chemical peripheral groups. Both functionalization abilities and solubility are tuned by branch-terminal chemical groups. Amine functions and carboxylic acids are often used as an initial platform.

The use of organocatalysts is often emphasized for their greener aspect since no metal leaching occurs during catalysis. Grafting organocatalysts on dendrimers was performed in several studies, in a very similar way as PEG-supported organocatalysts. However, a main difference stands on the fact that dendrimers, on the contrary to single PEG chain-supported catalysts, can render better catalysis performances than the unsupported organocatalyst. This effect is generally attributed to cooperativity between catalytic centers or local trapping of reactants and is called positive dendritic effect. However, in some cases, the steric hindrance of dendrimer branches tends to reduce the catalyst performance in term of yield or enantiomeric excess; this is designated as a negative dendritic effect. In general, dendrimers allow much more grafting points than single chain polymers for functionalization with organocatalysts, opening the route toward a broader variety of catalysis strategies. Consequently, oxidation, reduction, bond coupling, and cleavage reactions have all been catalyzed with dendrimer-supported organocatalysts [30]. In this chapter, we will distinguish two main approaches whether the organocatalyst moieties have been grafted peripherally or incorporated in the dendrimer core or backbone.

4.3.4.1 Peripheral grafting of organocatalysts

This synthesis approach is easier to implement since it "only" involves a final functionalization of pre-assembled dendrimers.

Conformational and structural changes of the dendrimer are thus limited. The resulting dendrimer-supported catalysts exhibit a large number of accessible external active centers. However, they often exhibit lower performances than core and backbone modified dendrimers. When the generation number of such peripherally modified dendrimer catalyst increases, the catalyst recovery process is eased, but the yield and the stereo-selectivity tend to meet a maximum at a low generation number (typically less than 5) before eventually decreasing.

Both oxidation and reduction reactions were investigated with peripherally modified dendrimer catalysts. In that case, the dendrimer backbone has to be designed to resist to the redox conditions used, for instance, by using ethylene glycol or benzylether moieties. Terminally grafted arylchalcogenides moieties containing selenium and tellurium have allowed catalyzing the oxidation of cyclohexene by hydrogen peroxide [31]. The most studied reduction reaction with dendrimer-supported organocatalyst is the conversion of ketones to alcohols. For instance, prolinol moieties coupled on arylether dendrimers have allowed catalyzing the borane reduction of ketones with an enantiomeric excess superior to 90% without performance fading after recycling five times [32].

Dendrimer-supported amino acids have been extensively studied for asymmetric catalysis. For instance, proline derivatives have been peripherally coupled to poly(propyleneimine) dendrimers for asymmetric reactions [33]. When the generation number increased from 1 to 5, the number of proline derivatives per dendrimer increased from 4 to 64 (Fig. 4.6). However, this did not result in an increase of the catalysis performance for aldol reaction: The yield obtained was comparable to the unsupported proline but the enantioselectivity dropped from 50–70% (first generation) to 20–40% (fifth generation). A similar decrease of stereoselectivity was observed in several studies where proline was peripherally grafted on dendrimers with a peptide-based [34], or pyridyl-based backbone [35]. These results suggest that both folding and high generation number induce a negative effect on stereoselectivity by steric hindrance. In a similar way as amino acids, immobilization of chiral amino alcohols on dendrimers has been tested for enantioselective catalysis. Chiral ephedrine derivatives were immobilized on the corona of carbosilane and polystyrene-based dendrimers [36–39]. The corresponding catalysts were tested for diethyl- or diisopropylzinc addition to aryl aldehydes and aryl N-diphenylphosphinylimine in toluene. Performances similar to the unsupported catalyst were achieved, which suggests the lack of any dendritic effect occurrence.

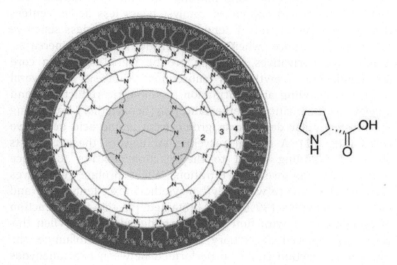

Figure 4.6 Structures of proline and of a generation 5 poly(propyleneimine) dendrimer. Peripheral functionalization is typically performed by amide bonding.

Peripherally functionalized dendrimers were also used for bond cleavage such as hydrolysis or decarboxylation reactions. Simple primary amine terminal polyamidoamine (PAMAM) dendrimers could efficiently catalyze the hydrolysis of p-nitrophenol [40–42]. Increasing the generation number of PAMAM from 1 to 4 also increased the positive dendritic effect, up to 28-fold better than the control [42]; however, this effect eventually decreased for higher generation numbers. Another example of bond cleavage catalyzed by PAMAM and poly(propyleneimine)-supported catalysts stands for the decarboxylation reaction [43–45]. In that case, the catalyst was assembled by peripheral grafting of alkyl and ether chains on dendrimers.

4.3.4.2 Incorporation organocatalysts in the core or backbone

This synthesis strategy can have a strong impact on the dendrimer structure and conformation. Therefore, most studies choose to use the catalyst as an initial core and then grow the dendrimer from it, in one direction, limiting structural problems. The resulting dendrimer-supported catalyst bears less active centers than peripherally modified dendrimers, but is often less sensitive to steric hindrance when the generation number increases. Amino acid derivatives including proline were used as a core for dendrimer growth and subsequent catalysis of several reactions including aldol formation, asymmetric reductions, and alkylations. Dendritic polyether and poly(benzylether) are among the most popular dendrimers grown from amino acid derivative cores (Fig. 4.7). A typical recycling strategy of these catalysts consists of adding MeOH to the reaction mixture to induce precipitation, followed by filtration. When proline derivatives were used as a core with poly(benzylether), excellent yields and enantioselectivities (99%) were obtained for the aldol reaction of cyclohexanone with functional benzaldehydes [46]. When this core was replaced by tertiary amines like triethanolamine, the nitro aldol reaction could be performed between benzaldehydes and nitroalkanes [47]. Interestingly, increasing the generation number of this kind of dendrimer-supported organocatalysts resulted in lower catalytic efficiency but a nearly constant stereoselectivity [48].

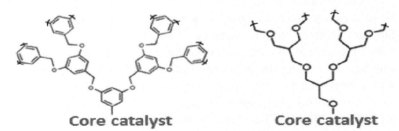

Figure 4.7 Incorporation of the catalyst in the core of poly(benzylether) and polyether dendrimers.

Dendrimers based on poly(benzylether) were broadly used for designing core-immobilized organocatalysts for reduction reactions [49–51]. For instance, prolinol derivatives were used as cores for performing enantioselective reduction of ketones, in presence of borane. In most cases, yields higher than 80% with enantiomeric excess higher than 90% were obtained [52]. A core-immobilized prolinol catalyst could be reused five times without any performance loss [53]. This type of core-modified poly(benzylether) is not restricted to reduction reaction catalysis. Modifying a core based on prolinol with trimethylsilylether allowed performing the enantioselective Michael addition of 3-methylbutanal with nitrostyrene, reaching 99% of enantiomeric excess (Fig. 4.7) [54].

Oxidation reactions with core-functionalized dendrimer-supported organocatalysts were catalyzed in presence of O_2 and peroxides. By using flavin as a core of a poly(benzylether) dendrimer, a mimic of flavoenzymes was created. In presence of O_2, this catalyst could outperform riboflavin for N-benzyl-1,4-dihydronicotinamide oxidation [55]. Singlet oxygen could also be generated by engineering dendrimers with photosensitizing cores such as tetraalkoxybenzophenone derivatives [56]. This strategy could be improved further by using porphyrin photosensitizers for both the core and the branches of the dendrimer. The corresponding photocatalyst could be recycled three times for converting 1-methyl-1-cyclohexene into a mixture of three peroxides [57]. Peptide derivatives incorporated on dendrimers exhibits also interesting catalytic properties for bonds cleavage. In that sense, they are promising enzyme mimic platforms. Several esters could be hydrolyzed by using multifunctional dendrimers almost entirely composed of aspartate, histidine and serine derivatives [58, 59]. Histidine derivatives were found to play a critical role in such a catalysis system. For instance, when a dendrimer constituted of only histidine-serine was assembled from generation 1 to 4, the catalysis efficiently increased 140,000-fold compared to the 4-methylimidazole catalyst [60].

4.3.4.3 Dendrimer-supported metal–ligand complexes

A majority of the most popular organometallic catalysts have been coupled to dendrimers including bispyridine, diphenylamine, phosphine, carbine, and binaphthol moieties. In presence of metal ions, these modified dendrimer-supported pre-catalysts are able to form metal–ligand complexes and subsequently catalyze reactions. Rhodium (Rh), palladium (Pd), ruthenium (Ru) and copper (Cu)-based ions are typically used. However, metal leaching is still a strong limiting point of this family of catalysts, restricting its application in the pharmaceutical industry. Recently, many efforts to address this problem have been performed by using phosphorus dendrimers. In this chapter, we will review the main types of metal–ligand couples currently used for dendrimer-supported organometallic catalysis.

Cu(II)-loaded dendrimers were obtained by peripheral grafting of bis(oxazoline) [61] or bipyridine [62] moieties on poly(benzylether) and subsequent incubation with Cu(II) triflate complexes. These dendrimer-supported organometallic catalysts exhibited Lewis acid character that allowed catalyzing aldol and Diels-Alder reactions. For instance, cyclopentadiene, 1,3-cyclohexadiene, and dimethylbutadiene could all react with activated dienophiles in dichloromethane with yields higher than 85% [62]. Recycling the catalyst was possible by changing the solvent to hexane, and catalytic performances were preserved after five cycles. Recently the need of Cu(I) ions-based catalysts has emerged due to the rise of the Cu(I) catalyzed alkyne-azide "click" reaction (CuAAC). PAMAM dendrimers offer ideal platforms for designing this type of catalyst, without any functionalization step, due to their ability to complex Cu(II) ions which can subsequently be converted into Cu(I) [63]. The resulting catalyst was reusable by ultrafiltration and allowed performing faster click couplings between azido-propanol and propargyl-alcohol in water. When the click catalyst was designed as a Cu(I) core surrounded by dendritic branches, catalysis performances increased significantly [64].

Rh(I) catalysts have been extensively used with dendrimer-supported diphosphines or monophosphines for hydrogenation reactions [35, 65, 66]. Poly(amidoamines), poly(ethyleneimine), and poly(benzylethers) have all been tested with that approach. For instance, pyrphos-Rh(norbornadiene) terminated dendrimers allowed catalyzing the asymmetric hydrogenation reaction of Z-methyl acetamidocinnamate, but exhibited a negative dendritic effect for both the activity and the stereoselectivity [67]. When the pyrphos moiety was immobilized in the core of the dendrimer, the dendrimer-supported Rh(I) catalysis of the reaction reached a constant stereoselectivity higher than 97%

for all reuse cycles, but the yield dropped from 94% to 55% after three cycles [68]. Rhodium complexes were also formed with phosphoramidite and 2,2'-bis(diphenylphosphino)-1-1'-binaphthyl ligands [62, 69, 70], exhibiting a positive dendritic effect for the hydrogenation of 2-acetamidocinnamate with up to 99% enantiomeric excess with high turnover frequencies.

Ru(II) catalysts are commonly obtained by complexation with diamine ligands such as diphenyldiamine and 2,2'-bis(diphenylphosphino)-1,1'-binaphthyl) derivatives. The corresponding catalysts were used for asymmetric hydrogenation of aryl alkyl ketones [71, 72]. The best results were obtained by using a chiral diamine dendrimer combined with 2,2'-bis(diphenylphosphino)-1,1'-binaphthyl) complexes of $RuCl_2$. Catalytic hydrogenation of acetonaphthone could be performed with a 99% yield and a 94% enantiomeric excess [73]. Even after recycling the catalyst twice by precipitation in methanol, this high enantioselectivity was preserved. However, some metal leaching occurred during the reaction that could be limited by increasing the dendrimer generation number.

A range of coupling reactions largely benefited from Pd-based catalysis during the past decades. Dendrimer bearing peripheral chiral phosphines were successful at allylic substitution reactions with various leaving groups [74]. The dendritic polymer was easily recovered by solvent precipitation, and reaction yields remained almost constant upon three cycles. On the contrary, the enantioselectivity of the reaction dropped significantly after the third cycle. The phosphine ligand stability becomes indeed an issue when the catalyst is to be reused. In an attempt to generate more stable catalysts, biarylphosphine-terminal dendrimers have been synthesized and tested for Pd-catalyzed Suzuki coupling [75]. High yields were obtained, similar to those given by the unsupported catalyst. However, performances began to fade from the fourth and fifth cycles, dropping down to no activity during the sixth cycle.

Metallocene moieties have been grafted on dendrimers and tested for catalysis. The most popular examples probably deal with ferrocenyl moieties that can easily be grafted on amine functions found in dendrimers like PAMAM [76] or on trietylamine cores. The resulting dendrimers possess redox properties. Recently, Ru and Pd complexes could be coupled on these ferrocene-grafted dendrimers, allowing designing redox switchable catalysts (Fig. 4.8) [77]. For instance, the conversion of 1-octen-3-ol to 3-octanone was largely catalyzed by dendrimer-supported (Fc, Ru) catalysts but was shut down by the oxidized form of this dendrimer. This process was reversible upon reduction of the functional dendrimer, gaining back 96% of its catalysis efficiency.

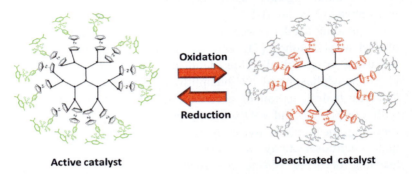

Figure 4.8 Redox-switchable dendrimer-supported catalyst. Reversible oxidation of ferrocenyl moieties reduces the catalysis activity of peripheral Ru complexes [77].

4.4 Polymers and Heterogeneous Catalysts

Heterogeneous catalysis is defined by a phase separation between the catalyst and the reactants. Typically, reactants are in liquid or gas phase while the catalyst is in solid phase. Intrinsically, this strategy possesses desirable advantages like an easy separation between the catalyst, reactants and products, an easier integration in industrial processes and better scaling-up capabilities than homogeneous catalysts. For these reasons, this catalysis strategy is largely preferred by industries. Because of the phase separation, the reaction occurs on the surface of the material, triggering the need of high specific surface area materials. Sometimes, the catalyst itself can be composed of highly porous materials, or else it is dispersed on adequate supports. A last approach consists of immobilizing homogeneous catalysts on heterogeneous supports. Although silica is probably the most popular material to support such catalysts, polymers play a growing role on all these strategies by helping on the synthesis of porous catalysts or by providing tunable supports.

4.4.1 Polymer for Immobilizing Homogeneous Catalysts

Immobilization of homogeneous catalysts on polymeric substrates is a very dynamic research field that has attracted considerable attention during the last decades. Generally speaking, it intends to facilitate the separation of products from the catalysts, ultimately allowing reusing them several times. Among other advantages, reduced toxicity, cleaner chemistry, and adaptability to continuous flow processes are commonly cited. In the literature, catalysts obtained by using these processes are depicted as mixed-phase, polymer-bound, immobilized, or heterogenized catalysts. A main argument motivating this research is the cost decrease of the catalyst use. To be of interest for industry, this cost must decrease enough to also cover the fees and risks induced by the heterogenization process. However, the immobilization is often performed at the price of lower catalysis performances such as lower yields, lower regioselectivity, longer reaction times and smaller turnover numbers. In fact, there are little examples of immobilized homogeneous catalysts that reached equal or higher performances than their homogeneous counterparts. For these reasons, it was estimated that heterogenized catalysts cost 1.5–3 times more than their homogeneous counterparts [78]. It was even recently suggested that almost all heterogenization methods invented over the years will never be used in large scale by industry [1]. Consequently, this type of catalyst occupies a niche-market in the pharmaceuticals industry, for instance for the purification of molecular libraries in solution. Far from discouraging scientists to investigate this field, this situation should stimulate research on resolving the phenomena limiting the applicability of heterogenized catalysts. Broadening the impact of this field to more industry processes is an immense and emergent challenge: there is plenty of room for exciting research.

4.4.1.1 Polystyrene-supported catalysts

Cross-linked polystyrene is probably the most popular polymeric support for heterogenized catalysts. For instance, the chloromethylated cross-linked polystyrene, also named Merrifield resin, offers a convenient functionalization point. Thus, several ligands for designing homogeneous catalysts, including carbenes, xyliphos, and phosphines ligands were covalently immobilized on cross-linked polystyrene matrices. The corresponding Ru, Ir, and Pd catalysts were synthesized and their activity was tested for hydrogenation, ring closing metathesis and coupling reactions. Heterogenized Ru-based Grubbs and Hoveyda catalysts could reach comparable turnover numbers as their homogeneous counterparts for catalyzing various ring closing metathesis reactions [79]. However, no recycling could be performed, due to the degradation of the catalyst. Immobilization of iridium-based xyliphos complexes on polystyrene allowed catalyzing the asymmetric hydrogenation of imines [80]. In that case, the heterogenization process resulted in much less active catalysts than homogenous competitors, but a similar enantioselectivity of around 75% was obtained. Although several attempts for immobilizing palladium complexes on polystyrene have been performed, the formation of palladium nanoparticles after one catalytic run is suspected [81–86]. This type of polymer-supported nanoparticles is discussed later in this chapter.

4.4.1.2 Azlactone-containing polymers

Azlactone functional groups have attracted much attention in recent years because they can react with a variety of nucleophilic species by ring opening reaction. Consequently a library of polymers bearing azlactone as pendant groups has been developed [87]. This emerging type of polymers constitutes promising support materials for protein-related catalysis, since azlactone moieties react rapidly and selectively with lysine amino acids [88]. For instance a porous copolymer monolith composed of poly(2-vinyl-4,4-dimethylazlactone-co-acrylamide-co-ethylene methacrylate) could successfully immobilize trypsin [89]. The resulting heterogenized catalyst retained a high hydrolysis activity for casein, ultimately allowing working under flow rate conditions (Fig. 4.9).

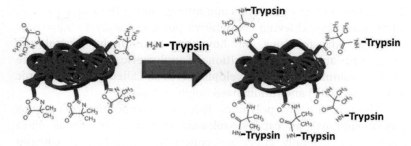

Figure 4.9 Covalent immobilization of proteins (trypsin for instance) on a polymer monolith containing azlactone moieties. The resulting functional monolith allows performing recyclable asymmetric catalysis.

4.4.1.3   Ion exchange approach

Polymer materials offer the advantage to trap counterions to ensure the electroneutrality of the system. This ability to act like an ion-exchange material offers an alternative immobilization strategy for the heterogenization of catalysts. For instance, negatively charged Nafion, bearing pendant sulfonate groups, has been extensively used for immobilizing cationic catalysts by electrostatic interactions. Because the inner-pores of the polymer matrix have generally a higher local proton concentration than the bulk, a special focus was brought to the heterogenization of rhodium-based complexes for asymmetric hydrogenation reactions of methyl 2-acetamidoacrylate [90–92]. This strategy allowed reaching similar catalysis performances compared with the homogeneous catalysts competitors [93]. However, up to now limited metal loading could be obtained (around 1%) and both the catalyst efficiency and its leaching were highly dependent on the type of solvent used, due to swelling of the polymeric matrix [44, 45].

4.4.1.4   Porous organic polymers and rationalized polymer design

Recently more attention has been paid to polymeric matrices to achieve a better cooperativity between the support and the catalyst [96]. Potential benefits of this approach include an enhanced activity of catalysts and attenuation of their deactivation. The quest for achieving cooperative polymer/catalyst composites triggers investigations on how one can finely control the structure and the composition of polymeric materials. Some great examples are the development of so-called porous organic polymers (POP) and rationalized polymer design.

Porous organic polymers face the competition of porous silica supports as catalysis-relevant materials. If porous silica possesses generally superior properties, its surface is covered with silanol groups that interact strongly with catalysts [97, 98]. Polymeric supports typically address this issue but offer poorly defined structures, resulting in diffusion limitation and polymer swelling during catalysis. This leads ultimately to limited performances. A way to limit these drawbacks consists of developing porous organic polymers composed of selected monomers that are able to generate well-controlled networks [99–103]. Catalysts can be incorporated into POP either by complexation or as part of the polymeric backbone. The most popular monomers for assembling POPs include porphyrins, triazine, bispyridine, and isocyanate building blocks [104].

Phthalocyanines-based POPs were developed for hosting various transition metals, including cobalt and iron. The obtained polymeric networks could reach high specific surface areas ranging from 120 m^2g^{-1} to 866 m^2g^{-1} [105, 106]. When tested for hydrogen peroxide decomposition, cobalt POP catalysts achieved rate constants two orders of magnitude higher than cobalt phthalocyanine monomers. A similar enhanced catalytic performance was obtained with iron POP catalysts for hydroquinone oxidation. This type of catalyst was typically reused three to four times without fading of performances [107, 108]. Other popular building blocks for designing POPs include nitrogen-containing ligands like triazine and bispyridine. These functional groups have high affinity for metal ions. Consequently, POPs able to coordinate iron, manganese, platinum, palladium and rhenium were developed [109–114]. Typical specific surface area of the obtained materials exceeded 500 m^2g^{-1}. This porosity and versatility allowed performing recyclable catalysis of a broad range of reactions including additions, oxidations, and C–C couplings.

The concept of rationalized polymer design arose from the necessity of developing heterogenized catalysts that are more stable and more active than their unsupported counterparts. These two goals are usually not fulfilled by traditional heterogenization approaches that rely on polymeric matrices. This alternative strategy is based on carefully controlling the polymer backbone, the catalyst site density and the type of bonds between the catalyst and its support. For instance, Salen type complexes containing transition metal ions have been heterogenized on polymers as pendant groups, interchain linkers, backbone units, dendritic units and chain terminal groups. With this respect, the following trends were found for catalysis with rationally designed polymer-supported salen complexes [115]:

- **Choice of polymer**: reagents have a better access inside polymer matrices with superior solubility and porosity (if the matrix morphology is maintained during the solvation process).
- **Grafting site**: catalysts linked as pendant groups are more active than catalysts immobilized in the polymer backbone or as inter-chain linkers.
- **Molecular rigidity**: flexible linkers between the catalyst and the matrix must be preferred, as long as inter-sites interactions are limited.

4.4.2 Polymers for Designing Heterogeneous Catalysts

Polymers and polymeric materials are widely used for forming and stabilizing catalytic nanoparticles, as well as limiting the leakage of catalytic nanoparticles during catalysis [116]. The polymeric material used is typically considered as an inactive matrix capable of providing tunable porosity and stabilization features by electrostatic and steric interactions. The resulting hybrid composites are anticipated to retain the properties of each component, with sometimes additional synergetic effects between the organic and inorganic phases. Polymer-supported nanoparticles (NPs) of metals and metal-oxides are used in various applications, such as sensors and heterogeneous catalysts. NPs are obtained either by in situ growth within the polymer matrix or by ex situ dispersion in the polymeric material. Both strategies are very popular since they allow using several kinds of nanoparticles or host polymers. However, the ex situ approach faces significant limitations for achieving homogeneous and well-dispersed nanoparticles.

On the contrary, in situ approaches generally yield well-dispersed nanoparticles with better size distribution and improved stability [117]. Considering the number of polymer/nanoparticle combinations possible, there is definitely plenty of room for development of this field; however, the synthesis of polymer-supported nanoparticle materials still needs more standardization. In recent years, intelligent polymer materials have been developed, providing new synthesis routes for heterogeneous catalysts where polymers play a more significant role as described in the "emergent materials" section.

4.4.2.1 Ex situ synthesis of polymer-supported catalysts

This synthesis approach relies on pre-formed NPs that will be physically encapsulated within a polymer matrix. The entrapment method usually leads to films, membranes, and monoliths. Because these approaches imply synthesizing NPs in a first step, it is confronted to the typical drawbacks of bulk NP synthesis that are agglomeration and polydispersity of NPs. However, once the encapsulation process is performed, the resulting NPs gain in stability. Encapsulation of NPs can be achieved by impregnating polymer matrices with NPs or by polymerization of the matrix in a colloidal NP dispersion.

Impregnation of polymeric materials with NPs is typically performed by mixing polymer blends with preformed NPs followed by casting and solvent evaporation. In that case, electrostatic interactions are often used for dispersing NPs in the polymer (polyvinyl alcohol and Nafion being the most popular). Sometimes the layer-by-layer film deposition method was also used [118]. These methods allow obtaining hybrid films, membranes, and monolithic materials with a large range of polymers and nanoparticles [119]. However, such materials were rarely studied for catalysis applications, probably because of the competition with in situ grown polymer-supported NPs. For instance, mixtures of gold nanoparticles and PNIPAM have been spin-coated, leading to functional films with plasmonic abilities [120–122]. In some cases, catalysis could be performed by using decorated polymeric matrices, in which NPs are distributed only on the outer-surface of the material. Gold NPs are typically used for performing this approach since they easily adsorb on surfaces containing thiol, amine and carboxylic groups [123, 124]. Recently Au NPs were adsorbed on a chitosan-coated polyurethane matrix, allowing to catalyze the reduction of 4-nitrophenol [125]. Recyclability was demonstrated by performing 10 catalysis cycles without activity loss.

Polymerization of monomers inside of NP dispersions is an attractive alternative for ex situ preparation of polymer-supported NP catalysts. This strategy generally yields good spatial distribution of NPs within the polymeric matrix. Several monomers, including acrylamide, styrene and urethane derivatives were chemically polymerized in situ in the presence of tin oxide, iron oxide, and barium titanate nanoparticles [126–128]. In a similar way, living polymerization techniques have been extensively studied to synthesize such composites as they allow including metal compounds in the polymer structure [129]. There are plenty of recent examples of NP/polymer composites obtained by in situ polymerization, including (respectively) TiO_2 NPs in poly(methyl methacrylate) [130], Pd NPs in poly(2-hydroxyethyl methacrylate) [131], and Au NPs in polyaniline [132]. These composites open broad catalysis perspectives including photocatalysis, organic catalysis and electrocatalysis.

4.4.2.2 In situ synthesis of polymer-supported catalysts

This strategy is by far the most popular for assembling polymer/metal NPs composites [133]. It consists of synthesizing the catalytic NPs inside a pre-formed polymeric framework. The polymer matrix acts as a confined medium for the synthesis, but it also plays a prominent role for stabilizing the obtained NPs and preventing their aggregation. This process is generally performed in two steps: a sorption or impregnation of the matrix with ionic precursors and their subsequent reduction or precipitation in situ by using a chemical of photochemical stimulus. Because the obtained polymer-supported nanoparticles with this approach are smaller (up to 1–3 nm diameter NPs) and better dispersed than those synthesized ex situ, a very large number of polymers and precursor ions have been investigated.

In this respect, ion exchange resins, hydrogels and polyelectrolytes [134] are among the most popular matrices [135–140]. However, in many cases metallic ions are reduced into NPs by the polymers themselves without using any additional reagent or stimulus [141–143]. This occurs especially in the case of gold, platinum, palladium, and silver ions in the presence of nitrogen and sulfur containing polymers. In some reports, both the monomer polymerization and the ions reduction occur simultaneously [144, 145]. In this chapter, we will focus on three recent trends: the use of polysaccharides as a polyelectrolyte matrix, the development of colloidal polymer-supported nanoparticles, and the emergence of dendrimer-based alternatives.

Polysaccharides such as chitosan, alginate, and carrageenans are broadly available bio-sourced polymers possessing high binding ability for metals due to the presence of hydroxyl, carboxyl and amino groups in their structures. They were recently investigated for preparing versatile polymeric aerogels characterized by a high specific surface area (up to 500 m^{2}g^{-1}) and a macroporous framework. These matrices grant the absence of diffusion limitation and provide a high density of functional groups, constituting ideal materials for immobilizing catalysts. As a result, Pd(0) catalysts were immobilized on alginate, carrageenans, and chitosan matrices, reaching turnover numbers of, respectively 500, 190, and 40 for catalyzing the substitution reaction of an allyl carbonate with morpholine [146]. This strong influence of the polymeric support on the catalysis efficiency was correlated to a stronger electrostatic attraction in alginate (one carboxylic group per monomer) than that in carrageenans (one sulfate group for two monomers) and chitosan. Alginate aerogels, obtained by complexation with calcium ions, were used for ionic exchange with Pd cations and their subsequent reduction by ethanol into NPs [147]. The size of Pd NPs was finely tuned by simple control of the palladium loading in the matrix. The resulting material exhibited good catalysis activity and recyclability for Suzuki C–C coupling reactions reaching up to 94% conversion rate when Pd NPs were tuned to 2.5–3.5 nm.

Colloidal dispersions of polymer-supported nanoparticles allow both catalyzing reactions in quasi-homogeneous conditions and easily recovering the catalyst by precipitation or filtration processes. A broad range of amphiphilic copolymers have been developed for assembling micelles that are able to complex metal ions and promote the in situ growth and stabilization of NPs (Fig. 4.10) [148, 149]. For instance, an amphiphilic polystyrene-polyethyleneoxide block copolymer allowed forming micelles in presence of a surfactant and was used as a reactor for growing Pd nanoparticles [150]. These catalytic colloids were as efficient as homogeneous Pd catalysts for Heck couplings and could be reused three times with yields around 90%. Microgels constitute another fruitful approach for synthesizing colloidal polymer-supported NPs [138, 151, 152]. Several water-soluble polymers, including poly(1-vinyl imidazole), polymethacrylate, and polyacrylamide were used for synthesizing micro and nano hydrogels, followed by subsequent growth of Fe, Cu, Co, and Ni NPs in situ [153]. The resulting colloidal dispersions were tested as catalysts for the reduction of 4-nitrophenol. Recovery of the catalyst by simple rinsing with water allowed performing four catalysis cycles, 72% of the initial activity being retained [153].

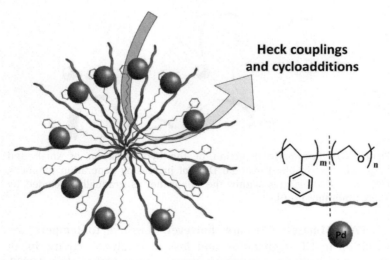

Figure 4.10 Polymeric micelles obtained from a polystyrene-polyethyleneoxide block copolymer in presence of the cetylpyridinium chloride surfactant for supporting Pd NPs.

Dendrimer-encapsulated nanoparticles (DEN) have been recently developed as an alternative strategy for stabilizing metallic nanoparticles that can be used as heterogeneous catalysts [154–157]. This concept consists of looking at dendrimers as monomolecular micelles whose size and properties can be tuned by the generation degree. The core of large dendrimers is able to stabilize NPs while its periphery protects these catalytic centers. These DEN allow preparing 1–4 nm sized NPs that are difficult to make and stabilize with other methods. DEN-based catalysts generally compare favorably with other supported NP methods in term of catalysis efficiency and recyclability [158, 159]. For instance, Pd(II) ions were coordinated by PAMAM and subsequently chemically reduced into Pd(0). This process resulted in the-growth of nanoparticles encapsulated inside the dendrimer for generation numbers higher than 4. The catalytic efficiency for alkene hydrogenation reactions was tuned by the generation number of PAMAM: Generation 4 allowed reactions of small and large alkenes, while only small alkenes could react with generation 6 and both small and large substrates were rejected by generation 8 (Fig. 4.11).

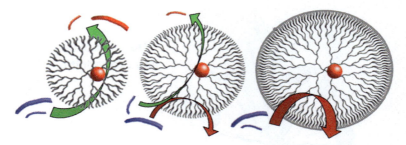

Figure 4.11 Schematic depiction of the hydrogenation of small and large alkenes (in blue) over Pd DEN of increasing generation numbers. Diffusion of substrates within the dendrimer is gradually excluded by steric hindrance [160].

PAMAM-based DEN are, however, sensitive to temperature, resulting in NP aggregation and loss of catalysis activity. In an attempt to perform catalysis at temperatures higher than 100°C, triazole containing dendrimers were prepared. These ligands exhibited great stabilization properties for Pd and Au NPs, allowing catalyze several C–C coupling reactions [160–162, 69]. Part of the good catalysis efficiency of DEN-based catalyst can be attributed to trapping of substrates inside the dendrimer, which are then acting as nanoreactors [163]. There is no doubt that this field will grow significantly in coming years; however, the price efficiency of DEN-based catalyst remains to be demonstrated.

4.5 Emerging Materials

For both homogeneous and heterogeneous catalysis fields, the tremendous development of polymers has recently opened up new opportunities. The tendency is clearly to stop considering polymer matrices as relatively inert supports, but rather trying to use polymers as intelligent materials capable of adding value to the catalyst. In this chapter, polymer-supported catalysts based on "emergent" strategies are described, including the use of stimuli-responsive polymers for recovering homogeneous catalysts, multifunctional polymers for tandem reactions, and nanostructured polymeric supports.

4.5.1 Thermoregulated Strategies for Homogeneous Catalysis

Thermoresponsive polymers attracted considerable attention during the last decades as they undergo an abrupt change of solubility with respect to a critical temperature. This phenomenon can occur in water or other solvents, with many polymers including polystyrene (in cyclohexane and butylacetate), polymethylmethacrylate (in 2-propanone), poly(vinylcaprolactame) (in water) and PEG (in water). Typically, solvated polymer chains adopt an expanded coil conformation that will collapse to form compact globules when a critical temperature is reached. If this phenomenon occurs upon heating up or cooling down the solution, these critical temperature values are called "lower critical solution temperature" (LCST) and "upper critical solution temperature" (UPST), respectively. A thermoresponsive polymer can present one or several critical temperatures (Fig. 4.12).

Thermoresponsive polymer-supported catalysts have been developed according to two LCST-based approaches:

A first strategy consists of performing a standard homogeneous catalysis below the LCST, generally at room temperature, and then heat up for recovering the catalyst by precipitation and filtration. This approach offers the advantage to be very similar to homogeneous catalysis; however, few studies actually investigated the number of times the catalysts could be recycled. The well-known Poly(N-isopropylacrylamide), (PNIPAM) has been used for this approach since its LCST in water is conveniently set up at 32°C. For instance, a Rh(I) hydrogenation catalyst was immobilized on PNIPAM, tested on 3-buten-1-ol, and reached a turnover frequency of 3000. The catalyst was recycled by heating up at 40°C and filtration [164].

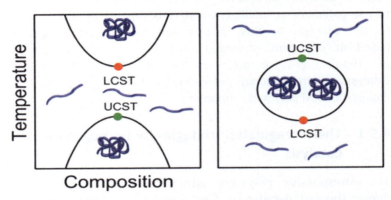

Figure 4.12 Schematic representation of the phase behavior of a thermoresponsive polymer possessing both a LCST and a UCST.

A second strategy relies on a biphasic system: Catalysts and substrates are in different phases, and heating up above the LCST will allow the catalyst to migrate to the next phase (typically the organic phase). These "thermoregulated phase transfer catalysts" have been extensively studied by using PEG as a stimulus-responsive oligomer. This approach is adapted to reactions in organic solvent and those requiring elevated temperatures (typically 100°C). However, the catalysis stereo-selectivity can be affected by the temperatures [165]. For instance, several pegylated triphenylphosphine ligands have been prepared for immobilizing rhodium and ruthenium. The corresponding catalysts were used for alkene hydroformylation and nitroarene reduction [165, 166]. Yields generally superior to 80% and very limited decrease of performances upon recycling were obtained with this strategy.

4.5.2 Poly(Ionic Liquid)-Supported Catalysts

Ionic liquids (ILs) are an extremely diverse family of molten salts formed of organic cations coordinated with inorganic or organic anions via electrostatic interactions [167]. These compounds have recently met a tremendous development, and there are now several examples of ILs used as stabilizing agents for NPs and as supports for reusable catalysts [168]. IL groups can be grafted on a polymer backbone (polyvinylic backbones for instance), leading to a new type of polyelectrolytes, called poly(ionic liquids) (PILs) [169]. These PILs benefit from the specific desirable properties of polymers and ionic liquids such as a high charge density, a high polarity, conductivity and a low flammability, making them attractive materials for catalysis (Fig. 4.13) [170]. Because of the polar environment, PILs can intrinsically trigger polar reactions. For instance, a fluoro-functionalized PIL recently exhibited intrinsic catalytic activity for CO_2 cycloaddition to epoxides, reaching yields higher than 90% even after five reuse cycles [171]. However, the fastest-growing application of PILs is as catalyst supports and stabilizers.

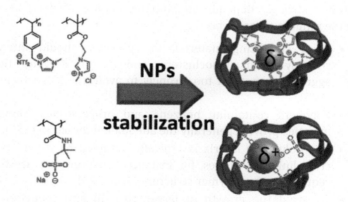

Figure 4.13 Examples of polystyrene, polyacrylic acid and polymethacrylic acid-based PILs and their stabilization effect on NPs.

Both in situ and ex situ NP synthesis strategies can be performed with PILs, owing to their polyelectrolyte nature. A broad range of NPs, including Pd, Pt, Au, Rh and Ag nanoparticles were immobilized in PILs [172–177]. In almost every case a homogeneous NP dispersion within the matrix was obtained;

however, the particle size varied from 1 nm up to 70 nm. These differences were often attributed to the type and the size of the organic cation used as IL group [170]. The resulting PIL-supported NP catalysts were tested for a variety of reactions including electrocatalysis, C–C couplings and hydrogenations. By tuning both the size of embedded NPs and the nature of the PILs used, optimized catalysts could be designed. Turnover frequencies higher than the corresponding non-supported catalyst were typically reported. However, ionic liquids are not yet widely used in industry due to their cost-intensive preparation. No doubt that this drawback will disappear with the emergence of more commercially available ILs.

4.5.3 Polymer-Supported Nanoclusters

Metal clusters, defined as particles smaller than 2 nm and composed of less than 100 atoms, are anticipated to provide new catalytic properties because of their unique size and electronic structure. At the moment, Pd, Au and Pt polymer-supported nanoclusters have been the most studied ones. Their in situ synthesis strongly depends on both the nature of the polymer and the reducer reagent chosen.

- As a stabilizing material, the polymer should interact strongly with nanoclusters enough to prevent their aggregation but not too strongly to avoid their deactivation [178].
- Polymers containing benzene and other cyclic aromatic moieties provide generally the adequate interaction [179].
- The polymeric matrix is typically obtained by reticulation: cross-linking moieties, for instance epoxy groups, are thus required in the polymer structure [180–183].
- To avoid the growth of larger NPs, in situ reduction of metal ions is performed with a strong reducing reagent (like $NaBH_4$) with an excess of polymeric coordination sites compared to metal ions [184].

Polymer-supported Pd nanoclusters were synthesized by using copolymers of polystyrene derivatives and tested for alkene hydrogenation in THF at room temperature. Yields higher than 90% were obtained with a variety of alkene compounds, and no activity fading was noticed upon recycling [185, 186]. This strategy seems thus to refrain both poisoning and leaching of Pd during catalysis. Gold nanoclusters were also extensively studied, in particular for catalyzing oxidation reactions (Fig. 4.14) [184]. Water-soluble polyvinylpyrrolidone (PVP) was found to both stabilize and modulate the electronic structure of Au clusters as well while allowing catalysis in aqueous dispersions [187]. A cooperative effect between the polymer and Au clusters enhanced the catalytic efficiency for CO aerobic oxidation, that could even be further improved by adding Ag dopant [188]. No doubt that such kind of rationally designed polymer-supported clusters for catalysis will meet a strong interest in future. For instance, multimetallic nanoclusters immobilized on polystyrene copolymer derivatives were recently developed for catalyzing the aerobic oxidation of alcohols [189].

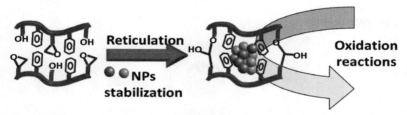

Figure 4.14 Design strategy of alloyed cluster catalysts (typically Au-Pt) by in situ stabilization and reticulation with a cross-linkable polystyrene-based copolymer.

4.5.4 Nanostructured Polymeric Materials

In recent years, a large number of studies have taken the problem of catalyst design from the polymer engineering point of view. The main idea is to control the polymeric matrix morphology at the nano level, allowing the emergence of synergetic, confinement and hierarchical effects for catalysis [190–192].

With this respect, the development of POPs matrices with a hierarchical structure is one of the most promising ways for obtaining nanostructured catalytic materials in scalable quantities. The size distribution of macropores (>50 nm), mesopores (2 nm) and micropores (<2 nm) within POPs was tuned by using hard-templating [193, 194], soft-templating [195] and template-free [196] strategies. The resulting matrices were used for immobilization of organocatalysts (prolinol [197], pyridine [194] and binol [198] derivatives), organometallics (porphyrin [105], bispyridine [113] and diphenylphosphine [199]-based complexes) and metal nanoparticles [200–202]. In several cases, improved catalytic performances and recyclability were found [194, 203, 204]. This enhancement was attributed to the hierarchical pore-size distribution in the POP. Micropores provide a high surface area and better catalysis selectivity, while meso/macropores allow superior diffusive properties.

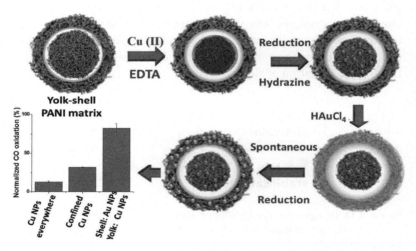

Figure 4.15 Example of self-structured polyaniline yolk-shell nanocapsules used for designing hierarchical gold-copper NPs catalyst. CO oxidation performances were improved by confinement of copper NPs in the yolk and cooperation with Au NPs in the shell [208].

Conductive conjugated polymers are another type of self-assembled matrices that have recently attracted great attention for designing nanostructured catalysts [205]. Most conductive polymers, including polyaniline, polypyrrole and polythiophene, undergo self-assembling processes during their polymerization that can be empirically controlled by adjusting the reaction parameters [206]. Because conducting polymers typically contain sulfur or nitrogen centers, they are able to easily coordinate transition metal ions for in situ growth of nanoparticles. At the same time, because of their intrinsic electrochemical potential, these polymers are able to directly reduce precious metal ions allowing forming Au, Ag or Pd nanoparticles without using any reagent. Consequently, colloidal dispersions and films of porous nano-objects, including nanofibers, nanotubes [207], and nanocapsules [208] were recently used as support for heterogeneous catalysis and electrocatalysis [116, 209–211]. For instance, confinement of active NPs as well as simultaneous trapping of different types of NPs in a nanostructured polyaniline matrix led recently to cooperative and confinement effects for CO oxidation (Fig. 4.15) [208].

References

1. Hübner, S., de Vries, J. G., and Farina, V. (2016). Why Does Industry Not Use Immobilized Transition Metal Complexes as Catalysts?, *Adv. Synth. Catal.*, **358**, pp. 3–25.
2. Glastrup, J. (1996). Degradation of Polyethylene Glycol. A Study of the Reaction Mechanism in a Model Molecule: Tetraethylene Glycol, *Polym. Degrad. Stab.*, **52**, pp. 217–222.
3. Hong, S. H., and Grubbs, R. H. (2006). Highly Active Water-Soluble Olefin Metathesis Catalyst, *J. Am. Chem. Soc.*, **128**, pp. 3508–3509.
4. Ferreira, P., Hayes, W., Phillips, E., Rippon, D., and Tsang, S. C. (2004). Polymer-Supported Nitroxyl Catalysts for Selective Oxidation of Alcohols, *Green Chem.*, **6**, pp. 310–312.
5. De Mico, A., Margarita, R., Parlanti, L., Vescovi, A., and Piancatelli, G. (1997). Versatile and Highly Selective Hypervalent Iodine (III)/2,2,6,6-Tetramethyl-1-piperidinyloxyl-Mediated Oxidation of Alcohols to Carbonyl Compounds, *J. Org. Chem.*, **62**, pp. 6974–6977.
6. Ferreira, P., Phillips, E., Rippon, D., Tsang, S. C., and Hayes, W. (2004). Poly(ethylene glycol)-Supported Nitroxyls: Branched Catalysts for the Selective Oxidation of Alcohols, *J. Org. Chem.*, **69**, pp. 6851–6859.
7. Benaglia, M., Cinquini, M., Cozzi, F., Puglisi, A., and Celentano, G. (2002). Poly(Ethylene Glycol)-Supported Proline: A Versatile Catalyst for the Enantioselective Aldol and Iminoaldol Reactions, *Adv. Synth. Catal.*, **344**, pp. 533–542.
8. Benaglia, M., Cinquini, M., Cozzi, F., Puglisi, A., and Celentano, G. (2003). Poly(Ethylene-Glycol)-Supported Proline: A Recyclable Aminocatalyst for the Enantioselective Synthesis of γ-Nitroketones by Conjugate Addition. *J. Mol. Catal. Chem.*, **204–205**, pp. 157–163.

9. Yamaguchi, M., Igarashi, Y., Reddy, R. S., Shiraishi, T., and Hirama, M. (1997). Asymmetric Michael Addition of Nitroalkanes to Prochiral Acceptors Catalyzed by Proline Rubidium Salts, *Tetrahedron*, **53**, pp. 11223–11236.
10. Kumaraswamy, G., Jena, N., Sastry, M. N. V., Rao, G. V., and Ankamma, K. (2005). Synthesis of 6,6′- and 6-MeO–PEG–BINOL-Ca Soluble Polymer Bound Ligands and their Application in Asymmetric Michael and Epoxidation Reactions, *J. Mol. Catal. Chem.*, **230**, pp. 59–67.
11. Benaglia, M., Danelli, T., and Pozzi, G. (2003). Synthesis of Poly(Ethylene Glycol)-Supported Manganese Porphyrins: Efficient, Recoverable and Recyclable Catalysts for Epoxidation of Alkenes, *Org. Biomol. Chem.*, **1**, pp. 454–456.
12. Li, X., Chen, W., Hems, W., King, F., and Xiao, J. (2003). Asymmetric Hydrogenation of Ketones with Polymer-Supported Chiral 1,2-Diphenylethylenediamine, *Org. Lett.*, **5**, pp. 4559–4561.
13. Bergbreiter, D. E., and Weatherford, D. A. (1989). Polyethylene-Bound Soluble Recoverable Palladium(0) Catalysts, *J. Org. Chem.*, **54**, pp. 2726–2730.
14. Leyva, A., García, H., and Corma, A. (2007). A Soluble Polyethyleneglycol-Anchored Phosphine as a Highly Active, Reusable Ligand for Pd-Catalyzed Couplings of Aryl Chlorides: Comparison with Cross and Non-Cross-Linked Polystyrene and Silica Supports, *Tetrahedron*, **63**, pp. 7097–7111.
15. Dickerson, T. J., Reed, N. N., and Janda, K. D. (2002). Soluble Polymers as Scaffolds for Recoverable Catalysts and Reagents, *Chem. Rev.*, **102**, pp. 3325–3344.
16. Farina, V., and Krishnan, B. (1991). Large Rate Accelerations in the Stille Reaction with Tri-2-Furylphosphine and Triphenylarsine as Palladium Ligands: Mechanistic and Synthetic Implications, *J. Am. Chem. Soc.*, **113**, pp. 9585–9595.
17. Lau, K. C. Y., He, H. S., Chiu, P., and Toy, P. H. (2004). Polystyrene-Supported Triphenylarsine Reagents and Their Use in Suzuki Cross-Coupling Reactions, *J. Comb. Chem.*, **6**, pp. 955–960.
18. Lau, K. C. Y., and Chiu, P. (2007). The Application of Non-Cross-Linked Polystyrene-Supported Triphenylarsine in Stille Coupling Reactions, *Tetrahedron Lett.*, **48**, pp. 1813–1816.
19. Luo, F.-T., Xue, C., Ko, S.-L., Shao, Y.-D., Wu, C.-J., and Kuo, Y.-M. (2005). Preparation of Polystyrene-Supported Soluble Palladacycle Catalyst for Heck and Suzuki reactions, *Tetrahedron*, **61**, pp. 6040–6045.

20. Herrmann, W. A., Brossmer, C., Reisinger, C.-P., Riermeier, T. H., Öfele, K., and Beller, M. (1997). Palladacycles: Efficient New Catalysts for the Heck Vinylation of Aryl Halides, *Chem. Eur. J.*, **3**, pp. 1357–1364.

21. Denmark, S. E., and Cullen, L. R. (2015). Development of a Phase-Transfer-Catalyzed, [2,3]-Wittig Rearrangement, *J. Org. Chem.*, **80**, pp. 11818–11848.

22. Liu, Y.-X., Sun, Y.-N., Tan, H.-H., and Tao, J.-C. (2007). Asymmetric Aldol Reaction Catalyzed by New Recyclable Polystyrene-supported L-proline in the Presence of Water, *Catal. Lett.*, **120**, pp. 281–287.

23. Nomura, K., and Kuromatsu, Y. (2006). Selective Hydrogen Transfer Reduction of Ketones by Recyclable Ruthenium Complex Catalysts Containing a 'ROMP Polymer-Attached' Ligand, *J. Mol. Catal. Chem.*, **245**, pp. 152–160.

24. Holbach, M., and Weck, M. (2006). Modular Approach for the Development of Supported, Monofunctionalized, Salen Catalysts, *J. Org. Chem.*, **71**, pp. 1825–1836.

25. Madhavan, N., and Weck, M. (2008). Highly Active Polymer-Supported (Salen)Al Catalysts for the Enantioselective Addition of Cyanide to α,β-Unsaturated Imides, *Adv. Synth. Catal.*, **350**, pp. 419–425.

26. Tanyeli, C., and Gümüş, A. (2003). Synthesis of Polymer-Supported TEMPO Catalysts and Their Application in the Oxidation of Various Alcohols, *Tetrahedron Lett.*, **44**, pp. 1639–1642.

27. Bastin, S., Eaves, R. J., Edwards, C. W., Ichihara, O., Whittaker, M., and Wills, M. (2004). A Soluble-Polymer System for the Asymmetric Transfer Hydrogenation of Ketones, *J. Org. Chem.*, **69**, pp. 5405–5412.

28. Pu, L., and Yu, H.-B. (2001). Catalytic Asymmetric Organozinc Additions to Carbonyl Compounds, *Chem. Rev.*, **101**, pp. 757–824.

29. Helms, B., and Fréchet, J. M. (2006). The Dendrimer Effect in Homogeneous Catalysis, *J. Adv. Synth. Catal.*, **348**, pp. 1125–1148.

30. Caminade, A.-M., Ouali, A., Keller, M., and Majoral, J.-P. (2012). Organocatalysis with dendrimers, *Chem. Soc. Rev.*, **41**, pp. 4113–4125.

31. Francavilla, C., Drake, M. D., Bright, F. V., and Detty, M. R. (2001). Dendrimeric Organochalcogen Catalysts for the Activation of Hydrogen Peroxide: Improved Catalytic Activity through Statistical Effects and Cooperativity in Successive Generations, *J. Am. Chem. Soc.*, **123**, pp. 57–67.

32. Niu, Y.-N., Yan, Z. Y., Li, G.-Q., Wei, H.-L., Gao, G.-L., Wu, L.-Y., and Liang, Y.-M. (2008). 1,2,3-Triazole-Linked Dendrimers as a Support

for Functionalized and Recoverable Catalysts for Asymmetric Borane Reduction of Prochiral Ketones, *Tetrahedron Asymm.*, **19**, pp. 912–920.

33. Bellis, E., and Kokotos, G. (2005). Proline-Modified Poly(Propyleneimine) Dendrimers as Catalysts for Asymmetric Aldol Reactions, *J. Mol. Catal. Chem.*, **241**, pp. 166–174.

34. Kofoed, J., Darbre, T., and Reymond, J.-L. (2006). Artificial Aldolases from Peptide Dendrimer Combinatorial Libraries, *Org. Biomol. Chem.*, **4**, pp. 3268–3281.

35. Mitsui, K., Hyatt, S. A., Turner, D. A., Hadad, C. M., and Parquette, J. R. (2009). Direct Aldol Reactions Catalyzed by Intramolecularly Folded Prolinamide Dendrons: Dendrimer Effects on Stereoselectivity, *Chem. Commun.*, pp. 3261–3263.

36. Sato, I., Hosoi, K., Kodaka, R., and Soai, K. (2002). Asymmetric Synthesis of N-(Diphenylphosphinyl)Amines Promoted by Chiral Carbosilane Dendritic Ligands in The Enantioselective Addition of Dialkylzinc Compounds to N-(Diphenylphosphinyl)Imines, *Eur. J. Org. Chem.*, **2002**, pp. 3115–3118.

37. Sato, I., Kodaka, R., Hosoi, K., and Soai, K. (2002). Highly Enantioselective Addition of Dialkylzincs to Aldehydes Using Dendritic Chiral Catalysts with Flexible Carbosilane Backbones, *Tetrahedron Asymm.*, **13**, pp. 805–808.

38. El-Shehawy, A. A. (2007). Synthesis of Novel Well-Defined Chain-End-Functionalized Polystyrenes with Dendritic Chiral Ephedrine Moieties and Their Applications as Highly Enantioselective Diethylzinc Addition to N-Diphenylphosphinoyl Arylimines, *Tetrahedron*, **63**, pp. 11754–11762.

39. El-Shehawy, A. A., Sugiyama, K., and Hirao, A. (2008). Application of Well-Defined Chain-End-Functionalized Polystyrenes with Dendritic Chiral Ephedrine Moieties as Reagents for Highly Catalytic Enantioselective Addition of Dialkylzincs to Aldehydes, *Tetrahedron Asymm.*, **19**, pp. 425–434.

40. Evans, D. J., Kanagasooriam, A., Williams, A., and Pryce, R. J. (1993). Aminolysis of Phenyl Esters by Microgel and Dendrimer Molecules Possessing Primary Amines, *J. Mol. Catal.*, **85**, pp. 21–32.

41. Martin, I. K., and Twyman, L. J. (2001). Acceleration of an Aminolysis Reaction Using a PAMAM Dendrimer with 64 Terminal Amine Groups, *Tetrahedron Lett.*, **42**, pp. 1123–1126.

42. Burnett, J., King, A. S. H., and Twyman, L. J. (2006). Probing the Onset of Dense Shell Packing by Measuring the Aminolysis Rates for

a Series Amine Terminated Dendrimers, *React. Funct. Polym.*, **66**, pp. 187–194.

43. Liu, L., and Breslow, R. (2004). Polymeric and Dendrimeric Pyridoxal Enzyme Mimics, *Bioorg. Med. Chem.*, **12**, pp. 3277–3287.

44. Murugan, E., Sherman, R. L., Spivey, H. O., and Ford, W. T. (2004). Catalysis by Hydrophobically Modified Poly(Propylenimine) Dendrimers Having Quaternary Ammonium and Tertiary Amine Functionality, *Langmuir*, **20**, pp. 8307–8312.

45. Pan, Y. J., and Ford, W. T. (2000). Amphiphilic Dendrimers with Both Octyl and Triethylenoxy Methyl Ether Chain Ends, *Macromolecules*, **33**, pp. 3731–3738.

46. Wu, Y., Zhang, Y., Yu, M., Zhao, G., and Wang, S. (2006). Highly Efficient and Reusable Dendritic Catalysts Derived from N-Prolylsulfonamide for the Asymmetric Direct Aldol Reaction in Water, *Org. Lett.*, **8**, pp. 4417–4420.

47. Morao, I., and Cossío, F. P. (1997). Dendritic Catalysts for the Nitroaldol (Henry) Reaction, *Tetrahedron Lett.*, **38**, pp. 6461–6464.

48. Zubia, A., Cossio, F. P., Morao, L. A., Rieumont, M., and Lopez, X. (2004). Quantitative Evaluation of the Catalytic Activity of Dendrimers with Only One Active Center at the Core: Application to the Nitroaldol (Henry) Reaction, *J. Am. Chem. Soc.*, **126**, pp. 5243–5252.

49. Imada, Y., Iida, H., Kitagawa, T., and Naota, T. (2011). Aerobic Reduction of Olefins by in situ Generation of Diimide with Synthetic Flavin Catalysts, *Chem. Eur. J.*, **17**, pp. 5908–5920.

50. Figlus, M., Caldwell, S. T., Walas, D., Yesibag, G., Cooke, G., Kočovský, P., Malkov, A. V., and Sanyal, A. (2010). Dendron-Anchored Organocatalysts: The Asymmetric Reduction of Imines with Trichlorosilane, Catalysed by an Amino Acid-Derived Formamide Appended to a Dendron, *Org. Biomol. Chem.*, **8**, pp. 137–141.

51. Zhang, X., Xu, H., Dong, Z., and Shen, J. (2004). Highly Efficient Dendrimer-Based Mimic of Glutathione Peroxidase, *J. Am. Chem. Soc.*, **126**, pp. 10556–10557.

52. Wang, G., Liu, X., and Zhao, G. (2006). Synthesis of Dendrimer-Supported Prolinols and Their Application in Enantioselective Reduction of Ketones, *Synlett*, **2006**, pp. 1150–1154.

53. Wang, G., Zheng, C., and Zhao, G. (2006). Asymmetric Reduction of Substituted Indanones and Tetralones Catalyzed by Chiral Dendrimer and Its Application to the Synthesis of (+)-Sertraline, *Tetrahedron Asymm.*, **17**, pp. 2074–2081.

54. Li, Y., Liu, X.-Y., and Zhao, G. (2006). Effective and Recyclable Dendritic Catalysts for the Direct Asymmetric Michael Addition of Aldehydes to Nitrostyrenes, *Tetrahedron Asymm.*, **17**, pp. 2034–2039.
55. Agasti, S. S., Caldwell, S. T., Cooke, G., Jordan, B. J., Kennedy, A., Kryvokhyzha, N., Rabani, G., Rana, S., Sanyal, A., and Rotello, V. M. (2008). Dendron-Based Model Systems for Flavoenzyme Activity: Towards a New Class of Synthetic Flavoenzyme, *Chem. Commun.*, pp. 4123–4125.
56. Hecht, S., and Frechet, J. M. J. (2001). Light-Driven Catalysis within Dendrimers: Designing Amphiphilic Singlet Oxygen Sensitizers, *J. Am. Chem. Soc.*, **123**, pp. 6959–6960.
57. Chavan, S. A., Maes, W., Gevers, L. E. M., Wahlen, J., Vankelecom, I. F. J., Jacobs, P. A., Dehaen, W., and De Vos, D. E. (2005). Porphyrin-Functionalized Dendrimers: Synthesis and Application as Recyclable Photocatalysts in a Nanofiltration Membrane Reactor, *Chem. Eur. J.*, **11**, pp. 6754–6762.
58. Esposito, A., Delort, E., Lagnoux, D., Djojo, F., and Reymond, J.-L. (2003). Catalytic Peptide Dendrimers, *Angew. Chem. Int. Ed.*, **42**, pp. 1381–1383.
59. Douat-Casassus, C., Darbre, T., and Reymond, J. L. (2004). Selective Catalysis with Peptide Dendrimers, *J. Am. Chem. Soc.*, **126**, pp. 7817–7826.
60. Delort, E., Darbre, T., and Reymond, J. L. (2004). A Strong Positive Dendritic Effect in a Peptide Dendrimer-Catalyzed Ester Hydrolysis Reaction, *J. Am. Chem. Soc.*, **126**, pp. 15642–15643.
61. Yang, B.-Y., Chen, X.-M., Deng, G.-J., Zhang, Y.-L., and Fan, Q.-H. (2003). Chiral Dendritic Bis(Oxazoline) Copper(II) Complexes as Lewis Acid Catalysts for Enantioselective Aldol Reactions in Aqueous Media, *Tetrahedron Lett.*, **44**, pp. 3535–3538.
62. Muraki, T., Fujita, K., and Kujime, M. (2007). Synthesis of Novel Dendritic 2,2′-Bipyridine Ligands and Their Application to Lewis Acid-Catalyzed Diels–Alder and Three-Component Condensation Reactions, *J. Org. Chem.*, **72**, pp. 7863–7870.
63. Martinez-Olid, F., Benito, J. M., Flores, J. C., and de Jesus, E. (2009). Polymetallic Carbosilane Dendrimers Containing N,N′-Iminopyridine Chelating Ligands: Applications in Catalysis, *Isr. J. Chem.*, **49**, pp. 99–108.
64. Iwasawa, T., Tokunaga, M., Obora, Y., and Tsuji, Y. (2004). Homogeneous Palladium Catalyst Suppressing Pd Black Formation in Air Oxidation of Alcohols, *J. Am. Chem. Soc.*, **126**, pp. 6554–6555.

65. Miyaura, N., and Suzuki, A. (1995). Palladium-Catalyzed Cross-Coupling Reactions of Organoboron Compounds, *Chem. Rev.*, **95**, pp. 2457–2483.
66. Mery, D., and Astruc, D. (2006). Dendritic Catalysis: Major Concepts and Recent Progress, *Coord. Chem. Rev.*, **250**, pp. 1965–1979.
67. Muller, C., Ackerman, L. J., Reek, J. N. H., Kamer, P. C. J., and van Leeuwen, P. (2004). Site-Isolation Effects in a Dendritic Nickel Catalyst for the Oligomerization of Ethylene, *J. Am. Chem. Soc.*, **126**, pp. 14960–14963.
68. Yi, B., Fan, Q.-H., Deng, G.-J., Li, Y.-M., Qiu, L.-Q., and Chan, A. S. C. (2004). Novel Chiral Dendritic Diphosphine Ligands for Rh(I)-Catalyzed Asymmetric Hydrogenation: Remarkable Structural Effects on Catalytic Properties, *Org. Lett.*, **6**, pp. 1361–1364.
69. Myers, V. S., Weir, M. G., Carino, E. V., Yancey, D. F., Pande, S., and Crooks, R. M. (2011). Dendrimer-Encapsulated Nanoparticles: New Synthetic and Characterization Methods and Catalytic Applications, *Chem. Sci.*, **2**, pp. 1632–1646.
70. Nakamula, I., Yamanoi, Y., Yonezawa, T., Imaoka, T., Yamamoto, K., and Nishihara, H. (2008). Nanocage Catalysts-Rhodium Nanoclusters Encapsulated with Dendrimers as Accessible and Stable Catalysts for Olefins and Nitroarene Hydrogenations, *Chem. Commun.*, pp. 5716–5718.
71. Chen, Y.-C., Wu, T.-F., Jiang, L., Deng, J.-G., Liu, H., Zhu, J., and Jiang, Y.-Z. (2005). Synthesis of Dendritic Catalysts and Application in Asymmetric Transfer Hydrogenation, *J. Org. Chem.*, **70**, pp. 1006–1010.
72. Chen, Y.-C., Wu, T.-F., Deng, J.-G., Liu, H., Cui, X., Zhu, J., Jiang, Y.-Z., Choi, M. C. K., and Chan, A. S. C. (2002). Multiple Dendritic Catalysts for Asymmetric Transfer Hydrogenation, *J. Org. Chem.*, **67**, pp. 5301–5306.
73. Liu, W., Cui, X., Cun, L., Wu, J., Zhu, J., Deng, J., and Fan, Q. (2005). Novel Dendritic Ligands of Chiral 1,2-Diamine and Their Application in Asymmetric Hydrogenation of Simple Aryl Ketones, *Synlett*, **2005**, pp. 1591–1595.
74. Laurent, R., Caminade, A.-M., and Majoral, J.-P. (2005). A Third Generation Chiral Phosphorus-Containing Dendrimer as Ligand in Pd-Catalyzed Asymmetric Allylic Alkylation, *Tetrahedron Lett.*, **46**, pp. 6503–6506.
75. Barder, T. E., and Buchwald, S. L. (2007). Rationale Behind the Resistance of Dialkylbiaryl Phosphines toward Oxidation by Molecular Oxygen, *J. Am. Chem. Soc.*, **129**, pp. 5096–5101.

76. Liu, X., Li, Y., Wang, G., Chai, Z., Wu, Y., and Zhao, G. (2006). Effective and Recyclable Dendritic Ligands for the Enantioselective Epoxidation of Enones, *Tetrahedron Asymm.*, **17**, pp. 750–755.

77. Neumann, P., Dib, H., Caminade, A.-M., and Hey-Hawkins, E. (2015). Redox Control of a Dendritic Ferrocenyl-Based Homogeneous Catalyst, *Angew. Chem. Int. Ed.*, **54**, pp. 311–314.

78. Ertl, G., Knozinger, H., Schüth, F., and Weitkamp, J. (2016). *Handbook of Heterogeneous Catalysis*, vol. 8, 2nd ed., Wiley (accessed: November 1).

79. Halbach, T. S., Mix, S., Fischer, D., Maechling, S., Krasue, J. O., Sievers, C., Blechert, S., Nuyken, O., and Buchmeiser, M. R. (2005). Novel Ruthenium-Based Metathesis Catalysts Containing Electron-Withdrawing Ligands: Synthesis, Immobilization, and Reactivity, *J. Org. Chem.*, **70**, pp. 4687–4694.

80. Pugin, B., and Blaser, H.-U. (2010). In *Heterogenized Homogeneous Catalysts for Fine Chemicals Production*, eds., Barbaro, P., and Liguori, F., Springer Netherlands, pp. 231–245.

81. Trzeciak, A. M., and Ziółkowski, J. J. (2007). Monomolecular, Nanosized and Heterogenized Palladium Catalysts for the Heck Reaction, *Coord. Chem. Rev.*, **251**, pp. 1281–1293.

82. de Vries, A. H. M., Parlevliet, F. J., Schmieder-van de Vondervoort, L., Mommers, J. H. M., Henderickx, H. J. W., Walet, M. A. M., and de Vries, J. G. (2002). A Practical Recycle of a Ligand-Free Palladium Catalyst for Heck Reactions, *Adv. Synth. Catal.*, **344**, pp. 996–1002.

83. Yao, F., Liu, J., and Cai, M. (2013). Heck Arylation of Conjugated Alkenes with Aryl Bromides or Chlorides Catalyzed by Immobilization of Palladium in MCM-41, *Catal. Lett.*, **143**, pp. 681–686.

84. Zhao, H., Yin, L., and Cai, M. (2013). A Phosphane-Free, Atom-Efficient Cross-Coupling Reaction of Triarylbismuths with Acyl Chlorides Catalyzed by MCM-41-Immobilized Palladium Complex, *Eur. J. Org. Chem.*, **2013**, pp. 1337–1345.

85. Rostamnia, S., Liu, X., and Zheng, D. (2014). Ordered Interface Mesoporous Immobilized Pd Pre-Catalyst: En/Pd Complexes Embedded Inside the SBA-15 as an Active, Reusable and Selective Phosphine-Free Hybrid Catalyst for the Water Medium Heck Coupling Process, *J. Colloid Interface Sci.*, **432**, pp. 86–91.

86. Veisi, H., Khazaei, A., Safaei, M., and Kordestani, D. (2014). Synthesis of Biguanide-Functionalized Single-Walled Carbon Nanotubes (SWCNTs) Hybrid Materials to Immobilized Palladium as New

Recyclable Heterogeneous Nanocatalyst for Suzuki–Miyaura Coupling Reaction, *J. Mol. Catal. Chem.*, **382**, pp. 106–113.

87. Buck, M. E., and Lynn, D. M. (2012). Azlactone-Functionalized Polymers as Reactive Platforms for the Design of Advanced Materials: Progress in the Last Ten Years, *Polym. Chem.*, **3**, pp. 66–80.

88. Heilmann, S. M., Rasmussen, J. K., and Krepski, L. R. (2001). Chemistry and Technology of 2-alkenyl Azlactones, *J. Polym. Sci. Part Polym. Chem.*, **39**, pp. 3655–3677.

89. Xie, S., Svec, F., and Fréchet, J. M. (1999). Design of Reactive Porous Polymer Supports for High Throughput Bioreactors: Poly(2-Vinyl-4,4-Dimethylazlactone-co-Acrylamide-co-Ethylene Dimethacrylate) Monoliths, *J. Biotechnol. Bioeng.*, **62**, pp. 30–35.

90. Tanielyan, S. K., Augustine, R. L., Marin, N., and Alvez, G. (2011). Anchored Wilkinson Catalyst, *ACS Catal.*, **1**, pp. 159–169.

91. Augustine, R., Tanielyan, S., Anderson, S., and Yang, H. (1999). A New Technique for Anchoring Homogeneous Catalysts, *Chem. Commun.*, pp. 1257–1258.

92. Augustine, R. L., Goel, P., Mahata, N., Reyes, C., and Tanielyan, S. K. (2004). Anchored Homogeneous Catalysts: High Turnover Number Applications, *J. Mol. Catal. Chem.*, **216**, pp. 189–197.

93. Augustine, R. L., Tanielyan, S. K., Mahata, N., Gao, Y., Zsigmond, A., and Yang, H. (2003). Anchored Homogeneous Catalysts: the Role of the Heteropoly Acid Anchoring Agent, *Appl. Catal. Gen.*, **256**, pp. 69–76.

94. Barbaro, P. (2006). Recycling Asymmetric Hydrogenation Catalysts by Their Immobilization onto Ion-Exchange Resins, *Chem. Eur. J.*, **12**, pp. 5666–5675.

95. Simons, C., Hanefeld, U., Arends, I. W. C. E., Maschmeyer, T., and Sheldon, R. A. (2006). Comparison of Supports for the Electrostatic Immobilisation of Asymmetric Homogeneous Catalysts, *J. Catal.*, **239**, pp. 212–219.

96. Notestein, J. M., and Katz, A. (2006). Enhancing Heterogeneous Catalysis through Cooperative Hybrid Organic–Inorganic Interfaces, *Chem. Eur. J.*, **12**, pp. 3954–3965.

97. Zhang, Y., Zhao, L., Lee, S. S., and Ying, J. Y. (2006). Enantioselective Catalysis over Chiral Imidazolidin-4-one Immobilized on Siliceous and Polymer-Coated Mesocellular Foams, *Adv. Synth. Catal.*, **348**, pp. 2027–2032.

98. Zhang, Y., Zhao, L., Patra, P. K., and Ying, J. Y. (2008). Synthesis and Catalytic Applications of Mesoporous Polymer Colloids in Olefin Hydrosilylation, *Adv. Synth. Catal.*, **350**, pp. 662–666.
99. McKeown, N. B. (2000). Phthalocyanine-Containing Polymers, *J. Mater. Chem.*, **10**, pp. 1979–1995.
100. Thomas, A. (2010). Functional Materials: From Hard to Soft Porous Frameworks, *Angew. Chem. Int. Ed.*, **49**, pp. 8328–8344.
101. McKeown, N. B., and Budd, P. M. (2006). Polymers of Intrinsic Microporosity (PIMs): Organic Materials for Membrane Separations, Heterogeneous Catalysis and Hydrogen Storage, *Chem. Soc. Rev.*, **35**, pp. 675–683.
102. McKeown, N. B. (2010). Nanoporous Molecular Crystals, *J. Mater. Chem.*, **20**, pp. 10588–10597.
103. Jiang, J.-X., and Cooper, A. I. (2009). *Functional Metal-Organic Frameworks: Gas Storage, Separation and Catalysis*, ed., Schröder, M., Springer Berlin Heidelberg, pp. 1–33.
104. Zhang, Y., and Riduan, S. N. (2012). Functional Porous Organic Polymers for Heterogeneous Catalysis, *Chem. Soc. Rev.*, **41**, pp. 2083–2094.
105. Mackintosh, H. J., Budd, P. M., and McKeown, N. B. (2008). Catalysis by Microporous Phthalocyanine and Porphyrin Network Polymers, *J. Mater. Chem.*, **18**, pp. 573–578.
106. Makhseed, S., Al-Kharafi, F., Samuel, J., and Ateya, B. (2009). Catalytic Oxidation of Sulphide Ions Using a Novel Microporous Cobalt Phthalocyanine Network Polymer in Aqueous Solution, *Catal. Commun.*, **10**, pp. 1284–1287.
107. Ray, S., and Vasudevan, S. (2003). Encapsulation of Cobalt Phthalocyanine in Zeolite-Y: Evidence for Nonplanar Geometry, *Inorg. Chem.*, **42**, pp. 1711–1719.
108. Chen, L., Yang, Y., and Jiang, D. (2010). CMPs as Scaffolds for Constructing Porous Catalytic Frameworks: A Built-in Heterogeneous Catalyst with High Activity and Selectivity Based on Nanoporous Metalloporphyrin Polymers, *J. Am. Chem. Soc.*, **132**, pp. 9138–9143.
109. Shultz, A. M., Farha, O. K., Hupp, J. T., and Nguyen, S. T. (2011). Synthesis of Catalytically Active Porous Organic Polymers from Metalloporphyrin Building Blocks, *Chem. Sci.*, **2**, pp. 686–689.
110. Kuhn, P., Antonietti, M., and Thomas, A. (2008). Porous, Covalent Triazine-Based Frameworks Prepared by Ionothermal Synthesis, *Angew. Chem. Int. Ed.*, **47**, pp. 3450–3453.

111. Palkovits, R., Antonietti, M., Kuhn, P., Thomas, A., and Schüth, F. (2009). Solid Catalysts for the Selective Low-Temperature Oxidation of Methane to Methanol, *Angew. Chem. Int. Ed.*, **48**, pp. 6909–6912.

112. Budd, P. M., Ghanem, B., Msayib, K., McKeown, N. B., and Tattershall, C. (2003). A Nanoporous Network Polymer Derived from Hexaazatrinaphthylene with Potential as an Adsorbent and Catalyst Support, *J. Mater. Chem.*, **13**, pp. 2721–2726.

113. Xie, Z., Wang, C., deKrafft, K. E., and Lin, W. (2011). Highly Stable and Porous Cross-Linked Polymers for Efficient Photocatalysis, *J. Am. Chem. Soc.*, **133**, pp. 2056–2059.

114. Kuhn, P., Forget, A., Su, D., Thomas, A., and Antonietti, M. (2008). From Microporous Regular Frameworks to Mesoporous Materials with Ultrahigh Surface Area: Dynamic Reorganization of Porous Polymer Networks, *J. Am. Chem. Soc.*, **130**, pp. 13333–13337.

115. Madhavan, N., Jones, C. W., and Weck, M. (2008). Rational Approach to Polymer-Supported Catalysts: Synergy between Catalytic Reaction Mechanism and Polymer Design, *Acc. Chem. Res.*, **41**, pp. 1153–1165.

116. Mahouche-Chergui, S., Guerrouache, M., Carbonnier, B., and Chehimi, M. M. (2013). Polymer-Immobilized Nanoparticles, *Colloids Surf. Physicochem. Eng. Asp.*, **439**, pp. 43–68.

117. Kharisov, B. I., Dias, H. V. R., Kharissova, O. V., and Vázquez, A. (2014). Ultrasmall Particles in the Catalysis, *J. Nanoparticle Res.*, **16**, pp. 2665.

118. Dhar, J., and Patil, S. (2012). Self-Assembly and Catalytic Activity of Metal Nanoparticles Immobilized in Polymer Membrane Prepared via Layer-by-Layer Approach, *ACS Appl. Mater. Interfaces*, **4**, pp. 1803–1812.

119. Sarkar, S., Guibal, E., Quignard, F., and SenGupta, A. K. (2012). Polymer-Supported Metals and Metal Oxide Nanoparticles: Synthesis, Characterization, and Applications, *J. Nanoparticle Res.*, **14**, p. 715.

120. Gupta, S., Agrawal, M., Uhlmann, P., Simon, F., Oertel, U., and Stamm, M. (2008). Gold Nanoparticles Immobilized on Stimuli Responsive Polymer Brushes as Nanosensors, *Macromolecules*, **41**, pp. 8152–8158.

121. Gehan, H., Fillaud, L., Chehimi, M. M., Aubard, J., Hohenau, A., Felidj, N., and Mangeney, C. (2010). Thermo-Induced Electromagnetic Coupling in Gold/Polymer Hybrid Plasmonic Structures Probed by Surface-Enhanced Raman Scattering, *ACS Nano*, **4**, pp. 6491–6500.

122. Gupta, S., Agrawal, M., Uhlmann, P., Simon, F., and Stamm, M. (2010). Poly(N-Isopropyl Acrylamide)–Gold Nanoassemblies on Macroscopic

Surfaces: Fabrication, Characterization, and Application, *Chem. Mater.*, **22**, pp. 504–509.

123. Guerrouache, M., Mahouche-Chergui, S., Chehimi, M. M., and Carbonnier, B. (2012). Site-Specific Immobilisation of Gold Nanoparticles on a Porous Monolith Surface by Using a Thiolyne Click Photo-patterning Approach, *Chem. Commun.*, **48**, pp. 7486–7488.

124. Ranoszek-Soliwoda, K., Girleanu, M., Tkacz-Szczęsna, B., Rosowski, M., Celichowski, G., Brinkmann, M., Ersen, O., and Grobelny, J. (2016). Versatile Phase Transfer Method for the Efficient Surface Functionalization of Gold Nanoparticles: Towards Controlled Nanoparticle Dispersion in a Polymer Matrix, *J. Nanomater. J. Nanomater.*, **2016**, p. e9058323.

125. Cheng, H.-H., Chen, F., Yu, J., and Guo, Z.-X. (2017). Gold-Nanoparticle-Decorated Thermoplastic Polyurethane Electrospun Fibers Prepared Through a Chitosan Linkage for Catalytic Applications, *J. Appl. Polym. Sci.*, **134**, p. 44336.

126. Manju, G. N., Anoop Krishnan, K., Vinod, V. P., and Anirudhan, T. S. (2002). An Investigation into the Sorption of Heavy Metals from Wastewaters by Polyacrylamide-Grafted Iron(III) Oxide, *J. Hazard. Mater.*, **91**, pp. 221–238.

127. Shubha, K. P., Raji, C., and Anirudhan, T. S. (2001). Immobilization of Heavy Metals from Aqueous Solutions Using Polyacrylamide Grafted Hydrous Tin (IV) Oxide Gel having Carboxylate Functional Groups, *Water Res.*, **35**, pp. 300–310.

128. Guo, Z., Lee, S.-E., Kim, Park, H. S., Hahn, H. T., Karki, A. B., and Young, D. P. (2009). Fabrication, Characterization and Microwave Properties of Polyurethane Nanocomposites Reinforced with Iron Oxide and Barium Titanate Nanoparticles, *Acta Mater.*, **57**, pp. 267–277.

129. Airaud, C., Ibarboure, E., Gaillard, C., and Heroguez, V. (2009). Simultaneous ROMP and ATRP in Aqueous Dispersed Media: A Straightforward Strategy to Prepare Polymer Composite Particles with Original Morphologies, *Macromol. Symp.*, **281**, pp. 31–38.

130. Cantarella, M., Sanz, R., Buccheri, M. A., Ruffino, F., Giancarlo, R., Scalese, S., Impellizzeri, G., Romano, L., and Privitera, V. (2016). Immobilization of Nanomaterials in PMMA Composites for Photocatalytic Removal of Dyes, Phenols and Bacteria from Water, *J. Photochem. Photobiol. Chem.*, **321**, pp. 1–11.

131. Kalbasi, R. J., and Mosaddegh, N. (2012). Palladium Nanoparticles Supported on Poly(2-Hydroxyethyl Methacrylate)/KIT-6 Composite

as an Efficient and Reusable Catalyst for Suzuki-Miyaura Reaction in Water, *J. Inorg. Organomet. Polym. Mater.*, **22**, pp. 404–414.

132. Mallick, K., Witcomb, M. J., Dinsmore, A., and Scurrell, M. S. (2005). Polymerization of Aniline by Auric Acid: Formation of Gold Decorated Polyaniline Nanoballs, *Macromol. Rapid Commun.*, **26**, pp. 232–235.

133. HaiZhu, S., and Bai, Y. (2008). In Situ Preparation of Nanoparticles/Polymer Composites, *Sci. China Ser. E Technol. Sci.*, **51**, pp. 1886–1901.

134. Landers, J., Colon-Ortiz, J., Zong, K., Goswami, A., Asefa, T., Vishnyakov, A., and Neimark, A. V. (2016). In Situ Growth and Characterization of Metal Oxide Nanoparticles within Polyelectrolyte Membranes, *Angew. Chem. Int. Ed.*, **55**, pp. 11522–11527.

135. Harish, S., Mathiyarasu, J., Phani, K. L. N., and Yegnaraman, V. (2008). Synthesis of Conducting Polymer Supported Pd Nanoparticles in Aqueous Medium and Catalytic Activity Towards 4-Nitrophenol Reduction, *Catal. Lett.*, **128**, p. 197.

136. Kuttiplavil Narayanan, R., Janardanan Devaki, S., and Prasada Rao, T. (2014). Robust Fibrillar Nanocatalysts Based on Silver Nanoparticle-Entrapped Polymeric Hydrogels, *Appl. Catal. Gen.*, **483**, pp. 31–40.

137. Ai, L., and Jiang, J. (2013). Catalytic Reduction of 4-Nitrophenol by Silver Nanoparticles Stabilized on Environmentally Benign Macroscopic Biopolymer Hydrogel, *Bioresour. Technol.*, **132**, pp. 374–377.

138. Ajmal, M., Siddiq, M., Al-Lohedan, H., and Sahiner, N. (2014). Highly Versatile p(MAc)–M (M: Cu, Co, Ni) Microgel Composite Catalyst for Individual and Simultaneous Catalytic Reduction of Nitro Compounds and Dyes, *RSC Adv.*, **4**, pp. 59562–59570.

139. Sahiner, N., Yildiz, S., and Al-Lohedan, H. (2015). The Resourcefulness of p(4-VP) Cryogels as Template for in situ Nanoparticle Preparation of Various Metals and their Use in H_2 Production, Nitro Compound Reduction and Dye Degradation, *Appl. Catal. B Environ.*, **166–167**, pp. 145–154.

140. Saha, S., Pal, A., Kundu, S., Basu, S., and Pal, T. (2010). Photochemical Green Synthesis of Calcium-Alginate-Stabilized Ag and Au Nanoparticles and Their Catalytic Application to 4-Nitrophenol Reduction, *Langmuir*, **26**, pp. 2885–2893.

141. Favier, I., Gómez, M., Muller, G., Picurelli, D., Nowicki, A., Roucoux, A., and Bou, J. J. (2007). Synthesis of New Functionalized Polymers and Their Use as Stabilizers of Pd, Pt, and Rh Nanoparticles. Preliminary Catalytic Studies, *Appl. Polym. Sci.*, **105**, pp. 2772–2782.

142. Washio, I., Xiong, Y., Yin, Y., and Xia, Y. (2006). Reduction by the End Groups of Poly(Vinyl Pyrrolidone): A New and Versatile Route to the Kinetically Controlled Synthesis of Ag Triangular Nanoplates, *Adv. Mater.*, **18**, pp. 1745–1749.

143. Hoppe, C. E., Lazzari, M., Pardiñas-Blanco, I., and López-Quintela, M. A. (2006). One-Step Synthesis of Gold and Silver Hydrosols Using Poly(N-Vinyl-2-Pyrrolidone) as a Reducing Agent, *Langmuir*, **22**, pp. 7027–7034.

144. Mallick, K., Witcomb, M. J., and Scurrell, M. S. (2006). In Situ Synthesis of Copper Nanoparticles and Poly(o-Toluidine): A Metal–Polymer Composite Material, *Eur. Polym. J.*, **42**, pp. 670–675.

145. Mallick, K., Mondal, K., Witcomb, M., Deshmukh, A., and Scurrell, M. (2008). Catalytic Activity of a Soft Composite Material: Nanoparticle Location–Activity Relationship, *Mater. Sci. Eng. B*, **150**, pp. 43–49.

146. Valentin, R., Molvinger, K., Viton, C., Domard, A., and Quignard, F. (2005). From Hydrocolloids to High Specific Surface Area Porous Supports for Catalysis, *Biomacromolecules*, **6**, pp. 2785–2792.

147. Primo, A., Liebel, M., and Quignard, F. (2009). Palladium Coordination Biopolymer: A Versatile Access to Highly Porous Dispersed Catalyst for Suzuki Reaction, *Chem. Mater.*, **21**, pp. 621–627.

148. Li, Y., and El-Sayed, M. A. (2001). The Effect of Stabilizers on the Catalytic Activity and Stability of Pd Colloidal Nanoparticles in the Suzuki Reactions in Aqueous Solution, *J. Phys. Chem. B*, **105**, pp. 8938–8943.

149. Klingelhöfer, S., Heitz, W., Greiner, A., Oesereich, S., Förster, S., and Antonietti, M. (1997). Preparation of Palladium Colloids in Block Copolymer Micelles and Their Use for the Catalysis of the Heck Reaction, *J. Am. Chem. Soc.*, **119**, pp. 10116–10120.

150. Semagina, N., Joannet, E., Parra, S., Sulman, E., Renken, A., and Kiwi-Minsker, L. (2005). Palladium Nanoparticles Stabilized in Block-Copolymer Micelles for Highly Selective 2-Butyne-1,4-Diol Partial Hydrogenation, *Appl. Catal. Gen.*, **280**, pp. 141–147.

151. Sahiner, N. (2013). Soft and Flexible Hydrogel Templates of Different Sizes and Various Functionalities for Metal Nanoparticle Preparation and Their Use in Catalysis, *Prog. Polym. Sci.*, **38**, pp. 1329–1356.

152. Sahiner, N., Butun, S., and Ilgin, P. (2011). Hydrogel Particles with Core Shell Morphology for Versatile Applications: Environmental, Biomedical and Catalysis, *Colloids Surf. Physicochem. Eng. Asp.*, **386**, pp. 16–24.

153. Sahiner, N., Butun, S., Ozay, O., and Dibek, B. (2012). Utilization of Smart Hydrogel–Metal Composites as Catalysis Media, *J. Colloid Interface Sci.*, **373**, 122–128.

154. Kassube, J. K., Wadepohl, H., and Gade, L. H. (2009). Immobilisation of the BINAP Ligand on Dendrimers and Hyperbranched Polymers: Dependence of the Catalytic Properties on the Linker Unit, *Adv. Synth. Catal.*, **351**, pp. 607–616.

155. Kavas, H., Durmus, Z., Tanriverdi, E., Şenel, M., Sozeri, H., and Baykal, A. (2011). Fabrication and Characterization of Dendrimer-Encapsulated Monometallic Co Nanoparticles, *J. Alloys Compd.*, **509**, pp. 5341–5348.

156. Knecht, M. R., Garcia-Martinez, J. C., and Crooks, R. M. (2006). Synthesis, Characterization, and Magnetic Properties of Dendrimer-Encapsulated Nickel Nanoparticles Containing < 150 Atoms, *Chem. Mater.*, **18**, pp. 5039–5044.

157. Kuchkina, N. V., Morgan, D. G., Kostopoulou, A., Lappas, A., Brintakis, K., Boris, B. S., Yuzik-Klimova, E. Yu., Stein, B. D., Svergun, D. I., Spilotros, A., Sulman, M. G., Nikoshvili, L. Zh, Sulman, E. M., Shifrina, Z. B., and Bronstein, L. M. (2014). Hydrophobic Periphery Tails of Polyphenylenepyridyl Dendrons Control Nanoparticle Formation and Catalytic Properties, *Chem. Mater.*, **26**, pp. 5654–5663.

158. Newkome, G. R., and Shreiner, C. (2010). Dendrimers Derived from 1 –> 3 Branching Motifs, *Chem. Rev.*, **110**, pp. 6338–6442.

159. Noh, J.-H., and Meijboom, R. (2014). Dendrimer-Templated Pd Nanoparticles and Pd Nanoparticles Synthesized by Reverse Microemulsions as Efficient Nanocatalysts for the Heck Reaction: A Comparative Study, *J. Colloid Interface Sci.*, **415**, pp. 57–69.

160. Li, N., Echeverria, M., Moya, S., Ruiz, J., and Astruc, D. (2014). 'Click' Synthesis of Nona-PEG-branched Triazole Dendrimers and Stabilization of Gold Nanoparticles That Efficiently Catalyze p-Nitrophenol Reduction, *Inorg. Chem.*, **53**, pp. 6954–6961.

161. Lu, A.-H., Salabas, E. L., and Schueth, F. (2007). Magnetic Nanoparticles: Synthesis, Protection, Functionalization, and Application, *Angew. Chem. Int. Ed.*, **46**, pp. 1222–1244.

162. Murugan, E., and Jebaranjitham, J. N. (2015). Dendrimer Grafted Core-Shell Fe_3O_4-Polymer Magnetic Nanocomposites Stabilized with AuNPs for Enhanced Catalytic Degradation of Rhodamine B-A kinetic study, *Chem. Eng. J.*, **259**, pp. 266–276.

163. Deraedt, C., and Astruc, D. (2016). Supramolecular Nanoreactors for Catalysis, *Coord. Chem. Rev.*, **324**, pp. 106–122.

164. Shaw, W. J., Chen, Y., Fulton, J., Linehan, J., Cutowska, A., and Bitterwolf, T. (2008). Structural Evolution of a Recoverable Rhodium Hydrogenation Catalyst, *J. Organomet. Chem.*, **693**, pp. 2111–2118.
165. Liu, C., Jiang, J., Wang, Y., Cheng, F., and Jin, Z. (2003). Thermoregulated Phase Transfer Ligands and Catalysis XVIII: Synthesis of N,N-Dipolyoxyethylene-Substituted-2-(diphenylphosphino)phenylamine (PEO-DPPPA) and the Catalytic Activity of Its Rhodium Complex in the Aqueous–Organic Biphasic Hydroformylation of 1-Decene, *J. Mol. Catal. Chem.*, **198**, pp. 23–27.
166. Jiang, J., Mei, J., Wang, Y., Wen, F., and Jin, Z. (2002). Thermoregulated Phase-Transfer Ligands and Catalysis: XV CO Selective Reduction of Nitroarenes Catalyzed by Ru$_3$(CO)$_9$(PEO-DPPSA)$_3$ in Two-Phasic System, *Appl. Catal. Gen.*, **224**, pp. 21–25.
167. Welton, T. (1999). Room-Temperature Ionic Liquids. Solvents for Synthesis and Catalysis, *Chem. Rev.*, **99**, pp. 2071–2084.
168. Olivier-Bourbigou, H., Magna, L., and Morvan, D. (2010). Ionic Liquids and Catalysis: Recent Progress from Knowledge to Applications, *Appl. Catal. Gen.*, **373**, pp. 1–56.
169. Mecerreyes, D. (2011). Polymeric Ionic Liquids: Broadening the Properties and Applications of Polyelectrolytes, *Prog. Polym. Sci.*, **36**, pp. 1629–1648.
170. Manojkumar, K., Sivaramakrishna, A., and Vijayakrishna, K. (2016). A Short Review on Stable Metal Nanoparticles Using Ionic Liquids, Supported Ionic Liquids, and Poly(ionic liquids), *J. Nanoparticle Res.*, **18**, p. 103.
171. Yang, Z.-Z., Zhao, Y., Ji, G., Zhang, H., Yu, B., Gao, X., and Liu, Z. (2014). Fluoro-Functionalized Polymeric Ionic Liquids: Highly Efficient Catalysts for CO$_2$ Cycloaddition to Cyclic Carbonates under Mild Conditions, *Green Chem.*, **16**, pp. 3724–3728.
172. Lowe, A. B., Sumerlin, B. S., Donovan, M. S., and McCormick, C. L. (2002). Facile Preparation of Transition Metal Nanoparticles Stabilized by Well-Defined (Co)polymers Synthesized via Aqueous Reversible Addition-Fragmentation Chain Transfer Polymerization, *J. Am. Chem. Soc.*, **124**, pp. 11562–11563.
173. Mu, X., Meng, J., Li, Z.-C., and Kou, Y. (2005). Rhodium Nanoparticles Stabilized by Ionic Copolymers in Ionic Liquids: Long Lifetime Nanocluster Catalysts for Benzene Hydrogenation, *J. Am. Chem. Soc.*, **127**, pp. 9694–9695.
174. Shang, L., Qin, C., Wang, T., Wang, M., Wang, L., and Dong, S. (2007). Fluorescent Conjugated Polymer-Stabilized Gold Nanoparticles for

Sensitive and Selective Detection of Cysteine, *J. Phys. Chem. C*, **111**, pp. 13414–13417.

175. Yang, X., Fei, Z., Zhao, D., Ang, W. H., Li, Y., and Dyson, P. (2008). Palladium Nanoparticles Stabilized by an Ionic Polymer and Ionic Liquid: A Versatile System for C–C Cross-Coupling Reactions, *J. Inorg. Chem.*, **47**, pp. 3292–3297.

176. Kvítek, L., Panáček, A., Soukupová, J., Kolář, M., Večeřová, R., Prucek, R., Holecová, M., and Zbořil, R. (2008). Effect of Surfactants and Polymers on Stability and Antibacterial Activity of Silver Nanoparticles (NPs), *J. Phys. Chem. C*, **112**, pp. 5825–5834.

177. Kim, J. W., and Choi, B. G. (2014). Polymeric Ionic Liquid-Promoted High Dispersion of Pt Nanoparticles on Graphene, *Mater. Lett.*, **132**, pp. 373–376.

178. Akiyama, R., and Kobayashi, S. (2001). Microencapsulated Palladium Catalysts: Allylic Substitution and Suzuki Coupling Using a Recoverable and Reusable Polymer-Supported Palladium Catalyst, *Angew. Chem. Int. Ed.*, **40**, pp. 3469–3471.

179. Kobayashi, S., and Miyamura, H. (2010). Polymer-Incarcerated metal(0) Cluster Catalysts, *Chem. Rec.*, **10**, pp. 271–290.

180. Akiyama, R., and Kobayashi, S. (2003). The polymer Incarcerated Method for the Preparation of Highly Active Heterogeneous Palladium Catalysts, *J. Am. Chem. Soc.*, **125**, pp. 3412–3413.

181. Hagio, H., Sugiura, M., and Kobayashi, S. (2006). Practical Preparation Method of Polymer-Incarcerated (PI) Palladium Catalysts Using Pd(II) Salts, *Org. Lett.*, **8**, pp. 375–378.

182. Hagio, H., Sugiura, M., and Kobayashi, S. (2005). Immobilization of a Platinum Catalyst Using the Polymer Incarcerated (PI) Method and Application to Catalytic Reactions, *Synlett*, pp. 813–816.

183. Miyazaki, Y., Hagio, H., Sugiura, M., and Kobayashi, S. (2006). Practical Access to the Polymer Incarcerated Platinum (PI Pt) Catalyst and Its Application to Hydrogenation, *Org. Biomol. Chem.*, **4**, pp. 3537–3537.

184. Yamazoe, S., Koyasu, K., and Tsukuda, T. (2014). Nonscalable Oxidation Catalysis of Gold Clusters, *Acc. Chem. Res.*, **47**, pp. 816–824.

185. Borowski, A. F., Sabo-Etienne, S., Donnadieu, B., and Chaudret, B. (2003). Reactivity of the Bis(dihydrogen) Complex [RuH2(eta(2)-H-2)(2)(PCy3)(2)] toward N-heteroaromatic Compounds. Regioselective Hydrogenation of Acridine to 1,2,3,4,5,6,7,8-Octahydroacridine, *Organometallics,* **22**, pp. 1630–1637.

186. Okamoto, K., Akiyama, R., and Kobayashi, S. (2004). Recoverable, Reusable, Highly Active, and Sulfur-Tolerant Polymer Incarcerated Palladium for Hydrogenation, *J. Org. Chem.*, **69**, pp. 2871–2873.

187. Okumura, M., Kitagawa, Y., Kawakami, T., and Haruta, M. (2008). Theoretical Investigation of the Hetero-Junction Effect in PVP-Stabilized Au13 Clusters. The role of PVP in Their Catalytic Activities, *Chem. Phys. Lett.*, **459**, pp. 133–136.

188. Chaki, N. K., Tsunoyama, H., Negishi, Y., Sakurai, H., and Tsukuda, T. (2007). Effect of Ag-doping on the Catalytic Activity of Polymer-Stabilized Au Clusters in Aerobic Oxidation of Alcohol, *J. Phys. Chem. C*, **111**, pp. 4885–4888.

189. Miyamura, H., and Kobayashi, S. (2014). Tandem Oxidative Processes Catalyzed by Polymer-Incarcerated Multimetallic Nanoclusters with Molecular Oxygen, *Acc. Chem. Res.*, **47**, pp. 1054–1066.

190. Jiang, S., Lv, L.-P., Landfester, K., and Crespy, D. (2016). Nanocontainers in and onto Nanofibers, *Acc. Chem. Res.*, **49**, pp. 816–823.

191. Sun, Q., Dai, Z., Meng, X., and Xiao, F.-S. (2015). Porous Polymer Catalysts with Hierarchical Structures, *Chem. Soc. Rev.*, **44**, pp. 6018–6034.

192. Zaera, F. (2013). Nanostructured Materials for Applications in Heterogeneous Catalysis, *Chem. Soc. Rev.*, **42**, pp. 2746–2762.

193. Johnson, S. A., Ollivier, P. J., and Mallouk, T. E. (1999). Ordered Mesoporous Polymers of Tunable Pore Size from Colloidal Silica Templates, *Science*, **283**, pp. 963–965.

194. Chakraborty, S., Colón, Y. J., Snurr, R. Q., and Nguyen, S. T. (2014). Hierarchically Porous Organic Polymers: Highly Enhanced Gas Uptake and Transport through Templated Synthesis, *Chem. Sci.*, **6**, pp. 384–389.

195. Sai, H., Tan, K. W., Hur, K., Asenath-Smith, E., Hovden, R., Jiang, Y., Riccio, M., Muller, D. A., Elser, V., Estroff, L. A., Gruner, S. M., and Wiesner, U. (2013). Hierarchical Porous Polymer Scaffolds from Block Copolymers, *Science*, **341**, pp. 530–534.

196. Zhang, Y., Wei, S., Liu, F., Du, Y., Liu, S., Jia, Y., Yokoi, T., Tatsumi, T., and Xiao, F.-S. (2009). Superhydrophobic Nanoporous Polymers as Efficient Adsorbents for Organic Compounds, *Nano Today*, **4**, pp. 135–142.

197. Wang, C. A., Zhang, Z. K., Yue, T., Sun, Y. L., Wang, L., Wang, W. D., Zhang, Y., Liu, C., and Wang, W. (2012). 'Bottom-Up' Embedding of the Jørgensen–Hayashi Catalyst into a Chiral Porous Polymer for Highly Efficient Heterogeneous Asymmetric Organocatalysis, *Chem. Eur. J.*, **18**, pp. 6718–6723.

198. Kundu, D. S., Schmidt, J., Bleschke, C., Thomas, A., and Blechert, S. (2012). A Microporous Binol-Derived Phosphoric Acid, *Angew. Chem. Int. Ed.*, **51**, pp. 5456–5459.

199. Sun, Q., Meng, X., Liu, X., Zhang, X., Yang, Y., Yang, Q., and Xiao, F.-S. (2012). Mesoporous Cross-Linked Polymer Copolymerized with Chiral BINAP Ligand Coordinated to a Ruthenium Species as an Efficient Heterogeneous Catalyst for Asymmetric Hydrogenation, *Chem. Commun.*, **48**, pp. 10505–10507.

200. Zhang, P., Weng, Z., Guo, J., and Wang, C. (2011). Solution-Dispersible, Colloidal, Conjugated Porous Polymer Networks with Entrapped Palladium Nanocrystals for Heterogeneous Catalysis of the Suzuki-Miyaura Coupling Reaction, *Chem. Mater.*, **23**, pp. 5243–5249.

201. Jiang, J.-X., Su, F., Trewin, A., Wood, C. D., Campbell, N. L., Niu, H., Dickinson, C. Ganin, A. Y., Rosseinsky, M. J., Khimyak, Y. Z., and Cooper, Andrew, I. (2007). Conjugated Microporous Poly(Aaryleneethynylene) Networks, *Angew. Chem. Int. Ed.*, **46**, pp. 8574–8578.

202. Wang, F., Mielby, J., Richter, F. H., Wang, G., Prieto, G., Kasama, T., Weidenthaler, C., Bongard, H.-J., Kegnæs, S. Fürstner, A., and Schüth, F. (2014). A Polyphenylene Support for Pd Catalysts with Exceptional Catalytic Activity, *Angew. Chem. Int. Ed.*, **53**, pp. 8645–8648.

203. Zhang, Y., Zhang, Y., Sun, Y. L., Du, X., Shi, J. Y., Wang, W. D., and Wang, W. (2012). 4-(N,N-Dimethylamino)Pyridine-Embedded Nanoporous Conjugated Polymer as a Highly Active Heterogeneous Organocatalyst, *Chem. Eur. J.*, **18**, pp. 6328–6334.

204. Su, B.-L., Sanchez, C., and Yang, X.-Y. (2011). *Hierarchically Structured Porous Materials: From Nanoscience to Catalysis, Separation, Optics, Energy, and Life Science*, Wiley-VCH Verlag Gmbh.

205. Nguyen, D. N., and Yoon, H. (2016). Recent Advances in Nanostructured Conducting Polymers: From Synthesis to Practical Applications, *Polymers*, **8**, p. 118.

206. Xu, P., Han, X., Zhang, B., Du, Y., and Wang, H.-L. (2014). Multifunctional Polymer-Metal Nanocomposites via Direct Chemical Reduction by Conjugated Polymers, *Chem. Soc. Rev.*, **43**, pp. 1349–1360.

207. Qiu, L., Peng,Y., Liu, B., Lin, B., Peng, Y., Malik, M. J., and Yan, F. (2012). Polypyrrole Nanotube-Supported Gold Nanoparticles: An Efficient Electrocatalyst for Oxygen Reduction and Catalytic Reduction of 4-Nitrophenol, *Appl. Catal. Gen.*, **413–414**, pp. 230–237.

208. M. Sanchez-Ballester, N., Rydzek, G., Pakdel, A., Oruganti, A., Hasegawa, K., Mitome, M., Golberg, D., Hill, J. P., Abe, H., and Ariga, K. (2016).

Nanostructured Polymeric Yolk–Shell Capsules: A Versatile Tool for Hierarchical Nanocatalyst Design, *J. Mater. Chem. A*, **4**, pp. 9850–9857.

209. Rydzek, G., Terentyeva, T. G., Pakdel, A., Golberg, D., Hill, J. P., and Ariga, K. (2014). Simultaneous Electropolymerization and Electro-Click Functionalization for Highly Versatile Surface Platforms, *ACS Nano*, **8**, pp. 5240–5248.

210. Selvaraj, V., and Alagar, M. (2007). Pt and Pt–Ru Nanoparticles Decorated Polypyrrole/Multiwalled Carbon Nanotubes and Their Catalytic Activity towards Methanol Oxidation, *Electrochem. Commun.*, **9**, pp. 1145–1153.

211. Zhou, Q., and Shi, G. (2016). Conducting Polymer-Based Catalysts, *J. Am. Chem. Soc.*, **138**, pp. 2868–2876.

Chapter 5

Carbons as Supports for Catalysts

Shenmin Zhu, Chengling Zhu, Yao Li, and Hui Pan

*State Key Laboratory of Metal Matrix Composites,
Shanghai Jiao Tong University, 800 Dongchuan Road,
Shanghai 200240, China*

smzhu@sjtu.edu.cn

5.1 Introduction

Carbon is the 15th most abundant element in the Earth's crust, and the fourth most abundant element in the universe by mass. The carbons play a key role in catalysis. Organic molecules in which carbon is one of the main composed element, are involved in most of catalytic processes. Organic carbon is often the main constituent of the organometallic catalysts or enzymes. Inorganic carbon materials also have attracted great interests for both acting as catalysts or catalyst supports. Due to their specific characteristics of acid/base resistance, controllable porosity, and surface chemistry, carbon materials have been used for a long time in heterogeneous catalysis, capable of satisfying most of

---

Soft Matters for Catalysts
Edited by Qingmin Ji and Harald Fuchs
Copyright © 2020 Jenny Stanford Publishing Pte. Ltd.
ISBN 978-981-4774-66-6 (Hardcover), 978-1-351-27284-1 (eBook)
www.jennystanford.com

the desirable properties required for a suitable support. The typically large surface area and high porosity of activated carbon catalysts favor the dispersion of the active phase over the support and increase its resistance on sintering at high metal loadings. The pore size distribution can be adjusted to suit the requirements of several reactions. Although carbon materials are normally hydrophobic, the surface chemistry of carbon materials can easily be modified to increase their hydrophilicity and favor ionic exchange.

Apart from an easily tailorable porous structure and surface chemistry, carbon materials present other advantages for catalysis [1]:

(i) Metals on the support can be easily reduced.
(ii) The structure is stable at high temperatures (even above 1023 K under inert atmosphere).
(iii) Porous carbon catalysts can be prepared in different physical forms, such as granules, cloth, fibers, pellets, films, etc.
(iv) The active phase can be easily recovered.
(v) The cost of conventional carbon supports is usually lower than that of other conventional supports, such as alumina and silica.

On the other hand, carbon supports also present some disadvantages. For example, they are difficult to use in high-temperature hydrogenation and oxidation reactions due to the easy gasification. Their reproducibility can also be relatively poor, as the same material may contain varying ash amounts. Activated carbon (AC) and carbon black (CB) are the most commonly used carbon supports. In the past decade, new carbon forms such as graphenes, carbon nanotubes (CNT) and various carbon nanostructures derived from molecular assembled morphologies have generated intense attention in the scientific community. Especially for CNT and graphenes, they have become one of the most active fields of nanoscience and nanotechnology due to their exceptional properties. These unique features make them as breakthough materials for many potential applications in storage, optics, electronics, and also catalysis. The recent availability of novel carbon nanomaterials offers new opportunities for the development of advanced metal-free materials with improved catalytic performance.

In this chapter, we discuss 1D carbon nanotubes, 2D graphene nanostructures, and 3D carbon nanostructures and briefly introduce the functionalization of carbon-based (nano)materials for their applications in catalysis. We present the specific characteristics of these carbon materials for enhancing catalytic activity and highlight the possible perspectives of using novel architectured carbon nanostructures.

5.2 1D Carbon Nanotubes for Catalysis

Since Iijima found the formation of carbon nanotubes (CNTs) during the synthesis of fullerenes [2, 3], CNTs have become one of the most active fields of nanoscience and nanotechnology due to their exceptional properties and potentials for many potential applications. CNTs represent an interesting alternative to conventional supports for a number of reasons, including [4]:

- high purity that eliminates self-poisoning;
- impressive mechanical properties, high electrical conductivity and thermal stability;
- high accessibility of the active phase and the absence of any microporosity, thus eliminating diffusion and intraparticle mass transfer in the reactions medium;
- possibility for macroscopic shaping of the support;
- possibility of tuning the specific metal-support interactions, which can directly affect the catalytic activity and selectivity;
- confinement effects in their inner cavity.

In addition, CNTs have a high flexibility on the dimensions of their inner cavity and surface properties. Their specific surface area can be modulated in a range of (50–1300 m^2g^{-1}). CNTs can be modified by chemical functionalizations or by elemental doping [5, 6]. Functionalizations facilitate fine-tuning the properties as the catalytic supports including electron donor-acceptor property, conductivity, acid-base property and interfacial property. The commonly used methods for functionalization include surface oxidation, grafting of organic groups or macromolecules, doping of heteroatoms, chemical or electrochemical coating of polymers or nanoparticles (NPs), irradiation functionalization, plasma activation, etc. [7].

To date, CNT-based catalysts have been used mainly for important liquid-(hydrogenation, hydroformylation) or gas-phase (Fischer–Tropsch process, ammonia decomposition) reactions, for photocatalysis and for electrocatalysis (fuel-cell electrodes) [4]. The confinement effect of CNTs also plays a large role on the performance of the catalysts. The encapsulation of small molecules, metal complexes, ionic liquids (ILs), or fullerenes, and NPs in the CNT cavity paves the way to interesting perspectives for performing chemistry in a nanoreactor and exploiting unique CNT properties [4].

5.2.1 Oxidized CNTs for Catalysis

Surface oxidation for CNTs can greatly change the hydrophilicity or wettability of CNT surfaces in polar solvents including water. Those oxygen-containing groups on surface of CNTs may also serve as anchoring sites for subsequent functionalization by covalent, electrostatic and hydrogen bonds. The extent of oxidation is determined by the strength of oxidants and oxidation conditions. Common oxidants include nitric acid, hydrogen peroxide, permanganate, hypochlorite, persulfate, hypochlorite, chlorates, dichromate, oxygen, ozone and nitric oxide [8–11].

Oxygen surface groups favor the preparation of carbon-supported catalysts. In the most common procedure, catalyst precursors were bound on/in the CNTs. For the example of Pt-based catalyst, pre-oxidized CNTs by the nitric acid treatment may retard the reduction of Pt nanoparticles supported on CNTs, where the surface oxygen-containing groups strongly stabilized the cationic platinum species. The later removal of oxygen-containing species from CNTs may suppress acid-catalyzed side reactions and expedite electron transfer between CNTs and platinum nanoparticles, consequently improving both the activity and selectivity in hydrogenation [12].

Biomass (residues from forestry, industry and agriculture) conversion into biofuels or synthetic fuels using Fischer–Tropsch process has become a key technology platform for the successful implementation of a sustainable energy production. Cobalt-based catalysts represent the optimal choice for the synthesis of middle distillate fuels and used in Fischer–Tropsch process for the production of liquids fuels. In contrast to conventional support materials, CNTs as catalytic supports remarkably reduce the duration of the induction period, improve the reducibility of the active phase with higher catalytic activity and stability in Fischer–Tropsch synthesis [13]. CNT surface functionalization of oxygen containing groups through acid treatment opens up the cap ends (Fig. 5.1), enlarges the BET surface area. The introduction of defects and acidic functional groups may facilitate to diminish the size of Co nanoparticles and their homogeneous distribution in CNTs [14]. In addition, the reducibility of the catalyst increases by 10% and 50% upon acid treatment at 25 and 100°C, respectively. The Fischer–Tropsch activity and CO conversion efficiency both strongly increase after acid treatment on the CNT support [15, 16].

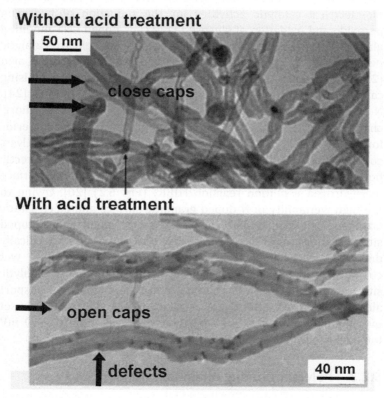

Figure 5.1 TEM images of CNT with close caps and CNT with defects and open caps after hot acid treatment [14].

5.2.2 Doping CNTs for Catalysis

Doping heteroatoms such as boron, nitrogen, sulfur and phosphine on CNTs is also an efficient approach to functionalize CNTs with fine-tuned structural and electronic properties. During the synthesis processes of CNTs, the mixing N resource (amines, ammonia), organic borane or organic phosphine may result into the formation of N-doped, B-doped or P-doped CNTs [17–19]. The heteroatoms can cause some defects and contribute additional electrons and fast electron transfer onto the surface of nanotubes, which may lead to greater reactivity than that of pristine CNTs [20].

The N-doped CNTs were utilized for immobilizing hemoglobin. The N-doped CNTs functionalized electrodes exhibited favorable bio-electrical catalytic activities for the reduction of hydrogen peroxide and oxidation of glucose [21, 22]. N-doped CNT modified glassy carbon electrodes showed promising electro-catalytic performance for the oxidation of dopamine and ascorbic acid [23]. The N-doped CNT surface coated with Pt presents promising catalytic performance for methanol electro-oxidation [24]. N-doped CNT-supported iron showed high selectivity for short-chain olefins, decreased chain growth probability, and superior long-term stability [25]. Kim et al. synthesized hybrid catalysts composed of amorphous molybdenum sulfide (MoS_x) layer directly bound at vertical N-doped carbon nanotube (NCNT) forest surface for hydrogen evolution reaction (HER) (Fig. 5.2) [26]. Owing to the high wettability of N-doped graphitic surface and electrostatic attraction between thiomolybdate precursor anion and N-doped sites, ~2 nm scale thick amorphous MoS_x layers are specifically deposited at N-doped CNT surface under low-temperature wet chemical process. The synergistic effect from the dense catalytic sites at amorphous MoS_x surface and fluent charge transport along NCNT forest attains the excellent HER catalysis with onset overpotential as low as ~75 mV and small potential of 110 mV for 10 mA/cm² current density.

5.2.3 Polymer Coating on CNTs

Pristine CNTs tend to form large bundles due to the strong van der Waals interaction between nanotubes. To obtain dispersed CNTs and favor the binding of metal particles on surface, the strategies for functionalizing the side wall of nanotubes and coating with surfactants or polymers are employed [27]. With wrapping CNTs with linear polymers, CNTs were shown to be solubilized in water. In compared to functionalize side wall or using surfactants, the coating with polymers is thought to be superior, as chemical modification of the side wall may damage the intrinsic property of CNTs, while use of surfactant raises the possibility of contamination of the electrocatalysts. The surface wettability, conductivity, electron capacitance, adsorption properties can be significantly tuned by coating polymers on CNT surfaces.

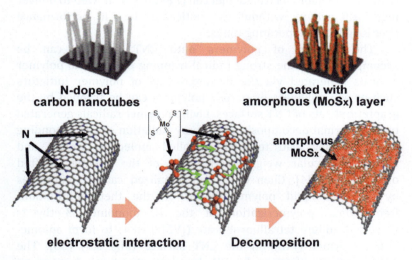

Figure 5.2 Schematic illustration for the synthesis of three-dimensional MoSx/N-doped CNT forest hybrid catalyst [26].

The covalent binding of polymers onto surfaces is mainly through carboxyl or phenolic hydroxyl groups at the corners and edges of the polycondensed aromatic rings or onto carbon atoms at defect positions. Poly(vinyl alcohol) (PVA)-coated CNTs were formed through a reaction between surface carboxyl groups and PVA along with a condensing agent such as N,N′-dicyclohexyl-carbodiimide (DCC) [28]. Through thionyl chloride treatment, the carboxyl groups on CNTs may also be transformed to the acyl chloride groups for introducing polymerization initiators on the surface. Polyethylene oxide (PEO)-grafted CNTs have been synthesized by reaction of acyl chlorides on CNT surfaces with the monoamine-terminated PEO [29]. The long alkyl chain grafted onto CNTs has been obtained by the reaction between acyl chloride groups and long chain aliphatic amine [30]. Ramaprabhu et al. modified CNTs with polypyrrole (PPV) by chemical oxidative polymerization of a monomer (pyrrole) on the surface of multi-walled CNTs (MWCNTs) [31]. The uniform coverage of polypyrrole (PPY) on MWCNT resulted into the homogeneously distribution of Co NPs on PPY-MWCNTs. The Co-PPY-MWCNT was used as an electrocatalyst for the oxygen reduction reaction (ORR) in polymer electrolyte fuel cell (PEMFC). It showed to deliver high ORR activity without any noticeable loss in performance over long PEMFC operating times.

The grafting of polymers onto CNT surfaces can be accomplished via the strong radical trapping effects. The polymer radicals generated via the decomposition of polymer initiators such as azo polymers or peroxy polymers can be trapped by the graphitic sheets of CNT surfaces. Living polymer radicals generated by the thermal decomposition or γ-ray radiation of corresponding polymers may lead to well-controlled molecular weight and narrow molecular weight distribution of the polymer grafted on CNTs [32–34]. Chen et al. functionalized carbon nanotubes by an ionic-liquid polymer based on the thermal-initiation-free radical polymerization of the IL monomer 3-ethyl-1-vinylimidazolium tetrafluoroborate ([VEIM]BF4) to form anionic-liquid polymer (PIL) on the CNT surface (Fig. 5.3) [35]. The process of modification by PIL has less structural damage on CNTs than the typical acid-oxidation treatment, because of the mild polymerization condition from the IL monomer. The uniform distribution of a large number of surface functional groups on the CNTs favors to anchor and grow metal nanoparticles on CNTs. PtRu and Pt nanoparticles were shown to be dispersed uniformly on the PIL-functionalized CNTs (CNTs-PIL). The obtained catalysts (PtRu/CNTs-PIL and Pt/CNTs-PIL) showed superb performance for direct electro-oxidation of methanol.

Giacalone et al. synthesized single-walled carbon nanotube (SWNT)-polyamidoamine dendrimers hybrids (SWCNT-PAMAM) via the direct reaction of cystamine-based PAMAM dendrimers (generations 2.5 and 3.0) with pristine SWCNTs in refluxing toluene (Fig. 5.4) [36]. The PAMAM dendrimers on SWCNTs surface offer binding sites for anchoring and immobilization of small Pd NPs, which showed to be homogeneously confined throughout the nanotube length. The SWCNT-PAMAM hybrids with Pd proved to be an efficient catalyst in Suzuki and Heck reactions, able to promote the above processes down to 0.002 mol% additional amount with a turnover number (TON) of 48,000 and a turnover frequency (TOF) of 566,000 h^{-1}. These catalysts could be recovered and recycled for up to 6 times. No leaching of the metal was detected during the reaction process.

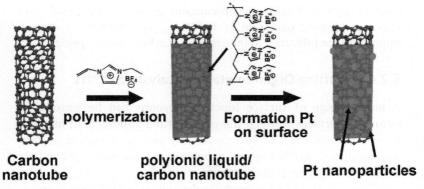

Figure 5.3 Scheme for the modification of carbon nanotubes (CNTs) with polyionic liquid (PIL) and the preparation of Pt/CNTs-PIL nanohybrids.

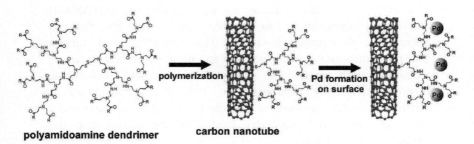

Figure 5.4 Synthesis of CNT–PAMAM Hybrids and Immobilization of PdNPs.

Electrochemical polymerization is also a readily available approach for polymer-functionalization of CNTs. For instance, polyaniline (PANI) covalently functionalized SWNTs can be obtained by the electrochemical polymerization of aniline directly onto the SWNT working electrode in an aqueous HCl electrolyte. Li and Cui et al. reported the fabrication of PANI/MWCNT composite by an electrochemical polymerization of aniline with airbrushed MWNTs on ITO substrates [37]. Their results indicated that the in situ polymerization process can effectively improve the electronic properties of the composites. Li et al. studied the effect of different loading levels of PANI on CNTs as Pt catalyst supports for methanol electro-oxidation [38]. It revealed that the loading level of PANI not only affects the size and distribution of Pt catalysts, but also the proportion of the different valencies of Pt components. The proper proportion of Pt(II) was essential for coordination of Pt with oxygen-containing ligands (OH$^-$, H$_2$O) and thus accelerating oxidation of Pt(CO)$_{ads}$ with release of CO$_2$ by improving the tolerance to intermediate carbonaceous species.

5.2.4 Grafting Organometallic Catalysts on CNTs

A broad range of surface functional groups can be introduced onto CNT surfaces through the post grafting approach. In many of these methods, the CNT surface already possesses certain surface functionalities, which can be further substituted by other catalytic organo- or organometallic catalysts.

Toma et al. have successfully developed an efficient oxygen-evolving anode using polyoxometalate clusters, M$_{10}$[Ru$_4$(H$_2$O)$_4$(μ-O)$_4$(μ-OH)$_2$(γ-SiW$_{10}$O$_{36}$)$_2$] (M = Cs or Li), and MWNTs [39]. The MWNTs were functionalized using polyamidoamine (PAMAM) ammonium dendrimers to anchor the negatively charged polyanionic Ru-containing cluster. The nanostructured carbon CNT-polyoxometalate was shown to perform efficient water oxidation, with appreciable catalytic current and remarkable turnover frequency numbers. The enhanced performance is attributed to the increase of the electrochemical active surface area between the redox-active center of the cluster and the surface of the electrode.

Álvarez and Jiménez et al. modified oxidized MWCNT with appropriate hydroxyl-ending imidazolium salts, which then found to favor the formation of nanohybrid materials containing iridium N-heterocyclic carbene (NHC)-type organometallic complexes (Fig. 5.5) [40]. The CNT-supported iridium-NHC materials were active in the heterogeneous iridium-catalyzed hydrogen-transfer reduction of cyclohexanone to cyclohexanol with 2-propanol/KOH as hydrogen source. The iridium hybrid materials are more efficient (with initial TOFs up to 5550 h^{-1}) than related homogeneous catalysts based on acetoxy-functionalized Ir-NHC complexes. Good recyclability without any loss of activity, and with well stability in air was also achieved.

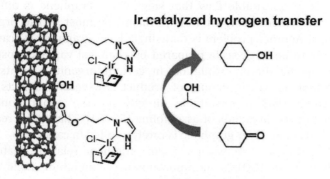

Figure 5.5 Enhanced hydrogen-transfer catalytic activity of iridium N-heterocyclic carbenes by covalent attachment on carbon nanotubes.

5.3 2D Graphenes for Catalysts

As a two-dimensional (2-D) allotrope of carbon and the basic unit of graphite, graphene possesses a single-atom-thick honey-comb lattice of sp^2-hybridized carbon atoms. The unique characteristics of graphene, such as ultra-high electrical conductivity of 1.00×10^8 S m^{-1}, thermal conductivity of ~5300 W m^{-1} K^{-1} [41], two-dimensional morphology, and extremely high theoretical specific surface area of 2630 m^2 g^{-1} [42], have made it and its related materials become one of the most extensively studied catalysts in recent ten years [43–48].

Up to now, thousands of research papers have been published on applying graphene and graphene-based materials as a certain kind of catalysts, including graphene itself and graphene oxide (GO), doped graphene, as well as graphene-based composite catalysts, in which graphene usually acts as a reinforcement or co-catalyst. The applications of graphene-based catalysts almost cover all the fields that a catalyst could be involved in, such as electrocatalysis, photocatalysis, and ORR.

Although graphene is strictly defined as a defect-free flat carbon monolayer in the field of physics [49], it is operationally referred to a broader range of related materials whose structures are similar (2D carbon forms) in the chemistry-oriented fields. Among them are the multilayer graphene, doped graphene, graphene oxide, and other functionalized graphene.

It is an inevitable flaw that single-layer graphene is difficult to be dispersed stably in water, ethanol or most of the organic solvents. Another problem is that single-layer graphene is scarcely possible to be large-scale prepared by chemical vapor deposition (CVD) or by direct exfoliation in certain organic solvents, let alone that the small amount of product is devoid of defects and impurities, thus is unsuitable directly for the applications as catalyst or the demands of assembling [50–54]. That is the reason why CVD single-layer graphene is rarely applied in catalysis.

The most extensively used graphene-related material, graphene oxide (GO), is an important succedaneum of single-layer non-defective graphene, which is free of the flaws mentioned above [55]. It has been an irresistible trend that most of the graphene and graphene-related materials in reported studies on catalyst and energy fields are prepared by applying GO as a precursor. Generally, Hummers' oxidation method is used for the preparation of GO [56], owing to its high output as well as controllability of defects and oxygenated functional groups. The product, conventionally referred to as graphite oxide, is prepared by oxidizing graphite with strong oxidizer such as sulfuric acid and phosphoric acid. During the process, abundant epoxy bridges, hydroxyl groups and carboxyl groups are created while preserving the layer structure of graphite. The groups expand the interlayer distance of graphite to be around ~0.7 nm between GO, making it easy to be exfoliated further into single layer or few layers structure. The ionized hydroxyl groups and carboxyl groups also give the exfoliated GO sheets good dispersity in water and some other solvents like dimethyl formamide [55]. This valuable property differs quite from the CVD graphene sheets without functionalization, which tend to stack back into graphite because of π-π stacking. Resulted from the estimable dispersity of GO in solvents and the resultant processability, it is studied not only as a catalyst itself, but also introduced as the precursor of various graphene-combined composite catalysts, e.g., noble metal/graphene catalysts, and transition metal oxide/graphene catalysts. Also benefited from the unique 2D structure of graphene, the graphene-based composite catalysts can be easily constructed into a variety of 3D porous structure. The formed hierarchical porous structure with micro-, meso-, and macro pores would provide transportation channels for the reactant and the resultant molecules, which is beneficial to the catalytic performances. When GO is reduced into the form lacking oxygen-contained functional groups, it will possess an atomic structure similar to the non-defective graphene, which is referred to as reduced graphene oxide (RGO) or indiscriminately called as graphene in the papers published earlier than 2010 [55, 57]. In the following discussion, the appellations are mostly followed as the ones in the reference papers, and in some cases the word graphene actually refers to RGO. Because of the lattice structure with fewer defects, RGO is closer to CVD graphene in electrical conductivity and chemical stability.

Moreover, GO can act not only as the precursor of graphene, but also as a high-efficiency and multifunctional catalyst itself. The groups on the surface of GO virtually make it a semiconductor, and capable for multiple applications [58–61].

5.3.1 Graphene-Based Materials as Photocatalyst

The fossil fuel all over the world is going to deplete, and so is the petrochemical industry. Photochemical reactions which make use of the inexhaustible solar energy are deemed as one of the most promising mode of sustainably producing macromolecule materials as well as clean energy like hydrogen. Since 1970s, the discovery of TiO_2 was reported to be able to split water by a photocatalysis process [62], great efforts have been made to the development of various semiconductor photocatalysts with filled valence band and empty conduction band [63–65].

It is proven that the photocatalytic activity of a semiconductor is benefitted from the electron–hole pairs appeared, when electrons are excited from the valence band into conduction band by the photons fitting to the band gap of the semiconductor. Normally, before the reaction occurs between the excited electron or the holes and catalysts, the electron–hole pairs can easily and ultrafast recombine together [64, 66]. In order to increase the photocatalytic efficiency, it is vital to increase the electron mobility in photocatalysts. Additionally, different from other kinds of catalysts, the absorption ability of light with wide-ranged wavelength is another determinant factor for the improvement of the photocatalytic efficiency [67].

It is justifiable that graphene-based (or graphene-reinforced) photocatalysts have attracted great attention. Owing to the multiple properties, i.e., good electrical conductivity, large surface area, and effective adsorption of light, graphene offers a promising application in improvement of the photocatalytic efficiency. Not only the electron mobility and the absorption ability of light can be promoted, but the effective contact area between the graphene-based photocatalysts and the catalysant is expanded to a large extent [68].

Soon after the first successful preparation of graphene in the world, it has been applied as a structural and conductive reinforcement in composite photocatalysts, such as TiO_2/graphene [68–70], WO_3/graphene [71–73], Bi_2WO_6/graphene [74–76], and $BiVO_4$/graphene [77–80], etc.

5.3.1.1 Photocatalysts for oxygen and hydrogen generation

In recent years, the utilization of solar energy to photodissociate water has gathered great interest and research enthusiasm, because solar photolysis of water is one of the cleanest ways of producing hydrogen and oxygen, which has great potential in solving energy problem. As a promising and clean fuel with the highest gravimetric heat of combusting (141.80 MJ kg^{-1}) and not producing greenhouse gas like CO_2 after combustion, hydrogen (H_2) has been increasingly utilized in hydrogen fuel cells as one of the substitutes of combustion engines [81, 82]. The most conventional method of producing hydrogen is electrolysis of water, which consumes electricity and thus is virtually not sustainable. Splitting water by photocatalytic reaction is a more direct and environmentally friendly way to obtain hydrogen.

As a semiconductor photocatalyst for spontaneous water splitting three main thermodynamic requirements should be fulfilled:

(1) Its band gap must be higher than water decomposition voltage (1.23 eV).

(2) Its band edge positions must straddle the hydrogen and oxygen redox potential.

(3) It must be stable against photo-corrosion during the catalytic reaction [83].

Accordingly, semiconductors including TiO_2, WO_3, Bi_2WO_6, ZnO, Bi_2O_3, CdS, etc., have been reported to be suitable in splitting water to produce hydrogen or oxygen [62, 65, 67, 71, 73, 75, 76, 79, 80, 84].

Commonly, when electrons in the photocatalysts like TiO_2, are excited by the photons with the energy larger than the bandgap ($hv \geq E_g$), the electron–hole pairs are produced. When the pairs come to the surface of the photocatalyst, they can react with O_2 molecules and OH^- ions to form singlet oxygen ($\bullet O_2^-$) and hydroxyl radicals ($\bullet OH$) [85]. Then, the electrons will react with H^+ to be reduced into hydrogen.

Photocatalytic splitting water for O_2 evolution is usually easier to achieve than H_2 evolution, and the former can be deemed as a foundation of the latter. In the development of O_2 and H_2 photocatalyst, various semiconductors have been combined with graphene materials to enhance or alter their performances.

Many approaches haven been developed for the fabrication of graphene-based photocatalyst. Guo et al. synthesized WO_3 nanoparticles on the surface of graphene sheets by a simple sonochemical method [71]. When the WO_3@graphene composite was used as a photocatalyst for water splitting, graphene sheets served as an acceptor for the conductive band (CB) electrons of WO_3 nanoparticles and effectively suppressed the recombination of electron–hole pairs in the composite. It thus leaves more positive charged holes on the WO_3 surface and promotes the production of oxygen. The amount of evolved O_2 from water by the WO_3@graphene composite was twice as much as that for pure WO_3.

Structure design is proved as an effective method to enhance the properties. Sun et al. applied an ultrasonication-assisted sol-gel synthesis route to controllably prepare $BiVO_4$@amorphous carbon core–shell nanoparticles on reduced graphene oxide (RGO-BVO@C) sheets (Fig. 5.6) [79]. The size of the nanoparticles can be controlled by adjusting the volume ratio of glycerol, which is also a precursor. The solid chemical bonds between $BiVO_4$ core and the amorphous carbon shell in the composites were demonstrated, which can promote charge transfer between BiVO$_4$ and the carbon shell, and further to RGO sheets. The light absorption ability improved by RGO sheets is another advantage in the RGO-based composites. As a result, the obtained RGO-BiVO$_4$@C nanocomposite showed a five times higher rate than that of pure BiVO$_4$ in O$_2$ evolution from water under visible-light irradiation.

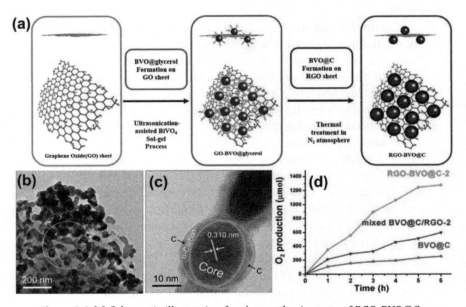

Figure 5.6 (a) Schematic illustration for the synthesis route of RGO-BVO@C. (b) TEM images and (c) high-resolution TEM image of RGO-BVO@C. (d) Photocatalytic O$_2$ generation from water of the samples under visible-light irradiation (λ > 420 nm) [79].

In the graphene-based photocatalysts for H$_2$ evolution, graphene materials play an even more important role to improve the performances. Xiang et al. introduced a microwave-hydrothermal synthesis method to combine GO and TiO$_2$ nanosheets into graphene/TiO$_2$ nanosheet composites with different weight ratio of the two components [86]. The graphene-based composites showed great photocatalytic H$_2$ production performance (736 μmol h^{-1} g^{-1}), which is 40 times more than that of TiO$_2$ nanosheets without graphene. The low performance of bare TiO$_2$ nanosheets was ascribed to the poor electron mobility in TiO_2, which leads to fast recombination of electron–hole pairs. Another reason is that the low CB of TiO_2 (−0.24 V vs. standard hydrogen electrode (SHE)) is not matchable with the H^+/H_2 potential (0 V vs. SHE). Unlike semiconductor e.g., TiO_2, graphene materials possess electrical properties more like conductor, which can enhance the electron mobility of TiO_2 with the surface-to-surface contact, and the higher potential of graphene/graphene- (−0.08 V vs. SHE) than the CB of TiO_2 makes graphene surface the real reaction sites for protons to capture electrons, and thus produce H_2 efficiently.

Despite TiO_2, some other semiconductor photocatalysts which are usually applied to produce O_2, e.g., Bi_2WO_6, can also be applied to photocatalytically produce H_2, through combined with graphene matrix. With the existence of the graphenes, the conduction band potential can be decreased and thus a more negative reduction potential than H^+/H_2 can be obtained. Sun et al. developed a mild sonochemical method to form Bi_2WO_6 nanoparticles in situ on the support of graphene sheets [75]. When used as a photocatalyst under visible-light irradiation, the O_2 production rate of the Bi_2WO_6-graphene composite was improved more than four times of that of bare Bi_2WO_6. More interestingly the composite showed a photocatalytic activity of H_2 production up to 159.2 µmol h^{-1}, which is completely distinct from the property of bare Bi_2WO_6. It is known that the electrons excited from the CB in Bi_2WO_6 are impossible to reduce H^+ to H_2. However, the electrons excited from the CB in Bi_2WO_6-graphene composite, which is −0.30 V and at least 0.39 V more negative than that of bare Bi_2WO_6 nanoparticles, are energetic enough to react with protons to produce H_2. The strong covalent bonding between graphene sheets and the Bi_2WO_6 nanoparticles facilitates the electron collection and transportation, at the same time inhibits the recombination of photo-generated charge carriers. As a result, a much higher production of O_2 and H_2 from the composite can be achieved in compared with those from a mixture of Bi_2WO_6 and graphene. In addition, the high content of graphene in the Bi_2WO_6-graphene composite overlaps the visible-light absorption, and an improved utilization of the solar energy is thus obtained. In another work of Sun et al., they used chemically reduced GO sheets as support and grew Bi_2WO_6 nanoneedles in situ on the support with the assistance of ultrasonic processing and subsequent thermal treatment [84]. The composite with unique morphology showed extremely high performance on photocatalytic production of H_2 (234.83 µmol h^{-1}) and O_2 (23.61 µmol h^{-1}) from water.

As a close related material and widely used precursor of graphene, GO itself can also act as a photocatalyst to split water into H_2. Because of the oxygen-contained functional groups, such as epoxy and hydroxyl, GO shows quite different electrical and electrochemical properties in compared with defectless graphene. Their properties can be tuned through various chemical treatments. Yeh et al. synthesized GO with bandgap of 2.4–4.3 eV by chemical oxidation-exfoliation method, and directly applied this GO as a dispersible photocatalyst to generate H_2 from aqueous methanol solution and pure water under visible or ultraviolet (UV) light irradiation, without any cocatalyst required [87]. This work revealed the possibility of developing a new generation of metal-free, environmentally friendly, and high-performance photocatalysts for the production of H_2.

5.3.1.2 Photocatalysts for degradation of pollutants

The utilization of photocatalysts to degrade diverse pollutants, especially the organic ones, is an effective approach to deal with the increasing environmental issues of global scale.

Since the very initial work of applying TiO_2 as photocatalyst in 1970s [62], TiO_2-based materials have been studied as one of the most promising photocatalysts for decontamination. From the point view of photo-conversion efficiency, TiO_2 can produce electron–hole pairs under the irradiation of UV light but deficient under visible light irradiation. Another problem is the serious recombination of the photo-induced electron–hole pairs because of the weak electron mobility inside TiO_2, which is similar for most of semiconductor photocatalysts and has been discussed in the above section. TiO_2/graphene composite for photodegradation was first reported by Zhang et al. in 2010 [88]. Commercial TiO_2 nanoparticles, P25, were applied to be loaded on graphene nanosheets through one-step hydrothermal reaction [68]. In the photodegradation of methylene blue, the as-prepared composite photocatalyst showed great adsorptivity of dyes, extended light absorption range, and efficient charge separation properties simultaneously. A significant enhancement on photocatalysis was achieved by the TiO$_2$/graphene composite.

Guo et al. exploited a controllable sonochemical method to grow TiO$_2$ nanoparticles with 4–5 nm in diameter homo-geneously on graphene sheets [69]. The Ti–O–C bonds formed between TiO$_2$ nanoparticles and graphene sheets were evidenced by FT-IR. The solid covalent bonds help electrons to transmit from TiO$_2$ to graphene and reduce the recombination of the photo-generated electron–hole pairs, and thus give the composite an excellent performance in photodegradation of methylene blue in aqueous solution.

In recent years, more metal-free photocatalysts have been developed for environmental friendly processes. The most intensively studied one is graphitic carbon nitride (γ-C$_3$N$_4$), a novel polymeric semiconductor with graphite-like layer lattice and mainly composed of C and N [89–92]. γ-C$_3$N$_4$ possesses suitable bandgap, and superior photochemical stability, but lacks electrical conductivity and specific surface area. Therefore, introducing graphene into the metal-free photocatalyst system can improve the pollutant adsorption and optical absorption in visible light region, and enable C$_3$N$_4$ to efficiently degrade pollutants under visible light. Ai et al. used one-pot calcination method to prepare γ-C$_3$N$_4$/graphene hybrids from GO and melamine, which possess specific surface area three times higher than that of the sample without GO [92]. When used as photocatalysts to oxidize methylene blue and phenol, the hybrid exhibited enhanced performances under various light irradiations. It is the benefits from graphene sheets, which also promote the excited electrons on the conduction band of γ-C$_3$N$_4$ to transport away quickly, and leave holes in γ-C$_3$N$_4$ to capture electrons from the target organic pollutants.

5.3.1.3 Graphene-based materials for CO$_2$ photocatalytic reduction

The growing unbalance of carbon circulation between the atmosphere and the crust surface is mainly resulted from the consumption of fossil fuels, which gives rise to the global warming and the unsustainability of petrochemical industry. Solar generation provides a clean and promising strategy to solve those energy problems, i.e., photocatalytically reducing CO_2, coverting the primary products from fuel combustion and the greenhouse gas into fuels like CO, CH_4, CH_3OH or raw monomer materials like HCOOH and HCHO [93–95]. This strategy can simultaneously decrease greenhouse gas and directly convert solar energy, the most abundant and renewable but low-grade energy source, to high-grade and processable chemical energy.

However, the photocatalytic reduction of CO_2 to hydrocarbon fuels is challenging and more complicated than photocatalytic H_2 production. The reduction of CO_2 involves several products, which selectivity is depended on many factors, such as reaction kinetics, redox potentials of photoinduced carriers, morphology, and the surface property of photocatalysts [93, 95, 96]. As shown in Fig. 5.7, the conduction band edge of the photocatalysts should be more negative than the redox potential of H_2/H_2O or CO/CO_2 (or hydrocarbons/CO_2). Meanwhile, the potential of the valance band edge should be more positive than the redox potential of O_2/H_2O.

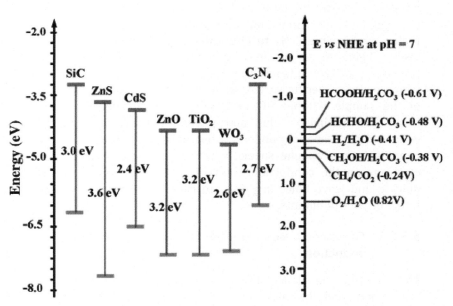

Figure 5.7 The comparison of band edge positions of selected semiconductor photocatalysts to the energy levels of various redox couples in neutral aqueous solution [94].

Toward the conventional amorphous-carbon- or carbon-nanotube-reinforced TiO_2 photocatalysts, the graphene-reinforced TiO_2 photocatalysis can have a more intimate interfacial contact between graphene sheets and TiO_2 particles and lead to higher CO_2 reduction performances, especially when TiO_2 also exhibits a 2D morphology. Liang et al. compared the 1D–2D carbon nanotube-TiO_2 nanosheet composites and the 2D–2D graphene–TiO_2 nanosheet composites for CO_2 photoreduction [97]. The 2D–2D graphene–TiO_2 nanosheet composites were found to yield superior electronic coupling and greater photoactivity enhancement under ultraviolet irradiation. The effect of the graphene ratio in the TiO_2/graphene photocatalysts on the conversion efficiency of CO_2 reduction into different products was studied by Tu et al. With increasing graphene content in the sandwich-like TiO_2/graphene structure, the production rate of CH_4 slowly decreases but the production rate of C_2H_6 noticeably increases [98]. This change is attributed to the balance between the absorption of the methyl radical (CH_3) intermediates on the surface of graphene by π-conjugation and the trend of excited-state electrons transferring to the graphene surface and reacting with CO_2.

Apart from graphene/TiO_2 composites, composites of graphene and other photocatalysts, such as graphene/CdS [99], graphene/ZnO [100], and graphene-modified NiO_x/Ta_2O_5 [101], have also been reported for the catalytic conversion of CO_2 into solar fuel. Despite the excellent electron-collecting/transferring capability of graphene, which promotes the multi-electron reduction of CO_2 and lead to high yields of products in the photocatalytic reaction, another role of graphene is to cause the shift of the CB potential of the semiconductor photocatalyst, which can reverse the intrinsic property of some semiconductors like WO_3 and make it capable to photocatalytically reduce CO_2 to CH_4 [102].

5.3.2 Graphene-Based Materials for Electrocatalyst

With the fossil fuels draining, there is also an increasing demand for the alternatives that can power vehicles and aircrafts. Among the various alternatives, fuel cells based on electrochemical energy conversion technology are of high power density, high energy density, and high stability, which thus is considered as one of the most promising technologies for the next generation of automotive power [103, 104].

In fuel cells, electrochemical oxidation reactions (i.e. HOR and AOR from hydrogen and alcohols) and reduction reactions (i.e. ORR from oxygen) occur respectively on the surface of anode and the cathode [105]. High-efficiency electrocatalysts are required in all the HOR, AOR and ORR reactions, which are determined the conversion efficiency of the fuel cells [106]. Pt-based materials are most generally used in electrochemical devices. Nevertheless, high cost and short durability of noble metals has hindered their commercial application [107].

Carbon materials of different phases and morphologies have been investigated for their potential application in electrocatalytic reactions, among which graphene is the most intensively studied one. When graphene works as a support for metal or ceramic nanoparticles, the planar 2D structure can offer high surface area for the attachment of nanocrystals with good distribution, and meanwhile act as a carrier of electrons due to the superior electrical conductivity [108]. Its excellent thermal and mechanical stability make it suitable for the electrochemical systems involving heat release [109]. With the relatively easy processibility and functionalization, GO-based composites are attractive candidate for acting as electrocatalysts.

Besides as a support substrate for nanocatalysts, graphene and graphene-based materials have also been intensively utilized as electrocatalyst themselves, mostly in the forms of doped or functionalized graphene. Heteroatoms such as nitrogen, boron, sulfur, phosphorus, and fluorine have been introduced to prepare doped graphene as electrocatalysts. The catalytic mechanisms of these graphene materials will be discussed as follows.

5.3.2.1 Graphene-based composite electrocatalysts

Noble metals and their alloys are still electrocatalysts with the highest performances nowadays [110]. However, the relatively low efficiency and high dosage limits their practical applications. To improve these drawbacks, two main strategies have been widely applied to modify noble-metal-based electrocatalysts: one is to construct nanostructure to increase the specific surface area, and the other is to supply supports for better dispersity of the noble metal particles. In comparison with traditional carbons such as active carbon, graphene materials, which possess higher specific surface area and superior conductivity, have attracted increasingly attentions as the supports for noble metal nanoparticles [111]. The graphene materials in these composite electrocatalysts not only can enhance the electrocatalytic activity with higher stability but also bring synergetic effects. The larger specific surface area can also save the dosage of the expensive noble metals [112].

Since 2004, many nanostructures and synthesis routes have been designed and developed to prepare Pt/graphene high-performance ORR electrocatalysts. Usually, Pt/graphene electrocatalysts are obtained by mixing GO and the precursors of Pt, followed by the reduction in the presence of H_2, N_2H_4, or $NaBH_4$, which are reductants [113]. The interaction between Pt and graphene, along with the morphology and crystallinity of Pt nanoparticles can play a key role in the electrocatalytic performances.

Yoo et al. found that the strong interaction between Pt and graphene can lead to Pt clusters (≤0.5 nm) stably anchoring on graphene sheets [114]. The specific electronic structures of the ultra-small Pt clusters result the Pt/graphene electrocatalyst ultra-high activity for methanol oxidation reaction (MOR). Xiao et al. developed the preparation route of in situ growing hollow Pt nanoparticles on graphene [115]. The hollow morphology of Pt nanoparticles and the synergetic effect from Pt and graphene were proven to enhance the electrocatalytic activity of MOR. Wang et al. synthesized ultrathin Pt nanowire arrays on S-doped graphenes. This composite have shown high electrocatalytic activity on both ORR and MOR, which efficiency is about 2–3 times higher than the commercial Pt/C catalysts [116].

As substitute materials for noble metal electrocatalysts, ceramic ones including metal oxides, sulfides, and nitrides have gained more and more interests in recent years [117]. Anchoring ceramic nanoparticles on graphene can not only greatly promote the electrical conductivity, but also retard the dissolution loss and agglomeration of the nanoparticles during usage [118, 119]. Dai et al. decorated Co_3O_4 nanoparticles on graphene and studied their electrocatalytic performance toward ORR and OER [120]. When N-doped graphene applied, a more uniform distribution and smaller size of Co_3O_4 nanoparticles can be obtained. The

Co$_3$O$_4$/graphene composite electrocatalyst possessed higher ORR performance than graphene or bare Co$_3$O$_4$ alone. The ORR activity of the composite monotonically increases with the content of Co$_3$O$_4$, which is in the range of 3~20%.

Many other transition metal oxides such as Fe$_3$O$_4$ [121], Cu$_2$O [122], and Ru$_2$O [123] combined with graphene have also been studied as electrocatalysts for ORR. Wu et al. presented the controllable growth of Fe$_3$O$_4$ nanoparticles on N-doped graphene aerogels. The aerogel composite has a 3D graphene framework with Fe$_3$O$_4$ nanoparticles uniformly depositing on graphene sheets [124]. The resultant sample showed considerable electrocatalytic activities for ORR, including lower ring current, higher electron transfer efficiency, and longer service life.

5.3.2.2 Doped graphene as electrocatalysts

Graphene without plenty of defects is electrocatalytically inert, but it tends to gain electrocatalytic activity after chemical functionalization or doping with various heteroatoms [125]. When the carbon atoms in the graphene lattice are replaced by heteroatoms, the lattice structure and the composition of graphene are changed, and its electrocatalytic activity can be thus promoted [126, 127].

Nitrogen-doped graphene (N-graphene) has been the most intensively studied electrocatalyst [128]. And significant progress has been made in recent years. As the next element to carbon in the periodic table, nitrogen possesses an electronegativity larger than carbon, and the introduction of nitrogen into graphene lattice has been demonstrated the great capability to modify the local electronic structure of graphene. Three main types of N formats can be introduced into graphene, including pyridinic N, pyrrolic N and graphitic N. Graphitic N is the N atoms in the hexagonal ring of graphene lattice with direct substitution of C, while pyridinic and pyrrolic N donate one and two p electrons to the π system, forming sp^2- and sp^3-hybridized bonds, respectively. Each of these three types affects the atomic charge distribution in graphene differently, and the atomic charge along with the spin density in graphene determines its catalytic activities, which has been revealed by theoretical studies on the electrocatalytic effect of N doping in graphene [129]. Therefore, it is important to explore the connection between the electrocatalytic activity of graphene and the nitrogen bonding types in it.

For this purpose, selective doping of graphene has been conducted to illuminate the ORR activity of different nitrogen species. Pyridinic N and quadrivalent nitrogen species are commonly considered as the active sites for ORR, while some doubts still exist. Yasuda et al. applied nitrogen-containing aromatic molecules to realize tunable pyridinic and quaternary nitrogen species in N-graphene for ORR performance [130]. The doping is based on thermal surface polymerization method. They found that the pyridinic and quaternary nitrogen species in the N-graphene can facilitate four- and two-electron reduction pathway, respectively. Lai et al. found that the onset potential of ORR can be promoted by pyridinic-N species in N-graphene, while the electrocatalytic activity is highly associated with the graphitic-N content [131]. From another point of view, Geng et al. further proved that a high ORR activity via a four-electron process can be obtained by a low nitrogen doping content of 2.8% in N-graphene [44]. Two different ORR pathways (associative and dissociative) have been proposed for N-graphene electrocatalysts, while the detailed mechanism should be further explored.

N-doping can be realized through post treatment of graphene or GO, e.g., hydrazine reduction and calcination in NH_3 [132]. Alternatively, it also can be directly achieved by the bottom-up synthesis of graphene by introducing NH_3 during the CVD growth process of graphene [133], or using nitrogen-containing molecules like lithium nitride or tetrachloromethane during solvothermal process [126]. Some graphite-like layer-structured compounds like γ-C_3N_4 has been recently introduced as a template for the fabrication of N-graphene electrocatalysts in ORR. Liao et al. developed a one-pot bottom-up synthesis route (Fig. 5.8) for the synthesis of N-graphene by calcinating the mixture of melamine and α-hydroxy acids [134]. The α-hydroxy acids, i.e., malic acid, tartaric acid, and citric acid were carbon sources, while melamine was the precursor of γ-C_3N_4. The N content in the resultant N-graphene can be controlled from 4.12 to 8.11 at% by choosing different α-hydroxy acids as carbon sources.

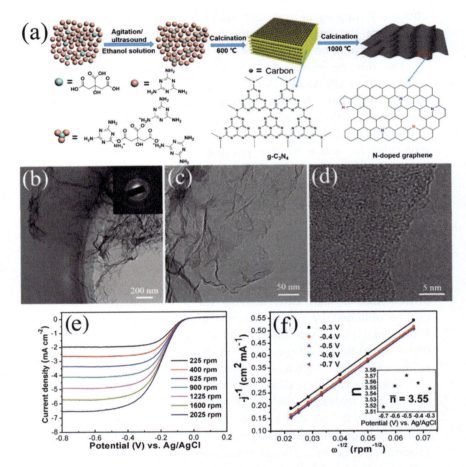

Figure 5.8 (a) Proposed synthetic protocol for the N-graphene. (b–d) High-magnification TEM image of N-graphene. (e) Electrocatalytic performances of N-graphene toward ORR [134].

To elevate the electrocatalytic performances, many other nonmetal elements have been experimentally doped into graphene lattice. The positive charge sites from the dopant atoms on graphene lattice act as the electrocatalytic active loci for O_2 reduction. Liu et al. synthesized phosphorus-doped graphene (P-graphene) with P content of 0.26 at% by copolymerize toluene and triphenylphosphine [135]. The P-graphene showed high electrocatalytic performance toward ORR and excellent durability.

Zhang et al. reported a higher P doping content up to 1.16 at% in P-graphene, which improved its electrocatalytic performances [136]. It was found that the ORR with P-graphene contains a four-electron process. Boron-doped graphene (B-graphene) is another alternate, owing to the strong electronegativity of boron. Sheng et al. prepared B-graphene through a thermal treatment of graphene in the presence of boron oxide, without the introduction of any catalyst [137]. The resultant B-graphene exhibited excellent electrocatalytic performance for ORR, which was attributed to its special electronic structure.

Doped with two elements with different electronegativities, dual-doped graphene materials are hoped to show better electrocatalytic activities than the sing-element-doped counterparts. Because of their unique electronic structure and the synergistic effect between two kinds of heteroatoms, dual-doped graphene materials are deemed capable to tailor the chemical properties of graphene for electrocatalysts [138, 139]. Wang et al. prepared B, N co-doped graphene with controllable constitute via thermal treatment of GO in the atmosphere containing H_3BO_3 and NH_3 [140]. The synthesized B, N co-doped graphene showed better electrocatalytic performances than commercial Pt/C electrocatalysts toward ORR. Zheng et al. designed a two-step doping route to synthesize B, N co-doped graphene [141]. The resultant B, N dual-doped graphene possessed high purity and well-defined doping sites, and demonstrated higher ORR electrocatalytic activity and better stability than the single doped graphene comparisons. The improved performance was analyzed to be derived from the synergistic coupling effect between boron and nitrogen atoms.

Doping of heteroatoms is a practicable approach to tune the chemical properties of graphene and create active sites on graphene sheets. Nevertheless, the electrocatalytic performances of these doped graphene materials are still far from being commercially applied into fuel cells, primarily because of insufficient doping content and the uncontrollable doping process. Tunable and selective matching of heteroatoms in graphene can conduce to increase the quantity of electrocatalytically active sites and improve their performance. The combination of experimental and theoretical studies is thus required to design synthesis routes for controllably doping of heteroatoms into specified sites, and to understand the inner mechanism in the electrocatalytic process of doped graphene.

5.3.3 Catalytic Oxidation Reaction

In 2010, Dreyer et al. revealed that GO can catalyze both the oxidation of alcohols and the hydration of alkynes with high yield and high selectivity [142]. A series of control experiments indicated that GO undergoes partial reduction during the catalytic oxidation reaction. The recyclability of the GO catalyst was also researched. Nevertheless, the recovered GO catalyst exhibited only approximately 5% conversion at low catalyst loadings. Dreyer and Bielawski have made great contributions to the studies of GO as catalysts in various catalytic reactions [143]. Their research works attracted great attention in the field, and many groups have made progress on exploring the performance of GO toward catalytic oxidation reactions.

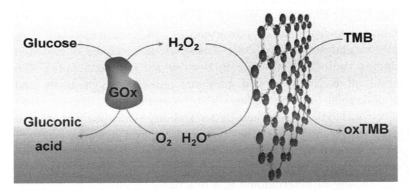

Figure 5.9 Schematic illustration of colorimetric detection of glucose by using glucose oxidase (GOx) and GO–COOH-catalyzed reactions [144].

Song et al. designed a colorimetric procedure for glucose detection as illustrated in Fig. 5.9 [144]. Sun et al. synthesized GO–COOH by adding NaOH and chloroacetic acid into GO solution to convert the –OH groups to –COOH via conjugation of acetic acid moieties [145]. Based on the catalytic performance of GO–COOH samples for the reaction of 3,3,5,5-tetramethylbenzidine oxidation in the presence of H_2O_2, GO–COOH has been demonstrated to exhibit intrinsic peroxidase-like activity in a blue color reaction following a ping-pong mechanism and with a maximum reaction rate. This method was simple, cheap, and highly selective. It was successfully employed for glucose detection in buffer solution, diluted food and fruit juice. These findings pave the way for the utilization of GO in medical diagnostics and biotechnology.

Long et al. prepared GO foam via freeze drying, which was found to serve both as a reactant affording reduced GO and as a catalyst to convert SO_2 to SO_3 [146]. Pan et al. reported the first example of using GO to facilitate the synthesis of organic compounds under visible light irradiation [147]. GO was employed as a cooperative catalyst to Rose Bengal, contributing to the photocatalytic oxidative C–H functionalization of tertiary amines to generate imines, which have great value in the synthesis of many industrially important materials and biologically active compounds. Huang et al. reported the first convenient metal-free catalytic process for an efficient imine synthesis from various amines under mild and neat conditions with molecular oxygen as the terminal oxidant [148]. Based on the oxidative coupling of amines to imines, Su et al. have probed the activity of GO and studied its catalytic mechanism. Through the base and acid treatment of GO, it was found that the carboxylic groups as well as the localized unpaired electrons facilitate synergistic intermolecular arrangements and contribute to enhance catalytic activity of oxidative coupling of various primary amines [149]. In order to understand the mechanism, theoretical simulations have also been employed to study the catalytic transition states and the oxidation process of GO catalysts. Boukhvalov et al. investigated the catalytic properties of GO based on the oxidation model of benzyl alcohol to benzaldehyde [150]. Based on density functional theory (DFT) calculations, it was revealed that the reaction proceeded via the transfer of hydrogen atoms from the organic molecule to the GO surface. The GO model with 12.5% of the carbon atoms can be covered by hydroxyl and epoxy groups initially. As the benzyl alcohol molecule moved close to the GO surface, one hydrogen atom would transfer from $PhCH_2OH$ to an epoxy group of GO, forming a diol on the GO sheet. The formation of dangling bonds on $PhCH_2OH$ or on GO was found to be energetically unfavorable, making this step endothermic. Then another hydrogen atom of benzyl alcohol molecule would migrate to one of the hydroxyl group on the GO surface. It finally results in the formation of one equivalent of water, which thus is exothermic step. These simulated results have depicted detailed images of the catalytic process of GO, which are in agreement with the above-mentioned experimental results of Bielawski and Dreyer, where the radical groups were found to disappear after catalytic courses [142].

Long et al. developed a heat treatment method to prepare N-graphene and studied its catalytic activity in selectively oxidizing benzylic alcohols [151]. The temperature of the heat treatment was reported to have a great influence on the contents and types of doped nitrogen. Graphitic N species in N-graphene were demonstrated to act as the active sites with preferable linear correlation with catalytic activity. The catalytic reaction process was proposed based on electron paramagnetic resonance characterization. It is suggested that the formed intermediate product of sp^2 $N-O_2$ has a high chemical reactivity towards alcohols.

What's more, graphene materials have been introduced as supports for other oxidation catalyst materials, such as noble metals and metal oxides. The roles played by graphene in these hybrid catalysts are similar to those in the photocatalysts and electrocatalysts mentioned in the previous sections. Lou et al. developed a facile hydrothermal method to synthesize Ag nanoprism on RGO with the assistance of sodium alginate [152]. Most of the Ag nanoparticles was found to have {111} external facets parallel to the (0001) direction of RGO sheets. When used as a catalyst in the oxidation of hydroquinone, the Ag/RGO composite catalyst showed a great catalytic rate constant.

5.3.4 Electrochemical Biosensing

Biosensing technology has been developed in the aim of improving the health quality of our life. Biosensors are desired to be able to detect various compounds sensitively and selectively, toward the applications in health care and environmental security. As a zero-gap, electroactive and transparent semiconductor material, graphene has been studied in a wide range of biosensing applications, such as enzymatic biosensing, DNA sensing, and immunosensing.

Glucose oxidation reaction is of great importance in bioscience. The glucose oxidase enzyme is generally utilized as a biorecognition unit. In the process, glucose is oxidized into gluconic acid with releasing electron to O_2 in the solution, and then producing H_2O_2. H_2O_2 is typically detected electrochemically. In some studies, it also revealed that the detection by direct electron transfer from enzyme is possible without needing O_2 [153].

Graphene has been employed in many studies for sensing glucose. Wang et al. introduced nitrogen plasma treatment to produce N-graphene, with the N content tunable from 0.11 to 1.35 wt% by changing the treating time [154]. The resultant N-graphene possessed good response and high sensitivity toward glucose oxidation. The influence of ascorbic acid and uric acid in the system was also investigated. Glucose with concentrations of 0.01 mM can be detected by the N-graphene-modified electrode, illustrating its great potential as biosensor. The reduction of H_2O_2 by N-graphene was studied by Wu et al., and excellent electrocatalytic activity in measuring H_2O_2 was obtained [155]. Furthermore, their calculations revealed the correlation between the activity and the types of the N-doped graphene: pyridinic N-graphene < pyrrolic N-graphene <graphitic N-graphene < pristine graphene, which was explained by comparing the electrostatic potential distributions in different types [156]. N-graphene has also been studied as an electrochemical sensor to determine ascorbic acid, dopamine, and uric acid.

Because the electrodes with enzyme are lack of operability, non-enzyme glucose biosensors have been developed, which mostly involve transition metals (including alloys and metal complexes) and metal oxide nanoparticles. However, poor electrical conductivity and poor dispersibility of the nanoparticles hinder their application in biosensors. The combination of graphene and the metal or metal oxide nanoparticles is a feasible approach to increase the conductivity, as well as the specific surface area of electrodes, and thus can promote their electrochemical catalytic activity. Up to now, graphene-based enzyme-free glucose biosensors have been studied by applying several transition metal oxides and noble metals, such as Cu_2O [157], Co_3O_4 [158], NiO [159], Pd [160], and Au [161]. By constructing 3D structure, graphene can provide more electrochemical surface area (ECSA) for sensing. Lou et al. developed a synthesis route (Fig. 5.10)

260 | *Carbons as Supports for Catalysts*

to prepare 3D RGO/Ag aerogel directly on the surface of glassy carbon electrodes (GCE) [162]. An ECSA 6 times high than the 2D-structured one was obtained for the 3D aerogel modified electrode. The composite demonstrated excellent electrocatalytic performance toward the detection of H_2O_2, i.e., a broad linear range of 0.016–27 mmol L^{-1}, and high sensitivity of 419.7 mA $mmol^{-1}$ L cm^{-2}.

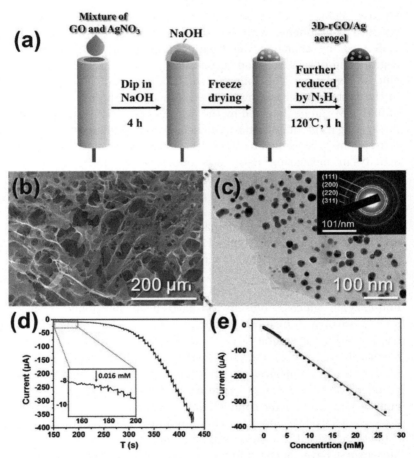

Figure 5.10 (a) Schematic for the fabrication of a GCE modified by 3D RGO/Ag aerogel. (b) SEM micrographs and (c) TEM micrograph of 3D RGO/Ag aerogel. (d), (e) The performances of GCE modified by 3D RGO/Ag aerogel toward the detection of H_2O_2 [162].

5.4 3D Architected Carbons for Catalyst

3D architected carbons are featured by high specific surface areas, stable physical and chemical properties and 3D interconnected porous structure. They can provide channels to efficiently transport protons, liquids, and gases. These materials can meet the demand required and have become the most significant materials which can be utilized as electrodes of lithium-ion battery (LIB), absorbent and catalyst. Porous carbon have become a hot topic from last decades.

According to the size of pore diameter, porous carbon can be divided into three groups: macro-porous (>50 nm), meso-porous (2–50 nm) and micro-porous (<2 nm) carbon. Macro-porous structure provides channels for molecule exchange; meco-porous structure is for ion exchange through capillary phenomenon, while micro-porous structure is where the reactions take place. Hierarchical porous structure can provide enough reaction sites under the prerequisite that the material exchanges can take place. The structure character of porous carbon could meet the demand of different catalytic reaction. This part focuses on the fabrication porous carbon-supported catalysts and investigation of their applications in energy and environment.

5.4.1 Classification of 3D Architected Carbons

5.4.1.1 Hard carbon

Hard carbon is a kind of carbon that is difficult to be graphitized (even above 3000°C) [163] and usually is fabricated from pyrolysis of polymers such as phenolic resins, epoxy resins and pitch. From the structural aspect, hard carbon is highly irregular and disordered, and primarily consists of single-layered carbon atoms that are closely and randomly connected [164].

5.4.1.2 Soft carbon

Soft carbon is a concept which is compared with hard carbon, easier to be graphitized and usually is fabricated from coke, carbonaceous mesophase spherules [165], carbon fiber [166], etc.,

In compared to hard carbons, soft carbons have better compatibility with ions and electron group.

5.4.1.3 Graphitized carbon

Generally, porous carbon is artificially synthesized through complicated processes. The pores in such artificially made materials are scattered in a single and homogeneous pattern. In most cases, they are made by aperiodic carbon with poor electrical conductivity [167]. However, amorphous carbon can be catalyzed to graphite layers by the introduction of catalytic particles. It improves not only the electronic conductivity but also other related properties because the formation of graphitic layer increases the exposure of surface area to the media. These features make porous carbons have great potential application in the catalyst field, especially for electrochemistry [168] and photocatalyst [169].

5.4.2 Synthesis of 3D Architectured Carbons

Normally, 3D architectured carbons are usually obtained via carbonization of precursors of different template and then followed by activation. To meet the requirements, a novel approach, the template carbonization method, has been proposed. Many researchers have prepared novel porous carbons based on this technique using a wide variety of porous templates. According to the templates used, they can be classified as hard template, soft template and natural template for the synthesis of 3D architecture carbons.

Replication, the process of filling the external and/or internal pores of a solid with a different material, physically or chemically separating the resulting material from the template, is a technique widely used for fabrication of microporous materials or 3D printing porous materials. Based on this technique, nanotubules and fibrils of replica polymers [170], metals [171], semiconductors [172], carbons, and other materials [173], can be obtained on the length scale of nanometers to microns. Some nanomaterials can also be prepared using nanoporous host matrixes. For instance, Q-state semiconductors are produced within the interlamellar region of a layered compound or layered materials [174]; semiconductors are formed in the spherical cavities of zeolite Y type template; carbon cadmium telluride nanoclusters are synthesized by vapor-phase deposition of elemental tellurium in Na-zeolite A as template [175].

Generally, there are five steps in the production of a templated porous carbon:

(i) synthesis of the inorganic template
(ii) impregnation of the template with an organic precursor (such as furfuryl alcohol, phenolformaldehyde, or acrylonitrile)
(iii) polymerization of the precursor
(iv) carbonization of the organic material
(v) leaching of the inorganic template

A variety of different methods are employed on each step, for example, leaching with HF, HCl, or NaOH, impregnation from liquid and gas phase, and carbonization at different temperatures.

5.4.2.1 Hard template method

Hard template method is usually been used to fabricate the 3D architecture carbons because of their solid nature and variety of structure. Porous silicon [176–178], zeolite [179], opals [180] and clays [181] are the most commonly used template for the fabrication of 3D architecture carbons.

As shown in Fig. 5.11, taking zeolite for example, the typical fabrication process of porous carbon is included pyrolytic carbon infiltration on the hard template and followed acid leaching process.

5.4.2.2 Soft template method

Porous carbon materials with ordered mesoporosity and narrow pore size distribution can be synthesized using a nanocasting approach that employs micelles of amphiphilic block-copolymers as templates. The most common copolymers are poly(ethylene oxide)-b-poly(propylene oxide)-b-poly(ethylene oxide) triblock-copolymers (PEO-b-PPO-b-PEO) from the pluronic family [184–188], polystyrene-b-poly(4-vinylpyridine) (PS-b-P4VP) or polystyrene-b-poly(ethylene oxide) (PS-b-PEO) [189]. Typically applied carbon precursors are small clusters of phenol–formaldehyde, or so-called "resol" (a soluble low-molecular weight polymer (M_w = 500–5000 g mol^{-1}) derived from acid-catalyzed or base-catalyzed polymerization of a mixture of phenol and formaldehyde) [188–191], so-called "RF resin" (resorcinol–formaldehyde) [179] and "PF resin" (phloroglucinol–formaldehyde) [186].

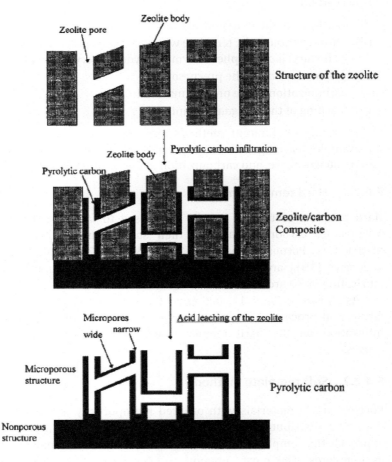

Figure 5.11 Scheme of pyrolytic carbon infiltration of zeolite [182, 183].

As shown in Fig. 5.12, typical synthesis procedures involve the mixing of a solvent (such as ethanol, water, tetrahydrofuran (THF), or mixtures of them), precursor (e.g., a resin of low molecular weight containing OH groups), and template (e.g., tri-block-copolymers of the type PEO-b-PPO-b-PEO). This mixture assembles into an ordered mesophase. The mesophase is stabilized via thermal or catalytic cross-linking. Finally, the template is removed, e.g., by thermal treatment, which results in a porous carbonaceous material. The thermal treatment also induces a compaction of the material, which leads to a shrinkage of the pore system.

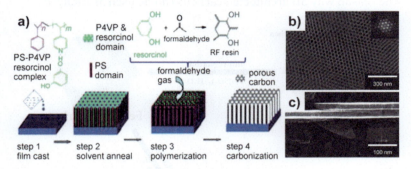

Figure 5.12 Synthesis of mesoporous carbon templated with micelles of PS-b-P4VP [186].

5.4.2.3 Nature template/precursor method

Natural materials are rich in resources, low in cost, renewable and in concord with the environment. It has many advantages to make porous carbon materials with hierarchical structure by natural biological templates. The biological templates can be both organic and inorganic, in which the organic natural polymers are the carbon source and inorganic components are used for creating pores after removal. To fabricate functional porous carbons from the biological templates, a typical process is always included procedures of carbonization, activation or functionalization and further treatment, as the example of Fe_2O_3/carbon composite shown in Fig. 5.13.

5.4.3 Applications of 3D Architected Carbons for Catalysts

Carbon has long been deemed a good catalyst support material because of its diverse porous structure, resistance to acidic and basic environments, low cost, easy accessibility, good recycling, low density, and amenability to synthesis variants or post-synthetically engineer traits using a wide range of manufacturing, activation and carbonization methods [187]. Beyond the advantage of carbon material, the 3D architected carbon have large surface area and interconnected porous structure which could provide the anchor sites and reaction room for electronic or chemical reaction, that is the reason why 3D architected carbons can be used as catalysts.

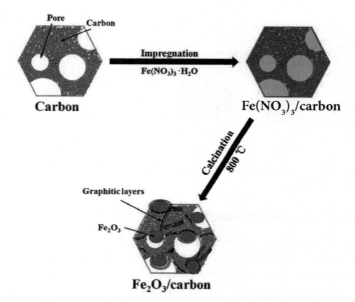

Figure 5.13 Illustration for the fabrication Fe_2O_3/carbon composite [169].

5.4.3.1 Porous carbons for heavy metal removal and reduction

Zhu et al. have reported that a designed N-doped 3D architected porous carbon with magnetic nanoparticles formed in situ (RHC-mag-CN) through simple impregnation followed by polymerization and calcination [168]. The doped nitrogen in RHC-mag-CN was in the form of graphite-type layers with the composition of C and N.

As described in Fig. 5.14, a precursor of rice husk was first used to fabricate porous carbon (RHC) with hierarchical structures via carbonization at 650°C followed by activation at 700°C wherein KOH was employed as an activator. Then granulated RHC (0.5 g) was mixed with 0.416 g iron chloride (FeCl$_3$) solution by milling and ultrasonication to produce a thick paste. The paste was exposed in pyrrole vapor at 50°C for 1 h, during which pyrroles were polymerized. Finally, the paste was calcinated at 850°C under nitrogen for graphitization and forming Fe$_3$O$_4$ nanoparticles in situ. The 3D porous composites denoted as RHC-mag-CN is then obtained.

This porous composite possesses high specific surface area of 1136 m^{2}g^{-1} with an average pore size of 3.8 nm and 18.5 wt% incorporation of magnetic nanoparticles. When used the RHC-mag-CN as an adsorbent for removal of Cr ions, a very quick adsorption on Cr(VI) ions was exhibited. 92% of Cr(VI) ions could be removed within 10 min for dilute solutions at 2 g L^{-1} adsorbent dose. The high adsorption capacity (16 mg g^{-1}) is related to the synergetic effects of physical adsorption from the surface area and chemical adsorption from complexation interactions between Cr(VI) and Fe$_3$O$_4$.

In addition, the basic nature of CNs in RHC-mag-CN composite may increase its negative charge density. It thus can strengthen the adsorption on metallic cations and also may facilitate the reduction of toxic Cr(VI) into low toxic Cr(III) ions in the solution. The magnetic nanoparticles in the composite also make a facial separation or collection of the samples after the adsorption of Cr ions from aqueous solutions.

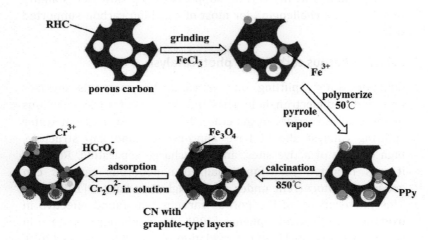

Figure 5.14 Schematic fabrication of RHC-mag-CN nanocomposite [168].

5.4.3.2 Porous carbons for nitrogen oxide and sulfur oxide removal

Nitrogen oxides (NO$_x$) are one type of the major contaminants emitted from high-temperature combustion process. They are not only associated with environmental problems such as acid rain, urban smog, ozone depletion but also bring health related issues such as bronchitis, pneumonia, immunodeficiency or even cancer [148]. Extensive research on selective catalytic reduction (SCR) of NO$_x$ has been conducted for many decades [189, 190]. One of the major sources for NO$_x$ is from internal combustion engines. Three-way catalysts (TWCs) are used in vehicles to reduce tailpipe emissions of NO$_x$ as well as CO and unburnt hydrocarbons [188, 191]. Different types of catalysts have been tested for the conversion of NO$_x$ conversion. Combinations of a number of metals (Pt, Pd, Rh, Au, Ag, Cu, Co, Fe, In, and Ga) with different supports (zeolites, Al$_2$O$_3$, SiO$_2$, and activated carbons) have been reported to produce active catalysts for the SCR of NO$_x$ reaction [192–195].

Like NO$_x$, SO$_x$ is another major pollutant, often accompanying NO$_x$ in combustion flue gas, which can also cause environmetal and health issues [196]. Activated carbons with acid treatments have proven to be able to strengthen the removal capability on SO$_x$ gases, which is due to the better immobilization effect on the catalytic metals in the carbon supports [197]. However, stability remains to be challenging for most of this kind carbon-supported catalysts.

5.4.3.3 Porous carbons for photocatalysis

Catalytic water splitting on semiconductor materials involves photoinduced electron–hole pairs and the phase transfer reactions of the charge carriers to generate O$_2$ and H$_2$. An efficient water splitting catalyst should have appropriate band gap structure, high density of active sites, and fast charge separation. Recently it was reported that functionalized nanoporous carbons can generate photocurrent under visible light irradiation [198, 199]. This photoactivity of the porous carbons was found effective in oxidation of dibenzothiophenes [200], electrical energy storage in supercapacitors [201], and in oxidation of water pollutants [202]. Moreover, through control the combination of heteroatoms in specific configurations with carbons, nanoporous carbons with photoluminescence can be obtained, which may emit light upon excitation [203].

In the usage of carbon materials in photoassisted reactions, Bandosz et al. showed the photoelectrochemical water splitting on a metal-free nanoporous carbon photoanode under visible light [204]. Compared to other reports on metal-free carbon photoanodes [205, 206], their material is a low cost electrode obtained from a commodity organic polymer. Based on an extensive surface characterization, they indicated that the photoelectrochemical performance of the nanoporous carbons on splitting water is related to their surface chemistry (O-, S-, and N-doping) and specific nanofeatures of carbon textures.

Another example for the fabrication of nanoporous carbons for photocatalysis is illustrated in Fig. 5.15. The nanoporous carbon materials is derived from poly(4-styrenesulfonic acid) ammonium salts and obtained based on a two-step heat-treatments (800°C under nitrogen for 40 min and 350°C under air for 3 h). The resulted carbon material named as CONS showed stable photocurrent with long illumination time (upon several hours) and reproducibility for several cycles. The production amount of O_2 detected after 3 h of illumination at 0.8 V (the lowest applied potential), was consistent with the value estimated from the accumulated charge passed, indicating a Faradaic efficiency close to 100%.

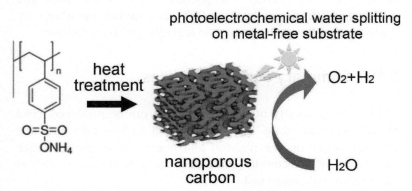

Figure 5.15 Visible light driven photoelectrochemical water splitting on metal-free nanoporous carbon.

However, the efficiency was rather small at short illumination time, which suggests that charge is also consumed in side reactions. Indeed, some changes become evident in the cyclic voltammogram (CV) of the carbon electrode after illumination, implying a photo- or electro-corrosion of the anode. The CV of the as-prepared electrode is typical for a nanoporous carbon with the electrical double layer capacitance (EDLC) [207], and a wide hump centered at 0.6 V, linked to the redox reactions of O, N, and S surface groups in various environments [208]. This hump disappears after exposure to visual light at high potentials. It implies the surface photosensitivity in the water splitting process, which should be due to some specific sites of the carbon electrode consumed the photogenerated charges [209]. After visual light exposure, the CV loses its rectangular shape, indicating charge transfer limitations. A broad hump appears on the cathodic sweep between 0.5 and 0.8 V after irradiation under anodic polarization. This indicates an irreversible change in carbon surface chemistry, which implies the essential role of surface functions for the photoelectrochemical activity.

Owing to the specificity of the precursor and synthesis conditions, surfaces of this porous carbon is rich in "photo-sensitizers" and conductive graphic units. Light exposure causes the photo-generation of charge carriers on the chromophore-like moieties and leads to the photoelectrochemical oxidation of water at a very low overpotential. The presence of well-developed porosity in this particular carbon is important for its performance and distinguished from graphene or CNT after various modifications. Furthermore, the synthesis method is simple and cost effective.

5.5 Summary and Perspectives

Today, a large variety of carbon materials can be produced, even at the industrial scale. The versatility of carbon materials and the complexity on the physical and chemistry properties make the design of structurally controlled catalysts based on carbon materials a challenging task.

As pristine carbons without modifications or functionalizations principally possess low catalytic activities, for design new novel catalysis systems based on carbons, a better understanding on the relationship between surface functions and bulk carbon structure is still required. The development of carbon-based catalysts reviewed above may profit significantly from the design and optimization of the active sites, whose architecture is necessary to be studied intensively, with the combination of experimental analysis and theoretical modeling. To fully understand the mechanisms of the doping and functionalization of carbon structures toward catalysis in the future, theoretical calculations are required apart from experimental analysis.

When carbon acts as a support for other catalytically active materials, such as organometallic clusters and noble metal nanoparticles, the size, shape, morphology and composition of the nanoparticles can be controlled and adjusted, and thus the catalytic performances for fuel oxidation, oxygen reduction and water splitting reactions can be considerably enhanced. However, the detachment, agglomeration, and dissolution in harsh service environments would badly influence their practical applications. On the other hand, the catalytic mechanisms of these carbon-based composites in energy-related systems still remain unclear, and further studies are required. More applications in catalysis also need to be explored in width and depth for the promising intrinsic properties of CNTs or graphenes and from the nanoarchitecturing of carbon structures.

Incontrovertibly, carbon-based catalysts are likely to be practically applied in various fields. Research efforts devoted to these directions will surely be recompensed in the near future.

References

1. Serp, P., and Machado, B. (2015). Carbon (nano)materials for catalysis, in *Nanostructured Carbon Materials for Catalysis*, Royal Society of Chemistry, Cambridge, UK, pp. 1–45.
2. Iijima, S. (1991). Helical microtubules of graphitic carbon, *Nature*, **354**, pp. 56–58.
3. Iijima, S., and Ichihashi, T. (1993). Single-shell carbon nanotubes of 1 nm diameter, *Nature,* **363**, pp. 603–605.
4. Serp, P., and Castillejos, E. (2010). Catalysis in carbon nanotubes, *ChemCatChem,* **2**, pp. 41–47.
5. Tasis, D., Tagmatarchis, N., Bianco, A., and Prato, M. (2006). Chemistry of carbon nanotubes, *Chem. Rev.,* **106**, pp. 1105–1136.

6. Castillejos, E., Debouttière, P. J., Roiban, L., Solhy, A., Martinez, V., and Kihn, Y. (2009). An efficient strategy to drive nanoparticles into carbon nanotubes and the remarkable effect of confinement on their catalytic performance, *Angew. Chem. Int. Ed.*, **48**, pp. 2529–2533.

7. Yan, Y., Miao, J., Yang, Z., Xiao, F.-X., Yang, H. B., Liu, B., and Yang, Y. (2015). Carbon nanotube catalysts: Recent advances in synthesis, characterization and applications, *Chem. Soc. Rev.*, **44**, pp. 3295–3346.

8. Kung, S. C., Hwang, K. C., and Lin, I. N. (2002). Oxygen and ozone oxidation-enhanced field emission of carbon nanotubes, *Appl. Phys. Lett.*, **80**, pp. 4819–4821.

9. Li, C. S., Wang, D. Z., Liang, T. X., Wang, X. F., Wu, J. J., Hu, X. Q., and Liang, J. (2004). Oxidation of multiwalled carbon nanotubes by air: Benefits for electric double layer capacitors, *Powder Technol.*, **142**, pp. 175–179.

10. Pradhan, B. K., and Sandle, N. K. (1999). Effect of different oxidizing agent treatments on the surface properties of activated carbons, *Carbon*, **37**, pp. 1323–1332.

11. Li, L. X., and Li, F. (2011). The effect of carbonyl, carboxyl and hydroxyl groups on the capacitance of carbon nanotubes, *New Carbon Mater.*, **26**, pp. 224–228.

12. Guo, Z., Chen, Y. T., Li, L. S., Wang, X. M., Haller, G. L., and Yang, Y. H. (2010). Carbon nanotube-supported Pt-based bimetallic catalysts prepared by a microwave-assisted polyol reduction method and their catalytic applications in the selective hydrogenation, *J. Catal.*, **276**, pp. 314–326.

13. Bai, S. L., Huang, C. D., Lv, J., and Li, Z. H. (2012). Comparison of induction behavior of Co/CNT and Co/SiO$_2$ catalysts for the Fischer-Tropsch synthesis, *Catal. Commun.*, **22**, pp. 24–27.

14. Kundu, S., Wang, Y. M., Xia, W., and Muhler, M. (2008). Thermal stability and reducibility of oxygen-containing functional groups on multiwalled carbon nanotube surfaces: A quantitative high-resolution XPS and TPD/TPR study, *J. Phys. Chem. C*, **112**, pp. 16869–16878.

15. Thiessen, J., Rose, A., Meyer, J., Jess, A., and Curulla-Ferré, D. (2012). Effects of manganese and reduction promoters on carbon nanotube supported cobalt catalysts in Fischer–Tropsch synthesis, *Microporous Mesoporous Mater.*, **164**, pp. 199–206.

16. Trepanier, M., Tavasoli, A., Dalai A. K., and Abatzoglou, N. (2009). Fischer–Tropsch synthesis over carbon nanotubes supported cobalt catalysts in a fixed bed reactor: Influence of acid treatment, *Fuel Process. Technol.*, **90**, pp. 367–374.

17. Adjizian, J. J., Leghrib, R., Koos, A. A., Suarez-Martinez, I., Crossley, A., Wagner, P., Grobert, N., Llobet E., and Ewels, C. P. (2014). Boron- and nitrogen-doped multi-wall carbon nanotubes for gas detection, *Carbon,* **66**, pp. 662–673.
18. Dong, J., Qu, X., Wang, L., Zhao, C., and Xu, J. (2008). Electrochemistry of nitrogen-doped carbon nanotubes (CNx) with different nitrogen content and its application in simultaneous determination of dihydroxybenzene isomers, *Electroanalysis,* **20**, pp. 1981–1986.
19. Cao, Y. H., Yu, H., Tan, J., Peng, F., Wang, H. J., Li, J., Zheng, W. X., and Wong, N. B. (2013). Nitrogen-, phosphorous- and boron-doped carbon nanotubes as catalysts for the aerobic oxidation of cyclohexane, *Carbon,* **57**, pp. 433–442.
20. Yang, Q. H., Hou, P. X., Unno, M., Yamauchi, S., Saito R., and Kyotani, T. (2005). Dual Raman features of double coaxial carbon nanotubes with N-doped and B-doped multiwalls, *Nano Lett.,* **5**, pp. 2465–2469.
21. Jia, N. Q., Wang, L. J., Liu, L., Zhou, Q., and Jiang, Z. Y. (2005). Bamboo- like CNx nanotubes for the immobilization of hemoglobin and its bioelectrochemistry, *Electrochem. Commun.,* **7**, pp. 349–354.
22. Jia, N. Q., Liu, L., Zhou, Q., Wang, L. J., Yan, M. M., and Jiang, Z. Y. (2005). Bioelectrochemistry and enzymatic activity of glucose oxidase immobilized onto the bamboo-shaped CNx nanotubes, *Electrochim. Acta,* **51**, pp. 611–618.
23. Dong, J. P., Qu, X. M., Wang, L. J., and Wang, T. L. (2007). Electrochemical behavior of nitrogen-doped carbon nanotube modified electrodes, *Acta Chim. Sin.,* **65**, pp. 2405–2410.
24. Du, H. Y., Wang, C. H., Hsu, H. C., Chang, S. T., Chen, U. S., Yen, S. C., Chen, L. C., Shih, H. C., and Chen, K. H. (2008). Electrochemical behavior of nitrogen-doped carbon nanotube modified electrodes, *Diamond Relat. Mater.,* **17**, pp. 535–541.
25. Schulte, H. J., Graf, B., Xia, W., and Muhler, M. (2012). Nitrogen- and oxygen-functionalized multiwalled carbon nanotubes used as support in iron-catalyzed, high-temperature Fischer–Tropsch Synthesis, *ChemCatChem,* **4**, pp. 350–355.
26. Li, D. J., Maiti, U. N., Lim, J., Choi, D. S., Lee, W. J., Oh, Y., Lee, G. Y., and Kim, S. O. (2014). Molybdenum sulfide/N-doped CNT forest hybrid catalysts for high-performance hydrogen evolution reaction, *Nano Lett.,* **14**, pp. 1228–1233.
27. O'Connell, M. J., Bachilo, S. M., Huffman, C. B., Moore, V. C., Strano, M. S., Haroz, E. H., Rialon, K. L., Boul, P. J., Noon, W. H., Kittrell, C., Ma,

J. P., Hauge, R. H., Weisman, R. B., and Smalley, R. E. (2002). Band gap fluorescence from individual single-walled carbon nanotubes, *Science,* **297**, pp. 593–596.

28. Lin, Y., Zhou, B., Fernando, K. A. S., Liu, P., Allard L. F., and Sun, Y. P. (2003). Polymeric carbon nanocomposites from carbon nanotubes functionalized with matrix polymer, *Macromolecules,* **36**, pp. 7199–7204.

29. Sano, M., Kamino, A., Okamura, J., and Shinkai, S. (2001). Self-organization of PEO-graft-single-walled carbon nanotubes in solutions and Langmuir–Blodgett Films, *Langmuir,* **17**, pp. 5125–5128.

30. Hamon, M. A., Chen, J., Hu, H., Chen, Y. S., Itkis, M. E., Rao, A. M., Eklund, P. C., and Haddon, R. C. (1999). Dissolution of single-walled carbon nanotubes, *Adv. Mater.,* **11**, pp. 834–840.

31. Reddy, A. L. M., Rajalakshmi, N., and Ramaprabhu, S. (2008). Cobalt-polypyrrole-multiwalled carbon nanotube catalysts for hydrogen and alcohol fuel cells, *Carbon,* **46**, pp. 2–11.

32. Tsubokawa, N. (2005). Preparation and properties of polymer-grafted carbon nanotubes and nanofibers, *Polym. J.,* **37**, pp. 637–655.

33. Lou, X. D., Detrembleur, C., Pagnoulle, C., Jerome, R., Bocharova, V., Kiriy, A., and Stamm, M. (2004). Surface modification of multiwalled carbon nanotubes by poly (2-vinylpyridine): Dispersion, selective deposition, and decoration of the nanotubes, *Adv. Mater.,* **16**, pp. 2123–2127.

34. Chen, J. H., Iwata, H., Tsubokawa, N., Maekawa, Y., and Yoshida, M. (2002). Novel vapor sensor from polymer-grafted carbon black: Effects of heat-treatment and γ-ray radiation-treatment on the response of sensor material in cyclohexane vapor, *Polymer,* **43**, pp. 2201–2206.

35. Wu, B., Hu, D., Kuang, Y., Liu, B., Zhang, X., and Chen, J. (2009). Functionalization of carbon nanotubes by an ionic-liquid polymer: Dispersion of Pt and PtRu nanoparticles on carbon nanotubes and their electrocatalytic oxidation of methanol, *Angew. Chem. Int. Ed.,* **48**, pp. 4751–4754.

36. Giacalone, F., Campisciano, V., Calabrese, C., La Parola, V., Syrgiannis, Z., Prato, M., and Gruttadauria, M. (2016). Single-walled carbon nanotube–polyamidoamine dendrimer hybrids for heterogeneous catalysis, *ACS Nano,* **10**, pp. 4627–4636.

37. Abdul Almohsin, S., Li, Z., Mohammed, M., Wu, K., and Cui, J. (2012). Electrodeposited polyaniline/multi-walled carbon nanotube composites for solar cell applications, *Synth. Met.,* **162**, pp. 931–935.

38. Li, X., Wei, J., Chai, Y., Zhang, S., and Zhou, M. (2015). Different polyaniline/carbon nanotube composites as Pt catalyst supports for methanol electro-oxidation, *J. Mater. Sci.*, **50**, pp. 1159–1168.
39. Toma, F. M., Sartorel, A., Iurlo, M., Carraro, M., Parisse, P., Maccato, C., Rapino, S., Gonzalez, B. R., Amenitsch, H., Da Ros, T., Casalis, L., Goldoni, A., Marcaccio, Scorrano, M. G., Scoles, G., Paolucci, F., Prato, M., and Bonchio, M. (2010). Efficient water oxidation at carbon nanotube-polyoxometalate electrocatalytic interfaces, *Nat. Chem.*, **2**, pp. 826–831.
40. Blanco, M., Álvarez, P., Blanco, C., Jiménez, M. V., Fernández-Tornos, J., Pérez-Torrente, J. J., Oro, L. A., and Menéndez, R. (2013). Enhanced hydrogen-transfer catalytic activity of iridium N-heterocyclic carbenes by covalent attachment on carbon nanotubes, *ACS Catal.*, **3**, pp. 1307–1317.
41. Raccichini, R., Varzi, A., Passerini, S., and Scrosati, B. (2015). The role of graphene for electrochemical energy storage, *Nat, Mater.*, **14**, pp. 271–279.
42. Zhu, Y., Murali, S., Stoller, M. D., Ganesh, K. J., Cai, W., Ferreira, P. J., Pirkle, A., Wallace, R. M., Cychosz, K. A., Thommes, M., Su, D., Stach, E. A., and Ruoff, R. S. (2011). Carbon-based supercapacitors produced by activation of graphene, *Science*, **332**, pp. 1537–1541.
43. Feng, W., Yang, L., Cao, N., Du, C., Dai, H., Luo, W., and Cheng, G. (2014). In situ facile synthesis of bimetallic CoNi catalyst supported on graphene for hydrolytic dehydrogenation of amine borane, *Int. J. Hydrogen Energ.*, **39**, pp. 3371–3380.
44. Geng, D., Chen, Y., Chen, Y., Li, Y., Li, R., Sun, X., Ye, S., and Knights S. (2011). High oxygen-reduction activity and durability of nitrogen-doped graphene, *Energy Environ. Sci.*, **4**, pp. 760–764.
45. Kou, R., Shao, Y., Wang, D., Engelhard, M. H., Kwak, J. H., Wang, J., Viswanathan, V. V., Wang, C., Lin, Y., Wang, Y., Aksay, I. A., and Liu, J. (2009). Enhanced activity and stability of Pt catalysts on functionalized graphene sheets for electrocatalytic oxygen reduction, *Electrochem. Commun.*, **11**, pp. 954–957.
46. Li, Y., Fan, X., Qi, J., Ji, J., Wang, S., Zhang, G., and Zhang, F. (2010). Gold nanoparticles–graphene hybrids as active catalysts for Suzuki reaction, *Mater. Res. Bull.*, **45**, pp. 1413–1418.
47. Sheng, Z.-H., Shao, L., Chen, J.-J., Bao, W.-J., Wang, F.-B., and Xia, X.-H. (2011). Catalyst-free synthesis of nitrogen-doped graphene via thermal annealing graphite oxide with melamine and its excellent electrocatalysis, *ACS Nano*, **5**, pp. 4350–4358.

48. Wang, X., Lian, J., and Wang, Y. (2014). The effect of Sn on platinum dispersion in Pt/graphene catalysts for the methanol oxidation reaction, *Int. J. Hydrogen Energ.*, **39**, pp. 14288–14295.
49. Novoselov, K. S., Geim, A. K., Morozov, S. V., Jiang, D., Katsnelson, M. I., Grigorieva, I. V., Dubonos, S. V., and Firsov, A. A. (2005). Two-dimensional gas of massless Dirac fermions in graphene, *Nature*, **438**, pp. 197–200.
50. Kuila, T., Bose, S., Mishra, A. K., Khanra, P., Kim, N. H., and Lee, J. H. (2012). Chemical functionalization of graphene and its applications, *Prog. Mater Sci.*, **57**, pp. 1061–1105.
51. Losurdo, M., Giangregorio, M. M., Capezzuto, P., and Bruno, G. (2011). Graphene CVD growth on copper and nickel: Role of hydrogen in kinetics and structure, *Phys. Chem. Chem. Phys.*, **13**, pp. 20836–20843.
52. Robertson, A. W., and Warner, J. H. (2014). Hexagonal single crystal domains of few-layer graphene on copper foils, *Nano Lett.*, **11**, pp. 1182–1189.
53. Wang, L., Zhang, X., Chan, H. L., Yan, F., and Ding, F. (2013). Formation and healing of vacancies in graphene chemical vapor deposition (CVD) growth, *J. Am. Chem. Soc.*, **135**, pp. 4476–4482.
54. Zhang, Y., Zhang, L., and Zhou, C. (2013). Review of chemical vapor deposition of graphene and related applications, *Acc. Chem. Res.*, **46**, pp. 2329–2339.
55. Dreyer, D. R. (2010). The chemistry of graphene oxide, *Chem. Soc. Rev.*, **43**, pp. 288–240.
56. Hummers Jr, W. S., and Offeman, R. E. (1958). Preparation of graphitic oxide, *J. Am. Chem. Soc.*, **80**, pp. 1339–1339.
57. Rao, C. N., Sood, A. K., Subrahmanyam, K. S., and Govindaraj, A. (2009). Graphene: The new two-dimensional nanomaterial, *Angew. Chem. Int. Ed.*, **48**, pp. 7752–7777.
58. Dreyer, D. R., Jarvis, K. A., Ferreira, P. J., and Bielawski, C. W. (2012). Graphite oxide as a carbocatalyst for the preparation of fullerene-reinforced polyester and polyamide nanocomposites, *Polym. Chem.*, **3**, pp. 757–766.
59. Dreyer, D. R., Jia, H. P., Todd, A. D., Geng, J., and Bielawski, C. W. (2011). Graphite oxide: A selective and highly efficient oxidant of thiols and sulfides, *Org. Biomol. Chem.*, **9**, pp. 7292–7295.
60. Jia, H. P., Dreyer, D. R., and Bielawski, C. W. (2011). Graphite oxide as an auto-tandem oxidation–hydration–aldol coupling catalyst, *Adv. Synth. Catal.*, **353**, pp. 528–532.

61. Jia, H.-P., Dreyer, D. R., and Bielawski, C. W. (2011). C–H oxidation using graphite oxide, *Tetrahedron,* **67**, pp. 4431–4434.
62. Fujishima, A. (1972). Electrochemical photolysis of water at a semiconductor electrode, *Nature,* **238**, pp. 37–38.
63. Chen, X., Shen, S., Guo, L., and Mao, S. S. (2010). Semiconductor-based photocatalytic hydrogen generation, *Chem. Rev.,* **110**, pp. 6503–6570.
64. Hoffmann, M. R., Martin, S. T., Choi, W., and Bahnemann, D. W. (1995). Environmental applications of semiconductor photocatalysis, *Chem. Rev.,* **95**, pp. 69–96.
65. Kudo, A., and Miseki, Y. (2009). Heterogeneous photocatalyst materials for water splitting, *Chem. Soc. Rev.,* **38**, pp. 253–278.
66. Woan, K., Pyrgiotakis, G., and Sigmund, W. (2009) Photocatalytic carbon-nanotube–TiO$_2$ composites, *Adv. Mater.,* **21**, pp. 2233–2239.
67. Malato, S., Fernández-Ibáñez, P., Maldonado, M. I., Blanco, J., and Gernjak, W. (2009). Decontamination and disinfection of water by solar photocatalysis: Recent overview and trends, *Catal. Today,* **147**, pp. 1–59.
68. Zhang, H., Lv, X., Li, Y., Wang, Y., and Li, J. (2009). P25-graphene composite as a high performance photocatalyst, *ACS Nano,* **4**, pp. 380–386.
69. Guo, J., Zhu, S., Chen, Z., Li, Y., Yu, Z., Liu, Q., Li, J., Feng, C., and Zhang, D. (2011). Sonochemical synthesis of TiO$_2$ nanoparticles on graphene for use as photocatalyst, *Ultrason. Sonochem.,* **18**, pp. 1082–1090.
70. Guo, J., Zhu, S., Chen, Z., Li, Y., Yu, Z., Liu, Q., Li, J., Feng, C., and Zhang, D. (2011). TiO$_2$/graphene composite from thermal reaction of graphene oxide and its photocatalytic activity in visible light, *J. Mater. Sci.,* **46**, pp. 2622–2626.
71. Guo, J., Li, Y., Zhu, S., Chen, Z., Liu, Q., Zhang, D., Moon, W. J., and Song, D. M. (2012). Synthesis of WO$_3$@ Graphene composite for enhanced photocatalytic oxygen evolution from water, *RSC Adv.,* **2**, pp. 1356–1363.
72. Qin, J., Cao, M., Li, N., and Hu, C. (2011). Graphene-wrapped WO$_3$ nanoparticles with improved performances in electrical conductivity and gas sensing properties, *J. Mater. Chem.,* **21**, pp. 17167–17174.
73. Weng, B., Wu, J., Zhang, N., and Xu, Y. J. (2014). Observing the role of graphene in boosting the two-electron reduction of oxygen in graphene–WO$_3$ nanorod photocatalysts, *Langmuir,* **30**, pp. 5574–5584.

74. Min, Y. L., Zhang, K., Chen, Y. C., and Zhang, Y. G. (2012). Enhanced photocatalytic performance of Bi_2WO_6 by graphene supporter as charge transfer channel, *Sep. Purif. Technol.*, **86**, pp. 98–105.
75. Sun, Z., Guo, J., Zhu, S., Mao, L., Ma, J., and Zhang, D. (2014). A high-performance Bi_2WO_6–graphene photocatalyst for visible light-induced H_2 and O_2 generation, *Nanoscale*, **6**, pp. 2186–2193.
76. Zhang, Y., Fei, L., Jiang, X., Pan, C., and Wang, Y. (2011). Engineering nanostructured Bi_2WO_6–TiO_2 toward effective utilization of natural light in photocatalysis, *J. Am. Ceram. Soc.*, **94**, pp. 4157–4161.
77. Fu, Y., Sun, X., and Xin, W. (2011). $BiVO_4$–graphene catalyst and its high photocatalytic performance under visible light irradiation, *Mater. Chem. Phys.*, **131**, pp. 325–330.
78. Li, Y., Sun, Z., Zhu, S., Liao, Y., Chen, Z., and Zhang, D. (2015). Fabrication of $BiVO_4$ nanoplates with active facets on graphene sheets for visible-light photocatalyst, *Carbon*, **94**, pp. 599–606.
79. Sun, Z., Li, C., Zhu, S., Cho, M., Chen, Z., Cho, K., Liao, Y., Yin, C., and Zhang, D. (2015). Synthesis of $BiVO_4$@ C core–shell structure on reduced graphene oxide with enhanced visible-light photocatalytic activity, *ChemSusChem*, **8**, pp. 2719–2726.
80. Wang, T., Li, C., Ji, J., Wei, Y., Zhang, P., Wang, S., Fan, X., and Gong, J. (2014). Reduced graphene oxide (rGO)/$BiVO_4$ composites with maximized interfacial coupling for visible light photocatalysis, *ACS Sustainable Chem. Eng.*, **2**, pp. 2253–2258.
81. Qi, L., Yu, J., and Jaroniec, M. (2011). Preparation and enhanced visible-light photocatalytic H_2-production activity of CdS-sensitized Pt/TiO_2 nanosheets with exposed (001) facets, *Phys. Chem. Chem. Phys.*, **13**, pp. 8915–8923.
82. Yu, J., Zhang, J., and Jaroniec, M. (2010). Preparation and enhanced visible-light photocatalytic H_2-production activity of CdS quantum dots-sensitized $Zn_{1-x}Cd_xS$ solid solution, *Green Chem.*, **12**, pp. 1611–1614.
83. Qu, Y., and Duan, X. (2013). Progress, challenge and perspective of heterogeneous photocatalysts, *Chem. Soc. Rev.*, **42**, pp. 2568–2580.
84. Sun, Z., Guo, J., Zhu, S., Ma, J., Liao, Y., and Zhang, D. (2014). Graphene-based photocatalysts for oxygen evolution from water, *RSC Adv.*, **4**, pp. 27963–27970.
85. Xiong, Z., Zhang, L. L., and Zhao, X. S. (2011). Visible-light-induced dye degradation over copper-modified reduced graphene oxide, *Chem. Eur. J.*, **17**, pp. 2428–2434.

86. Xiang, Q., Yu, J., and Jaroniec, M. (2011). Enhanced photocatalytic H_2-production activity of graphene-modified titania nanosheets, *Nanoscale,* **3**, pp. 3670–3678.
87. Yeh, T. F., Syu, J. M., Cheng, C., Chang, T. H., and Teng, H. (2010). Graphite oxide as a photocatalyst for hydrogen production from water, *Adv. Funct. Mater.,* **20**, pp. 2255–2262.
88. Zhang, X. Y., Li, H. P., Cui, X. L., and Lin, Y. (2010). Graphene/TiO_2 nanocomposites: Synthesis, characterization and application in hydrogen evolution from water photocatalytic splitting, *J. Mater. Chem.,* **20**, pp. 2801–2806.
89. Liao, Y., Zhu, S., Chen, Z., Lou, X., and Zhang, D. (2015). A facile method of activating graphitic carbon nitride for enhanced photocatalytic activity, *Phys. Chem. Chem. Phys.,* **17**, pp. 27826–27832.
90. Liao, Y., Zhu, S., Ma, J., Sun, Z., Yin, C., Zhu, C., Lou, X., and Zhang, D. (2014). Tailoring the morphology of g-C_3N_4 by self-assembly towards high photocatalytic performance, *ChemCatChem,* **6**, pp. 3419–3425.
91. Niu, P., Zhang, L., Liu, G., and Cheng, H. M. (2012). Graphene-like carbon nitride nanosheets for improved photocatalytic activities, *Adv. Funct. Mater.,* **22**, pp. 4763–4770.
92. Bing, A., Duan, X., Sun, H., Xiang, Q., and Wang, S. (2015). Metal-free graphene-carbon nitride hybrids for photodegradation of organic pollutants in water, *Catal. Today,* **258**, pp. 668–675.
93. Fan, W., Zhang, Q., and Wang, Y. (2013). Semiconductor-based nanocomposites for photocatalytic H_2 production and CO_2 conversion, *Phys. Chem. Chem. Phys.,* **15**, pp. 2632–2649.
94. Xiang, Q., Cheng, B., and Yu, J. (2015). Graphene-based photocatalysts for solar-fuel generation, *Angew. Chem. Int. Ed.,* **54**, pp. 11350–11366.
95. Tran, P. D., Wong, L. H., Barber, J., and Loo, J. S. C. (2012). Recent advances in hybrid photocatalysts for solar fuel production, *Energy Environ. Sci.,* **5**, pp. 5902–5918.
96. Cowan, A. J., and Durrant, J. R. (2013). Long-lived charge separated states in nanostructured semiconductor photoelectrodes for the production of solar fuels, *Chem. Soc. Rev.,* **42**, pp. 2281–2293.
97. Liang, Y. T., Vijayan, B. K., Lyandres, O., Gray, K. A., and Hersam, M. C. (2012). Effect of dimensionality on the photocatalytic behavior of carbon–titania nanosheet composites: Charge transfer at nanomaterial interfaces, *J. Phys. Chem. Lett.,* **3**, pp. 1760–1765.
98. Tu, W., Zhou, Y., Liu, Q., Tian, Z., Gao, J., Chen, X., Zhang, H., Liu, J., and Zou, Z. (2012). Robust hollow spheres consisting of alternating titania nanosheets and graphene nanosheets with high photo-

catalytic activity for CO_2 conversion into renewable fuels, *Adv. Funct. Mater.*, **22**, pp. 1215–1221.

99. Yu, J., Jin, J., Cheng, B., and Jaroniec, M. (2014). A noble metal-free reduced graphene oxide–CdS nanorod composite for the enhanced visible-light photocatalytic reduction of CO_2 to solar fuel, *J. Mater. Chem. A*, **2**, pp. 3407–3416.

100. Li, X., Wang, Q., Zhao, Y., Wu, W., Chen, J., and Meng, H. (2013). Green synthesis and photo-catalytic performances for ZnO-reduced graphene oxide nanocomposites, *J. Colloid Interface Sci.*, **411**, pp. 69–75.

101. Lv, X., Fu, W., Hu, C., Chen, Y., and Zhou, W. (2013). Photocatalytic reduction of CO_2 with H_2O over a graphene-modified NiO_x–Ta_2O_5 composite photocatalyst: Coupling yields of methanol and hydrogen, *RSC Adv.*, **3**, pp. 1753–1757.

102. Wang, P. Q., Bai, Y., Luo, P. Y., and Liu, J. Y. (2013). Graphene–WO_3 nanobelt composite: Elevated conduction band toward photo-catalytic reduction of CO_2 into hydrocarbon fuels, *Catal. Commun.*, **38**, pp. 82–85.

103. Xu, C., Xu, B., Gu, Y., Xiong, Z., Sun, J., and Zhao, X. S. (2013). Graphene-based electrodes for electrochemical energy storage, *Energy Environ Sci.*, **6**, pp. 1388–1414.

104. Cheng, F., and Chen, J. (2012). Metal–air batteries: From oxygen reduction electrochemistry to cathode catalysts, *Chem. Soc. Rev.*, **41**, pp. 2172–2192.

105. Zhao, X., Yin, M., Liang, M., Liang, L., Liu, C., Liao, J., Lu, T., and Wei, X. (2011). Recent advances in catalysts for direct methanol fuel cells, *Energy Environ Sci.*, **4**, pp. 2736–2753.

106. Yu, W., Porosoff, M. D., and Chen, J. G. (2012). Review of Pt-based bimetallic catalysis: From model surfaces to supported catalysts, *Chem. Rev.*, **112**, pp. 5780–5817.

107. Kamat, P. V. (2012). The magic of electrocatalysts, *J. Phys. Chem. Lett.*, **3**, pp. 3404–3404.

108. Wan, X., Huang, Y., and Chen, Y. (2012). Focusing on energy and optoelectronic applications: A journey for graphene and graphene oxide at large scale, *Acc. Chem. Res.*, **45**, pp. 598–607.

109. Pumera, M. (2010). Graphene-based nanomaterials and their electrochemistry, *Chem. Soc. Rev.*, **39**, pp. 4146–4157.

110. Xia, B. Y., Ng, W. T., Wu, H. B., Wang, X., and Lou, X. W. (2012). Self-supported interconnected Pt nanoassemblies as highly stable

electrocatalysts for low-temperature fuel cells, *Angew. Chem. Int. Ed.*, **124**, pp. 7325–7328.

111. Chen, D., Tang, L., and Li, J. (2010). Graphene-based materials in electrochemistry, *Chem. Soc. Rev.*, **39**, pp. 3157–3180.

112. Gao, L., Yue, W., Tao, S., and Fan, L. (2013). Novel strategy for preparation of graphene-Pd, Pt composite, and its enhanced electrocatalytic activity for alcohol oxidation, *Langmuir*, **29**, pp. 957–964.

113. Siburian, R., Kondo, T., and Nakamura, J. (2013). Size control to a sub-nanometer scale in platinum catalysts on graphene, *J. Phys. Chem. C*, **117**, pp. 3635–3645.

114. Yoo, E. J., Okata, T., Akita, T., Kohyama, M., Nakamura, J., and Honma, I. (2009). Enhanced electrocatalytic activity of Pt subnanoclusters on graphene nanosheet surface, *Nano Lett.*, **9**, pp. 2255–2259.

115. Xiao, Y. P., Wan, S., Zhang, X., Hu, J. S., Wei, Z. D., and Wan, L. J. (2012). Hanging Pt hollow nanocrystal assemblies on graphene resulting in an enhanced electrocatalyst, *Chem. Commun.*, **48**, pp. 10331–10333.

116. Wang, R., Higgins, D. C., Hoque, M. A., Lee, D. U., Hassan, F., and Chen, Z. (2013). Controlled growth of platinum nanowire arrays on sulfur doped graphene as high performance electrocatalyst, *Sci. Rep.*, **3**, p. 2431.

117. De, R. L., Cargnello, M., Gombac, V., Lorenzut, B., Montini, T., and Fornasiero, P. (2010). Embedded phases: A way to active and stable catalysts, *ChemSusChem*, **3**, pp. 24–42.

118. Cui, S., Mao, S., Lu, G., and Chen, J. (2013). Graphene coupled with nanocrystals: Opportunities and challenges for energy and sensing applications, *J. Phys. Chem. Lett.*, **4**, pp. 2441–2454.

119. Zheng, Y., Jiao, Y., Jaroniec, M., Jin, Y., and Qiao, S. Z. (2012). Nanostructured metal-free electrochemical catalysts for highly efficient oxygen reduction, *Small*, **8**, pp. 3550–3566.

120. Liang, Y., Li, Y., Wang, H., Zhou, J., Wang, J., Regier, T., and Dai, H. (2011). Co_3O_4 nanocrystals on graphene as a synergistic catalyst for oxygen reduction reaction, *Nat. Mater.*, **10**, pp. 780–786.

121. Xiao, J., Xu, G., Sun, S. G., and Yang, S. (2013). MFe_2O_4 and MFe@ oxide core–shell nanoparticles anchored on N-doped graphene sheets for synergistically enhancing lithium storage performance and electrocatalytic activity for oxygen reduction reactions, *Part. Part. Syst. Char.*, **30**, pp. 893–904.

122. Zhou, R., Zheng, Y., Hulicovajurcakova, D., and Qiao, S. Z. (2013). Enhanced electrochemical catalytic activity by copper oxide grown on nitrogen-doped reduced graphene oxide, *J. Mater. Chem. A*, **1**, pp. 3179–13185.

123. Jung, H. G., Jeong, Y. S., Park, J. B., Sun, Y. K., Scrosati, B., and Lee, Y. J. (2013). Ruthenium-based electrocatalysts supported on reduced graphene oxide for lithium-air batteries, *ACS Nano*, **7**, pp. 3532–3539.

124. Wu, Z. S., Yang, S., Sun, Y., Parvez, K., Feng, X., and Müllen, K. (2012). 3D Nitrogen-doped graphene aerogel-supported Fe_3O_4 nanoparticles as efficient electrocatalysts for the oxygen reduction reaction, *J. Am. Chem. Soc.*, **134**, pp. 9082–9085.

125. Wang, H., Maiyalagan, T., and Wang, X. (2012). Review on recent progress in nitrogen-doped graphene: Synthesis, characterization, and its potential applications, *ACS Catal.*, **2**, pp. 781–794.

126. Deng, D., Pan, X., Yu, L., Cui, Y., Jiang, Y., Qi, J., Li, W. X., Fu, Q., Ma, X., and Xue, Q. (2011). Toward N-doped graphene via solvothermal synthesis, *Chem. Mater.*, **23**, pp. 1188–1193.

127. Byon, H. R., Jin, S., and Yang, S. H. (2011). Graphene-based non-noble-metal catalysts for oxygen reduction reaction in acid, *Chem. Mater.*, **23**, pp. 3421–3428.

128. Yang, Z., Nie, H., Chen, X. A., Chen, X., and Huang, S. (2013). Recent progress in doped carbon nanomaterials as effective cathode catalysts for fuel cell oxygen reduction reaction, *J. Power Sources*, **236**, pp. 238–249.

129. Choi, C., Chung, M., Park, S., and Woo, S. (2013). Enhanced electrochemical oxygen reduction reaction by restacking of N-doped single graphene layers, *RSC Adv.*, **3**, pp. 4246–4253.

130. Yasuda, S., Yu, L., Kim, J., and Murakoshi, K. (2013). Selective nitrogen doping in graphene for oxygen reduction reactions, *Chem. Commun.*, **49**, pp. 9627–9629.

131. Lai, L., Potts, J. R., Zhan, D., Wang, L., Poh, C. K., Tang, C., Gong, H., Shen, Z., Lin, J., and Ruoff, R. S. (2012). Exploration of the active center structure of nitrogen-doped graphene-based catalysts for oxygen reduction reaction, *Energy Environ. Sci.*, **5**, pp. 7936–7942.

132. Wang, R., Wang, Y., Xu, C., Sun, J., and Gao, L. (2013). Facile one-step hydrazine-assisted solvothermal synthesis of nitrogen-doped reduced graphene oxide: Reduction effect and mechanisms, *RSC Adv.*, **3**, pp. 1194–1200.

133. Wei, D., Liu, Y., Wang, Y., Zhang, H., Huang, L., and Yu, G. (2009). Synthesis of N-doped graphene by chemical vapor deposition and its electrical properties, *Nano Lett.,* **9**, pp. 1752–1758.
134. Liao, Y., Yuan, G., Zhu, S., Zheng, J., Chen, Z., Chao, Y., Lou, X., and Di, Z. (2015). Facile fabrication of N-doped graphene as efficient electrocatalyst for oxygen reduction reaction, *ACS Appl. Mater. Interfaces,* **7**, pp. 19619–19625.
135. Liu, Z. W., Peng, F., Wang, H. J., Yu, H., Zheng, W. X., and Yang, J. (2011). Phosphorus-doped graphite layers with high electrocatalytic activity for the O_2 reduction in an alkaline medium, *Angew. Chem. Int. Ed.,* **123**, pp. 3315–3319.
136. Zhang, C., Mahmood, N., Yin, H., Liu, F., and Hou, Y. (2013). Synthesis of phosphorus-doped graphene and its multifunctional applications for oxygen reduction reaction and lithium ion batteries, *Adv. Mater.,* **25**, pp. 4932–4937.
137. Sheng, Z. H., Gao, H. L., Bao, W. J., Wang, F. B., and Xia, X. H. (2012). Synthesis of boron doped graphene for oxygen reduction reaction in fuel cells, *J. Mater. Chem,* **22**, pp. 390–395.
138. Zhao, Y., Yang, L., Chen, S., Wang, X., Ma, Y., Wu, Q., Jiang, Y., Qian, W., and Hu, Z. (2013). Can boron and nitrogen co-doping improve oxygen reduction reaction activity of carbon nanotubes?, *J. Am. Chem. Soc.,* **135**, pp. 1201–1204.
139. Choi, C., Chung, M., Kwon, H., Park, S., and Woo, S. (2013). B,N- and P,N-doped graphene as highly active catalysts for oxygen reduction reactions in acidic media, *J. Mater. Chem. A,* **1**, pp. 3694–3699.
140. Wang, S., Zhang, L., Xia, Z., Roy, A., Chang, D. W., Baek, J. B., and Dai, L. (2012). BCN graphene as efficient metal-free electrocatalyst for the oxygen reduction reaction, *Angew. Chem. Int. Ed.,* **51**, pp. 4209–4212.
141. Zheng, Y. (2013). Two-step boron and nitrogen doping in graphene for enhanced synergistic catalysis, *Angew. Chem. Int. Ed.,* **125**, pp. 3192–3198.
142. Dreyer, D. R., Jia, H. P., and Bielawski, C. W. (2010). Graphene oxide: A convenient carbocatalyst for facilitating oxidation and hydration reactions, *Angew. Chem. Int. Ed.,* **122**, pp. 6965–6968.
143. Dreyer, D. R., and Bielawski, C. W. (2011). Carbocatalysis: Heterogeneous carbons finding utility in synthetic chemistry, *Chem. Sci.,* **2**, pp. 1233–1240.

144. Song, Y., Qu, K., Zhao, C., Ren, J., and Qu, X. (2010). Graphene oxide: Intrinsic peroxidase catalytic activity and its application to glucose detection, *Adv. Mater.*, **22**, pp. 2206–2210.
145. Sun, X., Liu, Z., Welsher, K., Robinson, J. T., Goodwin, A., Zaric, S., and Dai, H. (2008). Nano-graphene oxide for cellular imaging and drug delivery, *Nano Res.*, **1**, pp. 203–212.
146. Long, Y., Zhang, C., Wang, X., Gao, J., Wang, W., and Liu, Y. (2011). Oxidation of SO_2 to SO_3 catalyzed by graphene oxide foams, *J. Mater. Chem.*, **21**, pp. 13934–13941.
147. Pan, Y., Wang, S., Kee, C. W., Dubuisson, E., Yang, Y., Loh, K. P., and Tan, C.-H. (2011). Graphene oxide and Rose Bengal: Oxidative C–H functionalisation of tertiary amines using visible light, *Green Chem.*, **13**, pp. 3341–3344.
148. Huang, H., Huang, J., Liu, Y. M., He, H. Y., Cao, Y., and Fan, K. N. (2012). Graphite oxide as an efficient and durable metal-free catalyst for aerobic oxidative coupling of amines to imines, *Green Chem.*, **14**, pp. 930–934.
149. Su, C., Acik, M., Takai, K., Lu, J., Hao, S. J., Zheng, Y., Wu, P., Bao, Q., Enoki, T., and Chabal, Y. J. (2012). Probing the catalytic activity of porous graphene oxide and the origin of this behaviour, *Nat. Commun.*, **3**, p. 1298.
150. Boukhvalov, P. D. D. W., Dreyer, D. R., and Son, P. D. Y.-W. (2012). A computational investigation of the catalytic properties of graphene oxide: Exploring mechanisms by using DFT methods, *ChemCatChem.*, **4**, pp. 1844–1849.
151. Long, J., Xie, X., Xu, J., Gu, Q., Chen, L., and Wang, X. (2012). Nitrogen-doped graphene nanosheets as metal-free catalysts for aerobic selective oxidation of benzylic alcohols, *ACS Catal.*, **2**, pp. 622–631.
152. Lou, X., Pan, H., Zhu, S., Zhu, C., Liao, Y., Li, Y., Zhang, D., and Chen, Z. (2015). Synthesis of silver nanoprisms on reduced graphene oxide for high-performance catalyst, *Catal. Commun.*, **69**, pp. 43–47.
153. Wilson, G. S., and Hu, Y. (2000). Enzyme-based biosensors for in vivo measurements, *Chem. Rev.*, **100**, pp. 2693–2704.
154. Wang, Y., Shao, Y., Matson, D. W., Li, J., and Lin, Y. (2010). Nitrogen-doped graphene and its application in electrochemical biosensing, *ACS Nano*, **4**, pp. 1790–1798.
155. Wu, P., Qian, Y., Du, P., Zhang, H., and Cai, C. (2012). Facile synthesis of nitrogen-doped graphene for measuring the releasing process of hydrogen peroxide from living cells, *J. Mater. Chem.*, **22**, pp. 6402–6412.

156. Wu, P., Du, P., Zhang, H., and Cai, C. (2013). Microscopic effects of the bonding configuration of nitrogen-doped graphene on its reactivity toward hydrogen peroxide reduction reaction, *Phys. Chem. Chem. Phys.,* **15**, pp. 6920–6928.

157. Liu, M., Liu, R., and Chen, W. (2013). Graphene wrapped Cu_2O nanocubes: Non-enzymatic electrochemical sensors for the detection of glucose and hydrogen peroxide with enhanced stability, *Biosens. Bioelectron.,* **45**, pp. 206–212.

158. Wang, X., Dong, X., Wen, Y., Li, C., Xiong, Q., and Chen, P. (2012). A graphene–cobalt oxide based needle electrode for non-enzymatic glucose detection in micro-droplets, *Chem. Commun.,* **48**, pp. 6490–6492.

159. Yuan, B., Xu, C., Deng, D., Xing, Y., Liu, L., Pang, H., and Zhang, D. (2013). Graphene oxide/nickel oxide modified glassy carbon electrode for supercapacitor and nonenzymatic glucose sensor, *Electrochim. Acta,* **88**, pp. 708–712.

160. Lu, L.-M., Li, H.-B., Qu, F., Zhang, X.-B., Shen, G.-L., and Yu, R.-Q. (2011). In situ synthesis of palladium nanoparticle–graphene nanohybrids and their application in nonenzymatic glucose biosensors, *Biosens. Bioelectron.,* **26**, pp. 3500–3504.

161. Kong, F.-Y., Li, X.-R., Zhao, W.-W., Xu, J.-J., and Chen, H.-Y. (2012). Graphene oxide–thionine–Au nanostructure composites: Preparation and applications in non-enzymatic glucose sensing, *Electrochem. Commun.,* **14**, pp. 59–62.

162. Lou, X., Zhu, C., Pan, H., Ma, J., Zhu, S., Zhang, D., and Jiang, X. (2016). Cost-effective three-dimensional graphene/Ag aerogel composite for high-performance sensing, *Electrochim. Acta,* **205**, pp. 70–76.

163. Yang, Y., Chiang, K., and Burke, N. (2011). Porous carbon-supported catalysts for energy and environmental applications: A short review, *Catal. Today,* **178**, pp. 197–205.

164. Zheng, H., Qu, Q., Zhang, L., Liu, G., and Battaglia, V. S. (2012). Hard carbon: A promising lithium-ion battery anode for high temperature applications with ionic electrolyte, *RSC Adv.,* **2**, pp. 4904–4912.

165. Lebedev, Y., Krekhov, A., Chuvyrov, A., and Gil'manova, N. K. (1986). Disclination structures of carbonaceous mesophase spherules, *Carbon,* **24**, pp. 719–724.

166. Gong, Y., Wei, Z., Wang, J., Zhang, P., Li, H., and Wang, Y. (2014). Design and fabrication of hierarchically porous carbon with a template-free method, *Sci. Rep.,* **4**, p. 6349.

167. Li, Y., Zhu, C., Lu, T., Guo, Z., Zhang, D., Ma, J., and Zhu, S. (2013). Simple fabrication of a Fe_2O_3/carbon composite for use in a high-performance lithium ion battery, *Carbon*, **52**, pp. 565–573.

168. Li, Y., Zhu, S., Liu, Q., Chen, Z., Gu, J., Zhu, C., Lu, T., Zhang, D., and Ma, J. (2013). N-doped porous carbon with magnetic particles formed in situ for enhanced Cr(VI) removal, *Water Res.*, **47**, pp. 4188–4197.

169. Guo, J., Zhu, S., Chen, Z., Li, Y., Yu, Z., Liu, Q., Li, J., Feng, C., and Zhang, D. (2011). Sonochemical synthesis of TiO_2 nanoparticles on graphene for use as photocatalyst, *Ultrason. Sonochem.*, **18**, pp. 1082–1090.

170. Parthasarathy, R. V., and Martin, C. R. (1994). Synthesis of polymeric microcapsule arrays and their use for enzyme immobilization, *Nature*, **369**, pp. 298–301.

171. Martin, C. R. (1991). Template synthesis of polymeric and metal microtubules, *Adv. Mater.*, **3**, pp. 457–459.

172. Klein, J. D., Herrick, R. D., Palmer, D., Sailor, M. J., Brumlik, C. J., and Martin, C. R. (1993). Electrochemical fabrication of cadmium chalcogenide microdiode arrays, *Chem. Mater.*, **5**, pp. 902–904.

173. Martin, C. R. (1996). Membrane-based synthesis of nanomaterials, *Chem. Mater.*, **8**, pp. 1739–1746.

174. Cao, G., Rabenberg, L., Nunn, C. M., and Mallouk, T. E. (1991). Formation of quantum-size semiconductor particles in a layered metal phosphonate host lattice, *Chem. Mater.*, **3**, pp. 149–156.

175. Brigham, E. S., Weisbecker, C. S., Rudzinski, W. E., and Mallouk, T. E. (1996). Stabilization of intrazeolitic cadmium telluride nanoclusters by ion exchange, *Chem. Mater.*, **8**, pp. 2121–2127.

176. Schuth, F., and Schmidt, W. (2002). Microporous and mesoporous materials, *Adv. Mater.*, **14**, pp. 629–638.

177. Schuth, F. (2003). Endo and exotemplating to create high-surface-area Inorganic materials, *Angew. Chem. Int. Ed.*, **42**, pp. 3604–3622.

178. Che, G., Lakshmi, B. B., Fisher, E. R., and Martin, C. R. (1998). Carbon nanotubule membranes for electrochemical energy storage and production, *Nature*, **393**, pp. 346–349.

179. Kim, K., Choi, M., and Ryoo, R. (2013). Ethanol-based synthesis of hierarchically porous carbon using nanocrystalline beta zeolite template for high-rate electrical double layer capacitor, *Carbon*, **60**, pp. 175–185.

180. Zakhidov, A. A., Baughman, R. H., Iqbal, Z., Cui, C., Khayrullin, I., Dantas, S. O., Marti, J., and Ralchenko, V. G. (1998). Carbon structures with

three-dimensional periodicity at optical wavelengths, *Science,* **282**, pp. 897–901.

181. Sonobe, N., Kyotani, T., and Tomita, A. (1988). Carbonization of polyacrylonitrile in a two-dimensional space between montmorillonite lamellae, *Carbon,* **26**, pp. 573–578.

182. Rodriguez-Mirasol, J., Cordero, T., Radovic, L., and Rodriguez, J. (1998). Structural and textural properties of pyrolytic carbon formed within a microporous zeolite template, *Chem. Mater.,* **10**, pp. 550–558.

183. Sakintuna, B., and Yürüm, Y. (2005). Templated porous carbons: A review article, *Ind. Eng. Chem. Res.,* **44**, pp. 2893–2902.

184. Liang, C., and Dai, S. (2006). Synthesis of mesoporous carbon materials via enhanced hydrogen-bonding interaction, *J. Am. Chem. Soc.,* **128**, pp. 5316–5317.

185. Liang, C., Hong, K., Guiochon, G. A., Mays, J. W., and Dai, S. (2004). Synthesis of a large-scale highly ordered porous carbon film by self-assembly of block copolymers, *Angew. Chem. Int. Ed.,* **43**, pp. 5785–5789.

186. Chuenchom, L., Kraehnert, R., and Smarsly, B. M. (2012). Recent progress in soft-templating of porous carbon materials, *Soft Matter,* **8**, pp. 10801–10812.

187. Yürüm, Y., Taralp, A., and Veziroglu, T. N. (2009). Storage of hydrogen in nanostructured carbon materials, *Int. J. Hydrogen Energy,* **34**, pp. 3784–3798.

188. Garcıa-Cortés, J., Pérez-Ramırez, J., Illán-Gómez, M., Kapteijn, F., Moulijn, J., and de Lecea, C. S.-M. (2001). Comparative study of Pt-based catalysts on different supports in the low-temperature de-NOx-SCR with propene, *Appl. Catal. B,* **30**, pp. 399–408.

189. Wei, Z., Zeng, G., Xie, Z., Ma, C., Liu, X., Sun, J., and Liu, L. (2011). Microwave catalytic NO_x and SO_2 removal using FeCu/zeolite as catalyst, *Fuel,* **90**, pp. 1599–1603.

190. Chen, G., Gao, J., Gao, J., Du, Q., Fu, X., Yin, Y., and Qin, Y. (2010). Simultaneous removal of SO_2 and NOx by calcium hydroxide at low temperature: Effect of SO_2 absorption on NO_2 removal, *Ind. Eng. Chem. Res.,* **49**, pp. 12140–12147.

191. Zhiming, L., Jiming, H., Lixin, F., Junhua, L., and Xiangyu, C. (2004). Advances in catalytic removal of NO_x under lean-burn conditions, *Chin. Sci. Bull.,* **49**, pp. 2231–2241.

192. Parvulescu, V., Grange, P., and Delmon, B. (1998). Catalytic removal of NO, *Catal. Today,* **46**, pp. 233–316.

193. Traa, Y., Burger, B., and Weitkamp, J. (1999). Zeolite-based materials for the selective catalytic reduction of NO_x with hydrocarbons, *Microporous Mesoporous Mater.*, **30**, pp. 3–41.
194. Garcıa-Cortés, J., Illán-Gómez, M., Solano, A. L., and de Lecea, C. S.-M. (2000). Low temperature selective catalytic reduction of NO(x) with C_3H_6 under lean-burn conditions on activated carbon-supported platinum, *Appl. Catal. B*, **25**, pp. 39–48.
195. Obuchi, A., Ohi, A., Nakamura, M., Ogata, A., Mizuno, K., and Ohuchi, H. (1993). Performance of platinum-group metal catalysts for the selective reduction of nitrogen oxides by hydrocarbons, *Appl. Catal. B*, **2**, pp. 71–80.
196. Macken, C., and Hodnett, B. (1998). Reductive regeneration of sulfated CuO/Al_2O_3 catalyst-sorbents in hydrogen, methane, and steam, *Ind. Eng. Chem. Res.*, **37**, pp. 2611–2617.
197. T. Tseng, H.-H., and Wey, M.-Y. (2006). Effects of acid treatments of activated carbon on its physiochemical structure as a support for copper oxide in $DeSO_2$ Reaction catalysts, *Chemosphere*, **62**, pp. 756–766.
198. Bandosz, T. J., Matos, J., Seredych, M., Islam, M., and Alfano, R. (2012). Photoactivity of S-doped nanoporous activated carbons: A new perspective for harvesting solar energy on carbon-based semiconductors, *Appl. Catal. A*, **445**, pp. 159–165.
199. Velasco, L. F., Lima, J. C., and Ania, C. (2014). Visible-light photochemical activity of nanoporous carbons under monochromatic light, *Angew. Chem. Int. Ed.*, **53**, pp. 4146–4148.
200. Seredych, M., and Bandosz, T. J. (2014). Effect of the graphene phase presence in nanoporous S-doped carbon on photoactivity in UV and visible light, *Appl. Catal. B*, **147**, pp. 842–850.
201. Seredych, M., and Bandosz, T. J. (2013). S-doped micro/mesoporous carbon-graphene composites as efficient supercapacitors in alkaline media, *J. Mater. Chem. A*, **1**, pp. 11717–11727.
202. Velasco, L. F., Fonseca, I. M., Parra, J. B., Lima, J. C., and Ania, C. O. (2012). Photochemical behaviour of activated carbons under UV irradiation, *Carbon*, **50**, pp. 249–258.
203. Bandosz, T. J., Rodriguez-Castellon, E., Montenegro, J. M., and Seredych, M. (2014). Photoluminescence of nanoporous carbons: Opening a new application route for old materials, *Carbon*, **77**, pp. 651–659.

204. Ania, C. O., Seredych, M., Rodríguez-Castellón, E., and Bandosz, T. J. (2014). Visible light driven photoelectrochemical water splitting on metal free nanoporous carbon promoted by chromophoric functional groups, *Carbon*, **79**, pp. 432–441.
205. Wang, X., Maeda, K., Thomas, A., Takanabe, K., Xin, G., Carlsson, J. M., Domen, K., and Antonietti, M. (2009). A metal-free polymeric photocatalyst for hydrogen production from water under visible light, *Nat. Mater.*, **8**, pp. 76–80.
206. Zhao, Y., Nakamura, R., Kamiya, K., Nakanishi, S., and Hashimoto, K. (2013). Nitrogen-doped carbon nanomaterials as non-metal electrocatalysts for water oxidation, *Nat. Commun.*, **4**, p. 2390.
207. Pandolfo, A., and Hollenkamp, A. (2006). Carbon properties and their role in supercapacitors, *J. Power Sources*, **157**, pp. 11–27.
208. Frackowiak, E., and Béguin, F. (2001). Carbon materials for the electrochemical storage of energy in capacitors, *Carbon*, **39**, pp. 937–950.
209. Wang, W.-S., Wang, D.-H., Qu, W.-G., Lu, L.-Q., and Xu, A.-W. (2012). Large ultrathin anatase TiO_2 nanosheets with exposed {001} facets on graphene for enhanced visible light photocatalytic activity, *J. Phys. Chem. C*, **116**, pp. 19893–19901.

Chapter 6

Porous Inorganic Nanoarchitectures for Catalysts

Qingmin Ji,[a] Jiao Sun,[a] and Shenmin Zhu[b]

[a]*Herbert Gleiter Institute of Nanoscience, Nanjing University of Science and Technology, 200 Xiaolingwei, Nanjing 210094, China*
[b]*State Key Laboratory of Metal Matrix Composites, Shanghai Jiao Tong University, 800 Dongchuan Rd, Minhang District, Shanghai, China*

jiqingmin@njust.edu.cn

6.1 Introduction

Heterogeneous catalysis has always been an inherently nanoscopic phenomenon with important technological and societal consequences for energy conversion and the production of chemicals. In catalytic processes, the effective molecular transport of reactants and production largely relies on the architectural design of catalysis systems at the nanoscale. The porous inorganic supports for catalysts provide new opportunities for improving catalytic performance through the confined effect from the void space and the surface multifunctionalities. The large surface areas

---

Soft Matters for Catalysts
Edited by Qingmin Ji and Harald Fuchs
Copyright © 2020 Jenny Stanford Publishing Pte. Ltd.
ISBN 978-981-4774-66-6 (Hardcover), 978-1-351-27284-1 (eBook)
www.jennystanford.com

may offer readily accessible platform to molecules. In addition, those nanostructures may be further assembled for special electrochemical catalysis or other catalytic requirements. The porous architectures open more easily operation pathways for various nanocatalysts and catalytic reactions and are promising for higher catalytic performance.

The design of novel porous catalytic nanoarchitectures for catalysts requires theoretical, synthetic, and analysis innovation on porous materials. New approaches are developed based on both computation and experimentation. The challenge might be to maintain the most active catalytic state of the system even under certain external conditions (temperature, pressure, electrical potential, photo energy, etc.), which would otherwise drive the restructuring, crystallization, or fusion of catalysts in the nanoscale dimension. It is essential to understand the relationship between porous structures and catalytic performance.

In the following sections, we focus on the achievements on the design of porous nanoarchitectures, especially of the non-carbon materials or structures for catalysis, and the application of porous nanostructures for biocatalysis, for gas-phase catalysis, and for electrochemical catalysis. We highlight the porous nanoarchitecturing on their effects to the loading and catalytic activity of various types of nanocatalysts, offer overviews on the conceptions for the development and the redesign of inorganic or hybrid catalytic nanoarchitectures.

6.2 Mesoporous Nanoarchitectures

Nanosized porous materials attract steady interest due to their potential to offer sustainable solutions to global issues such as energy, pollution, health improvement. Over the decades with intensive research and developments on chemistry, materials, nanotechnology, the family of mesoporous materials has grown rapidly. Nanoarchitectonics is expected to achieve an effective guidance in fabricating functional nanoporous materials [1, 2]. Advanced mesoporous nanoarchitectures are being created in which the pore and solid structural components are controlled at the nanoscale level.

Learning from nanoarchitectonic guidance in naturally occurring systems, various kinds of components in softly organized assemblies can operate synergistically to render incredibly high efficiency and selectivity [3]. Those soft-matters have great molecular flexibility and can form highly harmonized structures with specific functions, which are difficult to be gained for the robust inorganic materials. The structural transcript from soft-material assemblies into rigid inorganic materials has been treated as a breakthrough to create inorganic nanomaterials with complex morphologies [4–6].

The so-called "templated synthesis" has become more attractive since not only soft-materials, the characteristics of hard-materials may also be combined in the formation of multifunctional inorganic nanoarchitectures. According to the templates, various levels of porosity, which are always in the range of 2–50 nm, can be created in the solid. The template synthesis is thought of as the primary way to create mesoporous nanomaterials.

Nowadays a wide variety of synthesis techniques are utilized to produce porous materials. The types of templates reported so far can be the soft, hard, and sacrificial templates. Nearly all types of nanoarchitectured materials with different compositions (i.e., metal oxides, metals, semiconductors, and ceramics) and various structures/shapes (zero-dimensional (0D) nanoparticles, one-dimensional (1D) nanowires and two- and three-dimensional (2D and 3D) hierarchical nanostructures) can be created based on the template strategies [7].

Mesoporous nanoparticles have been one of the most studied porous nanostructures as supports for catalysts. As the catalysts may possess diverse dimension and molecular characters, the morphological requirement for porous supports is different. Both the nanostructures and the surface functions of porous materials need to be well manipulated to achieve the best match between nanofeatures of porous nanoarchitectures and the catalytic performance.

Cai et al. first reported factors affecting morphology control and particle size of mesoporous silica nanoparticles (MSNs) [8]. They synthesized mesoporous nanospheres and nanorods at extremely low surfactant concentration. They also proposed a detailed mechanism of the formation of mesoporous silica nanoparticles with different shapes and sizes. It is suggested that the synthesis of MSNs involves nucleation and co-self-assembly of silica precursor with the template. The morphology of the particles is determined by the type of catalyst used; for example, ammonia leads to a rod-like morphology, while NaOH forms the spherical particles.

In general, periodic porosity can be created for mesoporous nanostructures by surfactant-templated synthesis. However, the pore size by this strategy is limited, always smaller than 10 nm. To form porous materials with larger pores, co-template reagents or block polymer are used to enlarge the pore dimensions. Han et al. synthesized MSNs with a 3D cubic (Im3m) structure and large pore size by using cationic surfactants with lipophobic nature and a pore swelling agent (trimethylbenzene) [9]. They prepared nanoparticles with different pore sizes and mesostructures. The pore size of their MSNs can be reached to 20 nm. Through well adjusting the reaction systems with the combined templates, Fan et al. first created an ordered mesoporous material with an ultra large pore size of around 30 nm [10]; Qiu et al. synthesized MSNs with ultra large pores (20–40 nm) with interconnected channel structures (Fig. 6.1) [11]. The manipulation on the pore size may effectively improve the loading capacity of nanoscaled catalysts and provide optimal microenvironment for catalytic reactions.

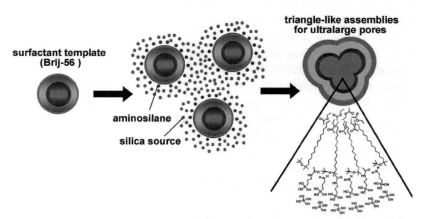

Figure 6.1 Illustration of a possible mechanism for the formation of mesoporous materials with ultra-large pores in a neutral pH system.

Besides mesoporous silica, other kinds of inorganic materials or their composites can also be synthesized by template methods. As some catalytic components can be directly formed into porous morphologies, synergism effect can be achieved for enhancing catalytic performance. Lyotropic liquid crystal (LLC) templating is often used to control porous nanostructures of non-siliceous materials. LLC is a stable binary system composed of amphiphilic molecules, such as surfactant molecules, and a polar solvent. LLC templating takes advantage of a self-assembled structure formed by the amphiphile to control the structure of an inorganic phase. Typically the inorganic phase grows in the hydrophilic domains of the liquid crystal, while the hydrophobic domains form pores. Saramat et al. reported the preparation of mesoporous Pt-Al_2O_3 composite and investigated the CO oxidation activity, showing a higher tolerance to CO poisoning compared to nonporous Pt [12]. Until now, a multitude of well-defined mesoporous metals with various compositions, such as Co, Ni, Rh, Sn, and Ge, have been synthesized by chemical or electrochemical deposition of metal ions in LLCs [13–18].

Yamauchi et al. reported the synthesis of mesoporous Pt with three types of mesostructures including 2D hexagonal, lamellar, and cage-type by changing the composition ratio between Pt salt and block copolymers [19]. The pore diameter and structure of the mesoporous metals can be easily controlled by modifying the type and composition of the LLCs (Fig. 6.2) [20]. The selective synthesis of large pores may favor the accessibility of guest species to inner parts of the mesoporous structure, and enhance the catalytic efficiency.

For the cases of porous nanostructures prepared from the template strategies, there exists the drawback of post-synthesis removal of the organic template, which can lead to the contamination of the resulting structures with remnants of organic components or may cause structural deformations and unexpected defects. Thus, the so-called template-free or self-templating methods have been developed as synthetic strategies [21, 22]. For the cases using a sacrificial template or self-templated strategy, inorganic template structures themselves will undergo a dissolution-regrowth process under certain reaction conditions. The nanofeatures of structures may change after the regrowth and lead to more complex structures in compared with the morphology of the original solid template.

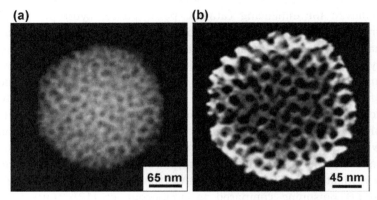

Figure 6.2 (a) Black field-STEM images of the mesoporous Pt and (b) the sliced structure at the cross section [20].

Ji et al. prepared cell-like flake-shell capsules from simple silica nanoparticles [23, 24]. Spontaneous formation of flake-shell capsules occurred during a hydrothermal process. Gradual dissolution of silica from the surface of nanoparticles and precipitation as silica nanosheets in the vicinity of the parent particle finally formed into hollow spherical capsules consisting of assembled silica nanosheets. Although inorganic structures, the silica nanosheet-assembled capsules also present flexibility somehow reminiscent of cells or other soft lipid assemblies. The nanosheet morphology provides more contact points within its network resulting in better control over shell flexibility. The morphology of the flake-shell hollow spheres is sensitive to heat and acidity, both of which can shrink the network of nanosheets or the morphology of the nanosheets upon application of some stimulus. The porous capsules showed advantages on the loading of various molecules (proteins and nanoparticles). The adsorbent molecules or NPs may be much more strongly attracted to the surfaces of nanosheets in the shell rather than penetrating into the confined nanosized channels. The "open" shell morphology and hollow interior should allow the substrates more easy access to the active sites and lead to high catalytic efficiency (Fig. 6.3).

Mesoporous Nanoarchitectures | **297**

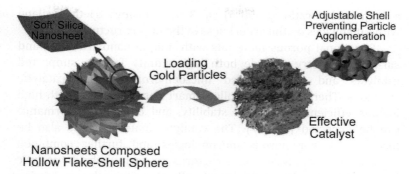

Figure 6.3 The scheme of the shell-adjustable hollow silica spheres as a support for gold nanoparticles.

The loading capacity, distribution, and fixation of catalytic components into the porous supports are key issues for the construction of porous catalyst systems. Since the porous supports have poor functionality in most cases, the surface modification may improve the affinity of porous supports to catalytic components. Controlling surface functionalities on porous supports may also bring smartness on the catalysis system for selective response to external stimuli.

Two main methods are typically carried out to organically functionalize the surface of porous support: post-treatment (grafting method) and co-condensation. For the post-synthesis grafting method, organic functionalities are added after removal of templates. This method offers the freedom on the generation of selective functional groups on the surface. However, it can also lead to a nonhomogenous surface coating. This is because the surface groups are more accessible than those inside the pore wall. In contrast to the grafting method, a homogeneous distribution of functional groups can be achieved by co-condensation, as the organic groups are directly incorporated into a silica framework. It thus avoids the occurrence of pore blocking in the grafting method. Due to the functional groups introduced during the formation process of porous materials, the co-condensation agents may affect the shape, the size, and the regularity of the pores. Thus, precise regulation on the cooperation of co-condensation agents is required. Weisner et al. successfully synthesized MSNs with a cubic Pm3n symmetry and a high molar ratio (> 50%) of 3-aminopropyl triethoxysilane without any loss on the orderliness of the mesostructure [25].

Functional porous materials with unique nanostructures and surface chemistry provide both high affinity for the supported catalysts and an optimized microenvironment for catalytic reactions. Therefore, a controlled smart immobilization with high loading efficiency, enhanced stability, and catalytic performance can be expected [26, 27]. The catalytic materials may also be formed into nanoporous morphologies and further combined with various catalytic nanostructures to form hybrid catalytic systems [28–31]. The multi-functionalization opens the application of porous catalysis systems for not only specific catalytic reaction, but also devices for sensing and fuel cell.

6.3 Inorganic Porous Nanoarchitectures for Enzymes

Although biological catalysts of enzymes have several unique advantages in contrast to conventional catalysts, like a high catalytic efficiency, substrate specificity selectivity, under mild reaction conditions etc., the use of enzymes still has limitations in large-scale production due to high cost and lack of long-term operational stability. The difficult separation of enzymes from the reaction system may not allow the recovery of enzymes and the repeated usage. It may also lead to the contamination of the final products. Under extreme temperatures or pH, enzymes are easily inactivated due to denaturation, either by changes of conformation or stereo chemical structure. These disadvantages can be largely overcome through immobilization in the solid supports. The immobilized enzymes are shown more stable and easier to recover than enzymes free in solution. Additionally, the catalytic activity of enzyme may also be improved by immobilization. There are several immobilization methods available, including binding or encapsulation of enzymes inside a hollow structures or cross-linking enzymes into large aggregates. The first strategy through noncovalent adsorption is the most simple immobilization method and also widely used in large-scale catalytic production.

Porous nanostructures are promising material as support for enzymes due to their large surface area and the adjustable pore size. Compared to smooth solids materials, the porous structure enables much larger encapsulation amount of enzymes. Moreover, porous inorganic materials are robust material and chemically stable over a broad pH and temperature range. Thus, the immobilized enzymes in porous nanostructures can produce recoverable and stable heterogeneous biocatalysts.

Porous nanostructures as supports for enzymes have generated vast interests ever since the successful synthesis of MSNs by Mobil researchers in 1992 [32]. Many reports have shown that the enzymes on MSNs improve the enzyme stability, catalytic activity, products specificity, and resistance to extreme environmental conditions [33, 34]. Besides silica, synthesis of other mesoporous materials can be achieved by adjusting templates or synthesis conditions. A broad range of porous materials have been applied for the immobilization of enzymes, including metals, oxides, and chalcogenides [35–37].

As enzymes are relatively large biological molecules, the pore dimension need to be tailored with high precision, which is essential for the loading capacity, catalytic activity, and reusability of the immobilized enzymes. Therefore, the exploitation of novel nanostructures that enable high enzyme loading and activity retention is the focus of research. Entrapment in the pores may protect the enzymes from the surrounding media. The use of immobilized enzymes with high enzyme loading and high volumetric capacity is also beneficial because high productivity and space-time yields [38]. The large pores in the support also facilitate transport of substrate and product [39]. In the porous supports, most enzymes in vivo may perform under a crowded environment where unfolding and aggregations are prevented. Due to the larger and controllable pore sizes, two or more different enzymes may also be encapsulated to perform parallel or sequential catalytic reactions.

The further surface design of porous materials may bring specific functions and provide favored microenvironment for the binding of enzymes and their catalytic reactions. The controlled release of bio-molecules from mesoporous supports may prevent biopharmaceuticals from degradation and decrease the non-specific release. Those mesoporous materials with biocompatibility, low cytotoxicity, large surface areas, and easy functionalization are excellent candidate as carriers not only for enzymes, but also genes and drug molecules. Biomimetic model compounds (artificial enzymes) have also been encapsulated in mesoporous supports to improve catalytic activity [40, 41]. It was recognized that the rigid structure and pore surface of functionalized mesoporous inorganic nanostructures can mimic protein skeleton, where the active sites are held stably with a fixed configuration to carry out the presumed enzymatic functions with good turnover efficiency (Fig. 6.4).

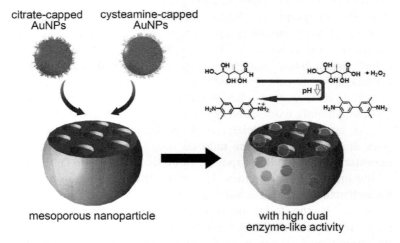

Figure 6.4 Design of mesoporous nanoparticle with Au nanoparticles as intelligent enzyme mimics for higher functions.

The enzymes can be immobilized to the porous material either by covalent binding, physical adsorption or electrostatic attraction. Covalent immobilization involves binding of reactive groups on the enzymes to a functional surface. The reactive groups may be amino, thiol or hydroxyl groups on amino acid side chains located on the enzyme surface. To covalently bind the enzyme to silica, the silica surface is generally modified with a silane coupling agent containing with nucleophilic (amine, thiol) or electrophilic groups (active carboxylic acid, alkyl chloride). Other bioconjugate techniques can further extend the covalent approach to greater possibility. For example, the addition of glutaraldehyde or glutardialdehyde for the covalent binding on the nanopore surface can provide a versatile linker with the peptide chains of enzymes. Covalent binding of the enzymes to the support provides stable fixation of enzymes, which make them can withstand elevated temperatures and can be reused without leaking out from the support. Li et al. reported the immobilization of triacylglycerol lipase (from porcine pancreas) on aldehyde modified mesoporous supports by covalent binding. They showed that the immobilized enzyme exhibited a high thermal stability (up to 65°C from 35°C) and reusability with high activity (over 80%) [42]. The major drawbacks of covalent bonding are reduced enzymatic activity due to conformational changes upon attachment and the pore blocking due to the cross-linking. Schlossbauer et al. developed the surface-based chemistry with azide-alkyne click reaction to covalently immobilize enzymes to the surface of mesopores (Fig. 6.5). This method gives very high surface loading of enzyme (trypsin, about 12%) without pore-blocking [43].

Physical adsorption is the most simple immobilization method and also frequently used in large-scale preparation. Only weak interactions of hydrogen bonding, electrostatic interactions or van der Waals forces are involved in the binding procedures. The pore features of supports can be largely maintained and the risk of enzyme denaturation by binding is also minimized. However, due to weak interaction, enzyme may leak out from the supports during the use. Therefore, it is important to tailor the pore size well match the enzymes [44, 45]. Ionic strength of the solution, pH and temperature may also affect the binding efficiency of enzymes in the porous supports. The control on the overall charge of the enzyme and support by solution conditions may improve the leakage of enzymes from porous supports [46].

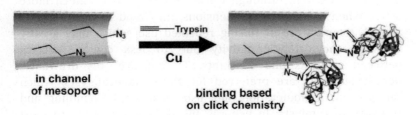

Figure 6.5 Covalent attachment of trypsin into the mesoporous system by click chemistry.

6.4 Porous Nanoarchitectures for Gas-Phase Catalysis

Heterogeneous catalysis involves systems in which catalyst and reactants form separate physical phases. Most of typical heterogeneous catalysts are inorganic solids such as metals, oxides, sulfides, and metal salts. The most common examples of heterogeneous catalysis in industry are involved the reactions of gases over the surface of solid catalysts. The catalytic reactions include the catalytic oxidation of volatile organic compounds (VOCs), preferential oxidation of CO, synthesis of ammonia, oxidation of HCl, partial oxidation of CH_4 and cracking of gas oil. The reactions proceed basically through several steps:

(i) Gaseous reactants are transferred on the surface of catalysts.
(ii) The reactants are adsorbed.
(iii) The gas molecules are interacted with atoms or ions on the surface of the catalysts.
(iv) Products are detached from the surface.

The enthalpy of chemisorption of gases on catalysts may reduce the activation energy barrier of the reaction, and promote the formation of products. A successful catalyst must allow the reaction to proceed at a suitable rate under conditions and is economically desirable. That means the catalysts need to have suitable good interactions with gases, should be chemically stable under temperature and pressure, should be easy separated after reactions, and possess sufficient mechanical strength.

Transition and noble metals such as Pt, Pd, Rh, Rd, Au, Ag, W, Cu, Co, Fe, and Ni are common catalysts used for the gas-phase reaction. Although those catalysts can be much more active when in nanoscale dimensions, they tend to fuse together at high temperature or during repeated use, which may greatly decrease the catalytic activity due to less contact surface between catalysts and reactants. To improve this drawback, porous supports are preferred for those nanocatalysts, which can prevent agglomeration and also make easier handling and recovery of catalysts. In general, the porous supports are inert materials. However, the surface chemistry of the porous supports may affect the distribution and the active phase of the catalytic components. Even the nanocatalysts are distributed in the porous supports homogeneously through proper design on the pore structures, nanocatalysts may still possible agglomerate in the pore channel and leach under catalytic reactions. The nanoarchitecturing on porous catalysis system can provide effective way to both homogeneous loading and maintain active phase of catalysts. The construction of multifunction catalytic systems also needs the design of nanoarchitectures of the porous catalysis systems.

As gases need to be easily transferred to the surface of catalysts, the post-grafting strategy may not be suitable. The post coverage of ligands on the porous supports (to prevent the leaking of catalysts from the pores) can shield partial of accessible surface of the catalysts, which may also weaken the catalytic efficiency. More advanced catalytic nanoarchitectures are built through introducing ligands which can stabilize the catalytic nanoparticles and also can be used as co-template agents.

Abe et al. reported the formation of silica supported Pt catalysis system by directly mixing dendrimers-covered Pt nanoparticles into a compartment-rich porous silica capsules [47]. The surface dendrimers (hydroxyl-terminated generation 6 polyamidoamine dendrimers) favor the encapsulation of Pt nanoparticles in the porous capsules. In contrast to channel-type mesoporous supports or solid supports, the compartment-rich silica capsules showed an enhanced loading capacity and confinement effect to prevent the sintering of Pt nanoparticle under high temperature (Fig. 6.6). The system showed superior catalytic activity in high-temperature CO oxidation. Budroni et al. prepared porous supported Pd catalysts by capped 1-dodecanethiol and 3-mercaptopropyltrimethoxysilane with Pd nanoparticles. The following co-condensation of silicate precursors with modified Pd nanoparticles resulted into a homogeneous incorporation of Pd nanoparticles into a sponge-like porous silica [48]. Both the pore and catalysis components can be controlled by the bridging ligand, the initial size and the necessary catalytic functionality of nanoparticles can be maintained into the porous network. The coassembly of the bifunctional components (template and catalyst) into the porous nanoarchitecture differs from the routes by post-impregnation of catalytic nanoparticles. This strategy allows more precise manipulation on the size of catalytic nanoparticles, pore structures of the supports, and the loading amount of catalytic nanoparticles within the porous support.

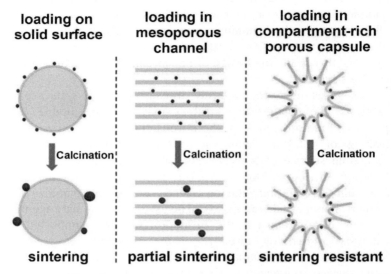

Figure 6.6 The illustration of nanoparticles dispersed on the solid supports, the porous supports, and the supports enriched with wide-mouthed compartments on the surface under calcination.

A general "encapsulation and etching" strategy is also developed for the fabrication of nanocatalyst systems in which catalyst nanoparticles are protected within porous shells [49]. The novelty of this approach lies in the use of chemical etching to assist the creation of mesopores in a protective oxide shell, which will then promote the efficient mass transfer to encapsulated metal nanoparticles, while the shells act as physical barriers against aggregation of the catalyst particles.

By using the surface-protected etching process, both yolk-shell and core-satellite type nanoreactors were synthesized and utilized in liquid- and gas-phase catalysis (Fig. 6.7). In the process, Pt nanoparticles were loaded on the surface-modified silica beads through the strong chemical affinity between the Pt and the primary amines. The loading of Pt nanoparticles can be well controlled by simply tuning the ratio of Pt precursor to silica beads used. In the presence of PVP, the composite colloids are over-coated with another layer of silica to fully encapsulate the Pt nanoparticles inside the silica matrix. Upon surface-protected etching with NaOH solution, the outer silica layer becomes mesoporous, exposing the Pt nanocatalysts to the outside chemical species. With carefully controlled conditions, it was possible to produce several-nanometer-thick SiO_2 shells (~3–4 nm). The porous shell can well prevent the Pt NPs from agglomerating and enhance the catalytic activity of Pt NPs appreciably [50, 51].

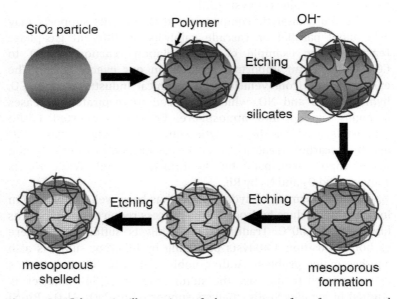

Figure 6.7 Schematic illustration of the concept of surface-protected etching for transforming solid structures into hollow structures with permeable shells.

To improve the distribution of certain catalytic nanoparticles and the catalytic efficiency, the supported catalysis systems with bimetallic or multimetallic catalytic components are also developed for multi-functional catalysts. A second metal may alter the electronic structure of a given supported catalyst, modify or increase the dispersion of active catalyst, and act as a "nanoglue" to improve the interface between the support and the active catalyst. Enache et al. added Au to Pd nanocrystals and formed Au-Pd nanocrystals with a Au-rich core and a Pd-rich shell. The Au can electronically influence the catalytic properties of Pd. After loaded into mesoporous TiO_2 support, the catalyst gives very high turnover frequencies (TOF) (up to 270,000 h^{-1}) for the oxidation of alcohols [52]. Chen et al. used mesoporous silica support SBA-16 for the immobilization of Au-Pd bimetallic nanocrystals. The "super-cage" morphology of SBA-16 allows the controlled formation of dispersed noble metals with different molar ratio in the channels. A higher TOF value can be achieved for bimetallic catalysts system in compared to the monometallic catalysts [53].

The multi-catalytic components in the porous supports may also allow parallel or cascade catalysis of different gas-phase reactions. For example of the oxidation reaction of CO to CO_2, which is an important reaction in our everyday life. The exhaust gases from vehicles and chemical industry contain CO, hydrocarbons and NO, which may lead to respiratory diseases if highly spread into atmosphere. Ceramic supported Pd-Rh catalyst is used for the catalytic conversion of those gases. CO and hydrocarbons react with the excess oxygen to form CO_2, and water vapor, principally by the catalysis of Pd. While NO is converted to O_2 and N_2 by Rh.

Steam reforming of hydrocarbons has been well established in industry to produce hydrogen. This process requires temperatures higher than 500°C. Traditional Ni catalysts suffer from severe carbon deposition. Catalysts poisoning by inherent sulfur is also another major problem. Adding noble metals to Ni catalysts has been reported to increase the sulfur resistance [54]. Song et al. studied bimetallic Ni-Rh catalysts supported in CeO_2-Al_2O_3. Rh-Ni is the active site for hydrocarbon activation, while the water molecule is activated on CeO_x. At high Ni loadings, strong interaction of Rh-Ni can be confirmed. As a sacrificial medium, Ni in the catalyst is demonstrated to be able to protect Rh against deactivating by sulfur poisoning. This supported bimetallic catalyst displayed dramatic improvement in sulfur tolerance, and >95% conversion was maintained for up to 28 h time-on-stream [55].

Design flexibility on the catalytic nanoarchitectures may bring new solution and access to improve the performance of catalysts from three viewpoints:

- The primary size of catalytic nanoparticles can be well stabilized. The size of nanocatalysts is one of the key factors

for the catalytic efficiency. Agglomeration of nanoparticles can be largely minimized in porous supports.
- The bicontinuous networks of the support and loaded catalysts provide a three dimensional molecular transport pathway for the catalytic reactions.
- The bifunctional catalysts can be developed. The supports can be not just a support substrate. They may be involved in the catalytic reaction, or improve the catalytic activity of guest catalysts by suitable interactions.

By the design of catalytic nanoarchitectures, catalytic activity can be fully exploited under less loading amount of noble metals, and multifunctional, long lasting, high-performance catalysts can become feasible.

6.5 Porous Nanoarchitectures for Electrocatalysis

To achieve electrocatalysts with high performance, it always requires multifunctionality on the nanostructures of catalysts, as the electrocatalysts need to possess high surface reactivity, electronic conductivity, ionic conductivity, and facile mass transport of molecules to enhance molecular conversion. In electrochemistry, the concentration of electron reactant scales with the electrified surface area. Therefore, mesoporous nanostructures with large specific area are preferred, which can provide more transport-accessible surfaces for the electrochemical processes. Mesoporous structures were originally obtained by a sol-gel route from inorganic alkoxides in the presence of surfactants. Later, various strategies were developed to successfully produce a range of mesoporous metallic systems [56-58]. Those mesoporous metals have attracted considerable interest due to their huge potentials for electrochemical applications.

Electrocatalyzed reactions range from synthetic (often using a molecular electron-transfer reactant) to destructive (in which pollutants are detoxified or mineralized) to energy conversion (turning fuel and oxidant to electricity, as in fuel cells). Electrocatalysis in fuel cells dominates the current research on electrocatalysis because of the societal interest in generating power without distributed pollution from the by-products of hydrocarbon combustion. High-performance fuel cell electrodes require nanoarchitecturing to establish nanoscopic interphases for reactions, to effectively integrate the catalytic surfaces, electrons, and protons, and to permit efficient molecular transport of gas- or liquid-phase reactants or products.

Pt-based electrocatalysts have been extensively studied and widely used for the application of fuel cells. The development of novel Pt electrocatalysts is also still an important subject of interest. To date, various Pt nanostructures, including nanospheres, nanodendrites, nanofibers/nanowires, nanocages, nanotubes, and nanosheets, have been successfully synthesized [59–66]. However, Pt nanostructures with very uniform mesopores and extremely high surface areas still cannot be easily obtained. Yamauchi et al. proposed a route to mesoporous metal films with various pore sizes by a simple electrodeposition method in an aqueous surfactant solution. Mesoporous Pt films were electrodeposited from an aqueous solution including K_2PtCl_4 and amphiphilic block copolymers (Pluronic F127, Brij 58 or PEO_{4500}-PPO_{3200}-PEO_{4500}). The mesopore sizes can be tuned in a wide range from 5 to 20 nm according to the chain length of surfactants. The mesoporous films are composed of connected Pt nanoparticles (around 3 nm). The Pt atomic crystallinity was shown coherently extending across over several Pt nanoparticles, providing a large number of atomic steps and defect sites, which facilitate the cleavage of the C–H and O–H bonds in methanol decomposition. As a result, the catalytic performance for methanol electro-oxidation is dramatically enhanced, compared to commercially available carbon-supported Pt nanoparticles and unsupported Pt nanoparticles [67].

Because of the high cost of Pt, optimizing both the size and shape of Pt nanostructures is necessary to reduce Pt loading. Recently, bimetallic core–shell Pt nanospheres with mesoporous structures have also been widely studied. In comparison to Pt alone, such core–shell nanostructures not only improve the utilization efficiency of catalysts but also reduce Pt usage in catalysts. In most cases for the preparation of core–shell nanostructures with a dendritic Pt shell, complex procedures are proposed, including at least two steps. High-temperature and high-pressure conditions sometimes are also required. The development of a facile and economic method for the one-step synthesis of core–shell nanostructures with a high Pt surface area is expected. As Pt shell thickness and homogeneous distribution of particles are vital for the structure-sensitive electrocatalytic reactions, the manipulations on the core–shell nanostructures are greatly associated with the catalytic activity and efficiency.

Wang and Yamauchi el al. report a facile synthetic route for bimetallic Au–Pt nanocatalysts with an Au core and a dendritic Pt shell (Fig. 6.8), by utilizing a surfactant-assisted process with an ultrasonic irradiation treatment [68]. Monodispersed Au–Pt nanocolloids with optimized Pt shell thicknesses can be formed, which exhibit enhanced activity as an electrocatalyst for a methanol oxidation reaction. Their strategy is promising for the large-scale production of nanocatalysts, because it is a one-step, inexpensive method for the synthesis of uniform nanoparticles with high Pt surface area.

Catalytic methanol oxidation is a complicated chemical reaction in fuel cells. Methanol molecules are first absorbed on Pt and disassembled as intermediate Pt-(CHO)$_{ads}$, which reacts with Pt-(OH)$_{ads}$ to give CO_2 and H^+. The reaction may suffer due to the slow oxidation kinetics of methanol and Pt catalyst "poisoning" by CO. However, the widely used carbon supports for Pt catalysts cannot prevent from being poisoned by CO. Therefore, the development for new support materials has also attracted great attention to improve catalytic performance of methanol oxidation.

Shi et al. synthesized a carbon-free catalyst, Pt loaded on mesoporous WO_3, by a template-replicating method [69]. The parent mesoporous silica with cubic Ia3d symmetry (designated as KIT-6) was used as a hard template for mesoporous WO_3. Controllable amount of Pt is loaded into the mesoporous WO_3. WO_3 can form a hydrogen tungsten bronzes compound (H_xWO_3) in acidic electrolyte. This tungsten bronzes compound can make the dehydrogenation of methanol molecules be adsorbed on Pt surface more effective, since the spillover of hydrogen onto the surface of hydrogen tungsten bronze can free these Pt sites for further chemisorption of methanol molecules [70]. In addition, the oxophilic nature of the oxide also helps in removing the adsorbed intermediates during the methanol oxidation [71]. This carbon-free mesostructrued Pt/WO_3 composite shows high electrocatalytic activity toward methanol oxidation and good electrochemical stability, and its overall activities are also significantly higher than that of commercial Pt/C catalysts with the same Pt loading amount. Both the assistant catalytic effect and mesoporous structure of WO_3 support play a large role for the enhanced electrocatalytic activity.

Figure 6.8 Bimetallic Au–Pt nanocatalysts with an Au core and a dendritic Pt shell.

6.6 Summary and Perspectives

Due to the intrinsic structural features, such as tunable pore diameter, huge surface area, and good flexibility to accommodate various functional groups and metals onto the surface, an inevitable linkage of porous materials with catalysis has been built up over the past few decades. Since a broad spectrum of techniques and concepts is now available for porous materials with pore dimensions ranging from micro- and meso- to macropores and compositions that include metals, metal oxides, carbons, semiconductors, or metal-organic frameworks, it brings more flexibility to design desired supports for the combination with various catalysts or improving the catalysis performance. Various routes to catalytic nanoarchitectures also offer flexibility in design and the compositional selection on nano-building blocks for catalysts.

New catalysis systems, with binary, ternary, and even quaternary metals and porous catalysts have been developed. We can expect that novel catalytic structures possessing multifunctionalities with optimized catalytic performance could be constructed through nanoarchitecturing processes.

References

1. Ariga, K., Vinu, A., Yamauchi, Y., Ji, Q., and Hill, J. P. (2012). Nanoarchitectonics for Mesoporous Materials, *Bull. Chem. Soc. Jpn.*, **85**, pp. 1–32.
2. Ariga, K., Ji, Q., McShane, M. J., Lvov, Y. M., Vinu, A., and Hill, J. P. (2012). Inorganic Nanoarchitectonics for Biological Applications, *Chem. Mater.*, **24**, pp. 728–737.
3. Ariga, K., Ji, Q., Mori, T., Naito, M., Yamauchi, Y., Abe, H., and Hill, J. P. (2013). Enzyme Nanoarchitectonics: Organization and Device Application, *Chem. Soc. Rev.*, **42**, pp. 6322–6345.
4. Ying, J. Y., Mehnert, C. P., and Wong, M. S. (1999). Synthesis and Applications of Supramolecular-Templated Mesoporous Materials, *Angew. Chem. Int. Ed.*, **38**, pp. 56–77.
5. Wang, Y., Angelatos, A. S., and Caruso, F. (2008). Template Synthesis of Nanostructured Materials via Layer-by-Layer Assembly, *Chem. Mater.*, **20**, pp. 848–858.
6. Huczko, A. (2000). Template-Based Synthesis of Nanomaterials, *Appl. Phys. A*, **70**, pp. 365–376.
7. Malgras, V., Ji, Q., Kamachi, Y., Mori, T., Shieh, F.-K., Wu, K. C.-W., Ariga, K., and Yamauchi, Y. (2015). Templated Synthesis for Nanoarchitectured Porous Materials, *Bull. Chem. Soc. Jpn.*, **88**, pp. 1171–1200.
8. Cai, Q., Luo, Z.-S., Pang, W.-Q., Fan, Y.-W., Chen, X.-H., and Cui, F.-Z. (2001). Dilute Solution Routes to Various Controllable Morphologies of MCM-41 Silica with a Basic Medium, *Chem. Mater.*, **13**, pp. 258–263.
9. Ji, X., Lee, K. T., Monjauze, M., and Nazar, L. F. (2008). Strategic Synthesis of SBA-15 Nanorods, *Chem. Commun.*, pp. 4288–4290.
10. Fan, J., Yu, C., Lei, J., Zhang, Q., Li, T., Tu, B., Zhou, W., and Zhao, D. (2005). Low-Temperature Strategy to Synthesize Highly Ordered Mesoporous Silicas with Very Large Pores, *J. Am. Chem. Soc.*, **127**, pp. 10794–10795.

11. Chen, L., Zhu, G., Zhang, D., Zhao, H., Guo, M., Shi, W., and Qiu, S. (2009). Novel Mesoporous Silica Spheres with Ultra-Large Pore Sizes and Their Application in Protein Separation, *J. Mater. Chem.*, **19**, pp. 2013–2017.
12. Saramat, A., Andersson, M., Hant, S., Thormählen, P., Skoglundh, M., Attard, G. S., and Palmqvist, A. E. C. (2007). Differences in Catalytic Properties between Mesoporous and Nanoparticulate Platinum, *Eur. Phys. J. D*, **43**, pp. 209–211.
13. Nelson, P. A., Elliott, J. M., Attard, G. S., and Owen, J. R. (2002). Mesoporous Nickel/Nickel Oxidea Nanoarchitectured Electrode, *Chem. Mater.*, **14**, pp. 524–529.
14. Bartlett, P. N., Gollas, B., Guerin, S., and Marwan, J. (2002). The Preparation and Characterisation of H1-e Palladium Films with a Regular Hexagonal Nanostructure Formed by Electrochemical Deposition from Lyotropic Liquid Crystalline Phases, *Phys. Chem. Chem. Phys.*, **4**, pp. 3835–3842.
15. Liu, Z., Li, M., Pu, F., Ren, J., Yang, X., and Qu, X. (2004). Fabrication of Magnetic Mesostructured Nickel–Cobalt Alloys from Lyotropic Liquid Crystalline Media by Electroless Deposition, *J. Mater. Chem.*, **14**, pp. 2935–2942.
16. Yamauchi, Y., Momma, T., Yokoshima, T., Kuroda K., and Osaka, T. (2005). Highly Ordered Mesostructured Ni Particles Prepared from Lyotropic Liquid Crystals by Electroless Deposition: the Effect of Reducing Agents on the Ordering of Mesostructured, *J. Mater. Chem.*, **15**, pp. 1987–1994.
17. Yamauchi, Y., Momma, T., Fuziwara, M., Nair, S. S., Ohsuna, T., Terasaki, O., Osaka, T., and Kuroda, K. (2005). Unique Microstructure of Mesoporous Pt (HI-Pt) Prepared via Direct Physical Casting in Lyotropic Liquid Crystalline Media, *Chem. Mater.*, **17**, pp. 6342–6348.
18. Nara, H., Fukuhara, Y., Takai, A., Komatsu, M., Mukaibo, H., Yamauchi, Y., Momma, T., Kuroda, K., and Osaka, T. (2008). Cycle and Rate Properties of Mesoporous Tin Anode for Lithium Ion Secondary Batteries, *Chem. Lett.*, **37**, pp. 142–143.
19. Takai, A., Yamauchi, Y., and Kuroda, K. (2010). Tailored Electrochemical Synthesis of 2D-Hexagonal, Lamellar, and Cage-Type Mesostructured Pt Thin Films with Extralarge Periodicity, *J. Am. Chem. Soc.*, **132**, pp. 208–214.
20. Yamauchi, Y., Sugiyama, A., Morimoto, R., Takai, A., and Kuroda, K. (2008). Mesoporous Platinum with Giant Mesocages Templated

from Lyotropic Liquid Crystals Consisting of Diblock Copolymers, *Angew. Chem. Int. Ed.*, **47**, pp. 5371–5373.

21. Pan, A., Wu, H. B., Yu, L., and Lou, X. W. (2013). Template-Free Synthesis of VO$_2$ Hollow Microspheres with Various Interiors and Their Conversion into V$_2$O$_5$ for Lithium-Ion Batteries, *Angew. Chem. Int. Ed.*, **52**, pp. 2226–2230.

22. Zhang, Q., Wang, W., Goebl, J., and Yin, Y. (2009). Self-Templated Synthesis of Hollow Nanostructures, *Nano Today*, **4**, pp. 494–507.

23. Ji, Q., Guo, C., Yu, X., Ochs, C. J., Hill, J. P., Caruso, F., Nakazawa, H., and Ariga, K. (2012). Flake-Shell Capsules: Adjustable Inorganic Structures, *Small*, **8**, pp. 2345–2349.

24. Ji, Q., Ishihara, S., Terentyeva, T. G., Deguchi, K., Ohki, S., Tansho, M., Shimizu, T., Hill, J. P., and Ariga, K. (2015). Manipulation of Shell Morphology of Silicate Spheres from Structural Evolution in a Purely Inorganic System, *Chem. Asian J.*, **10**, pp. 1379–1386.

25. Suteewong, T., Sai, H., Cohen, R., Wang, S., Bradbury, M., Baird, B., Gruner, S. M., and Wiesner, U. (2011). Highly Aminated Mesoporous Silica Nanoparticles with Cubic Pore Structure, *J. Am. Chem. Soc.*, **133**, pp. 172–175.

26. Lei, C., Shin, Y., Magnuson, J. K., Fryxell, G. E., Lasure, L. L., Elliott, D. C., Liu, J., and Ackerman, E. J. (2006). Characterization of Functionalized Nanoporous Supports for Protein Confinement, *Nanotechnology*, **17**, pp. 5531–5538.

27. Torres-Salas, P., del Monte-Martinez, A., Cutiño-Avila, B., Rodriguez-Colinas, B., Alcalde, M., Ballesteros, A. O., and Plou, F. J. (2011). Immobilized Biocatalysts: Novel Approaches and Tools for Binding Enzymes to Supports, *Adv. Mater.*, **23**, pp. 5275–5282.

28. Fujita, T., Guan, P., Mckenna, K. P., Lang, X., Hirata, A., Zhang, L., Tokunaga, T., Arai, S., Yamamoto, Y., Tanaka, N., Ishikawa, Y., Asao, N., Yamamoto, Y., Erlebacher, J., and Chen M. (2012). Atomic Origins of the High Catalytic Activity of Nanoporous Gold, *Nat. Mater.*, **11**, pp. 775–780.

29. Fujita, T., Abe, H., Tanabe, T., Ito, Y., Tokunaga, T., Arai, S., Yamamoto, Y., Hirata, A., and Chen, M. (2016). Earth-Abundant and Durable Nanoporous Catalyst for Exhaust-Gas Conversion, *Adv. Funct. Mater.*, **26**, pp. 1609–1616.

30. Pozogonzalo, C., Kartachova, O., Torriero, A. A. J., Howlett, P. C., Glushenkov, A. M., Fabijanic, D., Chen, Y., Poissonnet, S., and Forsyth, M. (2013). Nanoporous Transition Metal Oxynitrides as Catalysts

for the Oxygen Reduction Reaction, *Electrochim. Acta*, **103**, pp. 151–160.

31. Zhang, J., and Li, C. M. (2012). Nanoporous Metals: Fabrication Strategies and Advanced Electrochemical Applications in Catalysis, Sensing and Energy Systems, *Chem. Soc. Rev.*, **41**, pp. 7016–7031.

32. Kresge, C. T., Leonowicz, M. E., Roth, W. J., Vartuli, J. C., and Beck, J. S. (1992). Ordered Mesoporous Molecular Sieves Synthesized by a Liquid-Crystal Template Mechanism, *Nature*, **359**, pp. 710–712.

33. Wang, Y., and Caruso, F. (2005). Mesoporous Silica Spheres as Supports for Enzyme Immobilization and Encapsulation, *Chem. Mater.*, **17**, pp. 953–961.

34. Lei, C., Soares, T. A., Shin, Y., Liu, J., and Ackerman, E. J. (2008). Enzyme Specific Activity in Functionalized Nanoporous Supports, *Nanotechnology*, **19**, p. 125102.

35. Hartmann, M., and Jung, D. (2010). Biocatalysis with Enzymes Immobilized on Mesoporous Hosts: the Status Quo and Future Trends, *J. Mater. Chem.*, **20**, pp. 844–857.

36. Zucca, P., and Sanjust, E. (2014). Inorganic Materials as Supports for Covalent Enzyme Immobilization: Methods and Mechanisms, *Molecules*, **19**, pp. 14139–14194.

37. Zhou, Z., and Hartmann, M. (2012). Recent Progress in Biocatalysis with Enzyme Immobilized on Mesoporous Hosts, *Top. Catal.*, **55**, pp. 1081–1100.

38. Cao, L., and Schmid, R. D. (2005). *Carrier-Bound Immobilized Enzymes: Principles, Application and Design*, Wiley-VCH Verlag GmbH & Co., Weinheim, Germany.

39. Chong, A. S. M., and Zhao, X. S. (2004). Design of Large-Pore Mesoporous Materials for Immobilization of Penicillin G Acylase Biocatalyst, *Catal. Today*, **93–95**, pp. 293–299.

40. Lin, Y., Li, Z., Chen, Z., Ren, J., and Qu, X. (2013). Mesoporous Silica-Encapsulated Gold Nanoparticles as Artificial Enzymes for Self-Activated Cascade Catalysis, *Biomaterials*, **34**, pp. 2600–2610.

41. Li, X., Zhang, Z., and Li, Y. (2014). Artificial Enzyme Mimics for Catalysis and Double Natural Enzyme Co-immobilization, *Appl. Biochem. Biotechnol.*, **172**, pp. 1859–1865.

42. Bai, Y.-X., Li, Y.-F., Yang, Y., and Yi, L.-X. (2006). Covalent Immobilization of Triacylglycerol Lipase onto Functionalized Novel Mesoporous Silica Supports, *J. Biotechnol.*, **125**, pp. 574–582.

43. Schlossbauer, A., Schaffert, D., Kecht, J., Wagner, E., and Bein, T. (2008). Click Chemistry for High-Density Biofunctionalization of Mesoporous Silica, *J. Am. Chem. Soc.*, **130**, pp. 12558–12559.
44. Deere, J., Magner, E., Wall, J. G., and Hodnett, B. K. (2002). Mechanistic and Structural Features of Protein Adsorption onto Mesoporous Silicates, *J. Phys. Chem. B*, **106**, pp. 7340–7347.
45. Kisler, J. M., Dähler, A., Stevens, G. W., and O'Connor A. J. (2001). Separation of Biological Molecules Using Mesoporous Molecular Sieves, *Micropor. Mesopor. Mater.*, **44–45**, pp. 769–774.
46. Hudson, S., Magner, E., Cooney, J., and Hodnett, B. K. (2005). Methodology for the Immobilization of Enzymes onto Mesoporous Materials, *J. Phys. Chem. B*, **109**, pp. 19496–19506.
47. Liu, J., Ji, Q., Imai, T., Ariga, K., and Abe, H. (2017). Sintering-Resistant Nanoparticles in Wide-Mouthed Compartments for Sustained Catalytic Performance, *Sci. Rep.*, **7**, p. 41773.
48. Budroni, G., and Corma, A. (2006). Gold–Organic–Inorganic High-Surface-Area Materials as Precursors of Highly Active Catalysts, *Angew. Chem. Int. Ed.*, **45**, pp. 3328–3331.
49. Zhang, Q., Lee, I., Ge, J., Zaera, F., and Yin, Y. (2010). Surface-Protected Etching of Mesoporous Oxide Shells for the Stabilization of Metal Nanocatalysts, *Adv. Funct. Mater.*, **20**, pp. 2201–2214.
50. Lee, I., and Zaera, F. (2005). Selectivity in Platinum-Catalyzed cis-trans Carbon–Carbon Double-Bond Isomerization, *J. Am. Chem. Soc.*, **127**, pp. 12174–12175.
51. Lee, I., and Zaera, F. (2010). Catalytic Conversion of Olefins on Supported Cubic Platinum Nanoparticles: Selectivity of (100) Versus (111) Surfaces, *J. Catal.*, **269**, pp. 359–366.
52. Enache, D. I., Edwards, J. K., Landon, P., Solsona-Espriu, B., Carley, A. F., Herzing, A. A., Watanabe, M., Kiely, C. J., Knight, D. W., and Hutchings, G. J. (2006). Solvent-Free Oxidation of Primary Alcohols to Aldehydes Using Au-Pd/TiO$_2$ Catalysts, *Science*, **311**, pp. 362–365.
53. Chen, Y., Lim, H., Tang, Q., Gao, Y., Sun, T., Yan, Q., and Yang Y. (2010). Solvent-Free Aerobic Oxidation of Benzyl Alcohol over Pd Monometallic and Au–Pd Bimetallic Catalysts Supported on SBA-16 Mesoporous Molecular Sieves, *Appl. Catal. A*, **380**, pp. 55–65.
54. Hulteberg, C. (2012). Sulphur-Tolerant Catalysts in Small-Scale Hydrogen Production, *Int. J. Hydrogen Energy*, **37**, pp. 3978–3992.
55. Strohm, J. J., Zheng, J., and Song, C. S. (2006). Low-Temperature Steam Reforming of Jet Fuel in the Absence and Presence of Sulfur over Rh and Rh–Ni Catalysts for Fuel Cells, *J. Catal.*, **238**, pp. 309–320.

56. Guo, L., Cui, X., Li, Y., He, Q., Zhang, L., Bu, W., and Shi, J. (2009). Hollow Mesoporous Carbon Spheres with Magnetic Cores and Their Performance as Separable Bilirubin Adsorbents, *Chem. Asian J.*, **4**, pp. 1480–1485.
57. Corma, A. (2016). Porous Catalysts: Separate to Accumulate, *Nat. Mater.*, **15**, pp. 134–136.
58. Malgras, V., Ataee-Esfahani, H., Wang, H., Jiang, B., Li, C., Wu, K. C. W., Kim, J. H., and Yamauchi, Y. (2016). Nanoarchitectures for Mesoporous Metals, *Adv. Mater.*, **28**, pp. 993–1010.
59. Song, Y. J., Yang, Y., Medforth, C. J., Pereira, E., Singh, A. K., Xu, H. F., Jiang, Y. B., Brinker, C. J., van Swol, F., and Shelnutt, J. A. (2004). Controlled Synthesis of 2-D and 3-D Dendritic Platinum Nanostructures, *J. Am. Chem. Soc.*, **126**, pp. 635–645.
60. Wang, L., and Yamauchi, Y. (2009). Facile Synthesis of Three-Dimensional Dendritic Platinum Nanoelectrocatalyst, *Chem. Mater.*, **21**, pp. 3562–3569.
61. Wang, L., Wang, H. J., Nemoto, Y., and Yamauchi, Y. (2010). Rapid and Efficient Synthesis of Platinum Nanodendrites with High Surface Area by Chemical Reduction with Formic Acid, *Chem. Mater.*, **22**, pp. 2835–2841.
62. Yamauchi, Y., Takai, A., Nagaura, T., Inoue, S., and Kuroda, K. (2008). Pt Fibers with Stacked Donut-Like Mesospace by Assembling Pt Nanoparticles: Guided Deposition in Physically Confined Self-Assembly of Surfactants, *J. Am. Chem. Soc.*, **130**, pp. 5426–5427.
63. Takai, A., Yamauchi, Y., and Kuroda, K. (2009). Facile Formation of Single Crystalline Pt Nanowires on a Substrate Utilising Lyotropic Liquid Crystals Consisting of Cationic Surfactants, *J. Mater. Chem.*, **19**, pp. 4205–4210.
64. Wattanakit, C., Come, Y. B. S., Lapeyre, V., Bopp, P. A., Heim, M., Yadnum, S., Nokbin, S., Warakulwit, C., Limtrakul, J., and Kuhn, A. (2014). Enantioselective Recognition at Mesoporous Chiral Metal Surfaces, *Nat. Commun.*, **5**, p. 3325.
65. Song, Y. J., Garcia, R. M., Dorin, R. M., Wang, H. R., Qiu, Y., and Shelnutt, J. A. (2006). Synthesis of Platinum Nanocages by Using Liposomes Containing Photocatalyst Molecules, *Angew. Chem. Int. Ed.*, **45**, pp. 8126–8130.
66. Song, Y. J., Steen, W. A., Peña, D., Jiang, Y. B., Medforth, C. J., Huo, Q. S., Pincus, J. L., Qiu, Y., Sasaki, D. Y., Miller, J. E., and Shelnutt, J. A. (2006). Foamlike Nanostructures Created from Dendritic Platinum Sheets on Liposomes, *Chem. Mater.*, **18**, pp. 2335–2346.

67. Wang, H., Wang, L., Sato, T., Sakamoto, Y., Tominaka, S., Miyasaka, K., Miyamoto, N., Nemoto, Y., Terasaki, O., and Yamauchi, Y. (2012). Synthesis of Mesoporous Pt Films with Tunable Pore Sizes from Aqueous Surfactant Solutions, *Chem. Mater.*, **24**, pp. 1591–1598.

68. Ataee-Esfahani, H., Wang, L., Nemoto, Y., and Yamauchi, Y. (2010). Synthesis of Bimetallic Au@Pt Nanoparticles with Au Core and Nanostructured Pt Shell toward Highly Active Electrocatalysts, *Chem. Mater.*, **22**, pp. 6310–6318.

69. Cui, X., Shi, J., Chen, H., Zhang, L., Guo, L., Gao J., and Li, J. (2008). Platinum/Mesoporous WO_3 as a Carbon-Free Electrocatalyst with Enhanced Electrochemical Activity for Methanol Oxidation, *J. Phys. Chem. B*, **112**, pp. 12024–12031.

70. Hobbs, B. S., and Tseung, A. C. C. (1969). High Performance, Platinum Activated Tungsten Oxide Fuel Cell Electrodes, *Nature*, **222**, pp. 556–558.

71. Tseung, A. C. C., and Chen, K. Y. (1997). Hydrogen Spill-Over Effect on Pt/WO_3 Anode Catalysts, *Catal. Today*, **38**, pp. 439–443.

Chapter 7

Catalytic Reactions on Solid Surfaces

Huihui Kong,[a] Xinbang Liu,[a] and Harald Fuchs[a,b]

[a]*Herbert Gleiter Institute of Nanoscience,*
Nanjing University of Science and Technology,
Nanjing 210094, P. R. China
[b]*Center for Nanotechnology, University of Münster,*
D-48149 Münster, Germany

konghuihui@njust.edu.cn

7.1 Introduction

Recently, on-surface chemical reaction has been gradually becoming a rapidly blooming research field due to its potential applications on nanoscience and nanotechnology. Solid surfaces, especially metal surfaces, have been found to be capable of serving as catalysts and facilitate the synthesis of specific covalent nanostructures, even some polymers which are rather difficult to be achieved by conventional organic synthesis processes. Moreover, confinement of chemical reaction into two dimensions gives access to entirely new reaction pathways, such as stabilizing specific conformers. For such studies, scanning tunneling microscopy (STM) has proven to be the technique of choice since it allows direct characterization of the morphologies

---

Soft Matters for Catalysts
Edited by Qingmin Ji and Harald Fuchs
Copyright © 2020 Jenny Stanford Publishing Pte. Ltd.
ISBN 978-981-4774-66-6 (Hardcover), 978-1-351-27284-1 (eBook)
www.jennystanford.com

of precursor molecules, intermediates, and products with submolecular resolution in real space.

On-surface chemical reactions of organic molecules provide possibilities for in situ fabrication of robust molecular nanostructures with specific patterns and various dimensions through forming covalent bonds among organic molecules. Such nanostructures bear well-controlled properties, such as higher thermal stability and superior electron transport, and are considered to be potential candidates for the potential applications of the construction of functional devices.

In view of the importance and advantages of catalytic synthesis by metals, various types of organic reactions on metal surfaces have been widely studied based on STM techniques recently. Those reactions are including the common Ullmann-type reaction [1], Glaser reaction [2], condensation reaction of boronic acids [3], alkane polymerization [4], Bergman cyclization [5], and click reaction [6]. By manipulation on the reaction pathways of organic molecules on surfaces, functional nanoarchitectures such as one-dimensional nanowires, nanoribbons and two-dimensional networks can be precisely fabricated.

In this review, we present several important examples of on-surface catalytic reactions and summarize recent developments from the following aspects: (1) dehalogenative coupling reactions of organic molecules assisted by metal surfaces; (2) dehydrogenative coupling reactions of organic molecule assisted by metal surfaces.

7.2 Dehalogenative Coupling Reactions Assisted by Metal Surfaces

Among a large number of surface-assisted reactions, the surface-assisted dehalogenative coupling reaction has been most frequently used for constructing different robust covalent-bonded nanostructures and more importantly the fabrication of graphene nanoribbons with specific widths. In this section, we will mainly introduce the Ullmann-type reaction, which was discovered by Fritz Ullmann in 1901. Fritz Ullmann found that biphenyl compounds could be synthesized by heating halogenated aromatic compounds with copper powder, which is regarded as catalyst [7]. In 1992, the Ullmann reaction was demonstrated for the first time on the Cu(111) surface [8]. After that in 2000,

Hla et al. achieved Ullmann reactions of iodobenzene on Cu(111) by STM manipulation [9]. Since then, the surface-assisted Ullmann type reaction was extensively studied and is in most cases performed on copper surfaces, such as Cu(111) and Cu(110), which act as catalysts during the reactions. In general, the surface reactions proceed via a two-step process. Before the formation of the final products, organometallic intermediates are observed which verifies that the copper atoms play an important role in activating the on-surface Ullmann reactions. For example, Wang et al. have reported a C–Cu–C organometallic intermediates formed in the Ullmann coupling reaction of Br-(ph)^3-Br on Cu(111) when annealing the sample at 300 K (Fig. 7.1a–d). After further annealing at 473 K, it resulted in the final formation of polymeric chains [10] as shown in Fig. 7.1e,f. Similar C–Ag–C organometallic intermediate was also observed on Ag(111) surface [11]. It should be noted that the organometallic intermediate is not obligatory for Ullmann type reactions. In 2007, Grill et al. have reported the covalent bonded nanostructures on Au(111) by employing Ullmann type reactions where no organometallic intermediates are found [12].

In recent years, a lot of covalent-bonded nanostructures with controlled shapes and dimensions have been fabricated based on the surface-assisted Ullmann-type reactions. By respective deposition of porphyrins with different numbers of bromine substituents onto Au(111) and followed by subsequent annealing at 400 K to induce the Ullmann reactions, Grill et al. have controllably fabricated covalent bonded 0D dimers, 1D chains, and 2D networks of porphyrins [12]. By carefully choosing the molecular precursors, the graphene nanoribbons with different widths and edges also have been fabricated as shown in Fig. 7.2b. In 2010, Cai et al. for the first time fabricated seven-armchair graphene nanoribbons (AGNR) on Au(111) by employing on-surface Ullmann type reactions, as shown in the left panel of Fig. 7.2b [13]. In 2013, Chen et al. utilized the Ullmann reactions to fabricate 13-AGNR on Au(111) as shown in the middle panel of Fig. 7.2b, and later they seamlessly connected the 7-AGNR and 13-AGNR together by the Ullmann reaction, which finally results in the synthesis of 7–13 GNR heterojunctions as shown in the right panel of Fig. 7.2b. All the above examples exhibit the potential applications of on-surface Ullmann reactions on the construction of covalent-bonded nanostructures with special shapes and dimensions.

Besides the well-known Ullmann coupling of aryl halides which is related to C–C coupling after dehalogenation, other kinds of dehalogenative coupling are also studied and achieved on surfaces, like dehalogenated coupling of alkyl and alkynyl halides. Dehalogenative coupling of alkyl halides has been studied on several different metal surfaces including Cu(110), Ag(110) and Au(111) by Sun et al., as shown in Fig. 7.3. Such reactions have been achieved on all these surfaces by employing BMBP as a precursor molecule [14]. Interestingly, it was found that the debromination process of such reactions was more facile to occur on the metal surfaces, but no metal-mediated intermediate was observed. The temperature for triggering the C–C coupling after debromination was shown in the order of Cu(110)>Ag(110)>Au(111), which is distinctly different from that of Ullmann reaction.

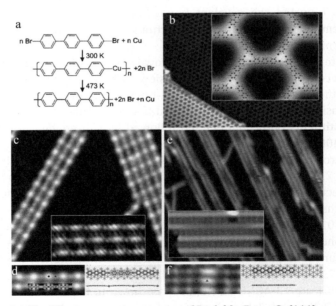

Figure 7.1 The Ullmann reaction process of Br-(ph)$_3$-Br on Cu(111) surface. (a) The schematic illustration of the Ullmann reactions. (b) Self-assembly nanostructure of Br-(ph)$_3$-Br molecules on Cu(111) after deposition. (c, d) Organometallic intermediates after annealing at 300 K. (e, f) Covalent bonded polymeric chains after annealing at 473 K [10].

Dehalogenative Coupling Reactions Assisted by Metal Surfaces | 323

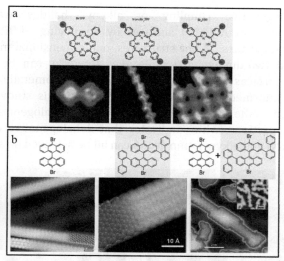

Figure 7.2 (a) Nanostructures with different dimensions formed through the Ullmann type reactions of substituted porphyrins on Au(111) [12]. (b) 7-AGNR, 13-AGNG and 7–13 heterojunction formed through Ullmann coupling of different molecular precursors [13].

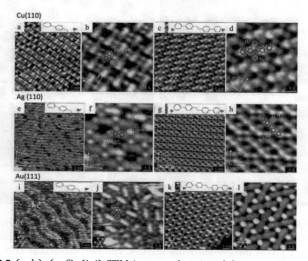

Figure 7.3 (a–b), (e–f), (i–j) STM images showing debromination of BMBP molecules after deposition on Cu(110), Ag(110) and Au(111) below room temperature. (c–d), (g–h), (k–l) STM images showing homocoupling of BMBP* molecules after separately annealing the samples to ~450, 420 and 350 K. A scaled model of a BMBP dimer is overlaid on the STM images in b, f, j, and detached Br atoms are indicated by dashed circles.

Dehalogenative coupling of alkynyl halides was also demonstrated on Au(111) surfaces by employing BEBP, bBEBP, and tBEP molecules. Dimer structures, one-dimensional molecular wires and two-dimensional molecular networks can be formed on the surfaces [15], where C–Au–C organometallic motifs serve as intermediates as shown in Fig. 7.4. This study further supplements the database of on-surface dehalogenative C–C coupling reactions, which indicates that dehalogenative coupling of alkyl, aryl and alkynylation group can all be achieved on surface.

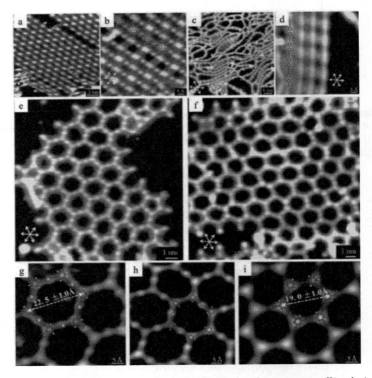

Figure 7.4 (a, b) STM images showing the C–Au–C organometallic chains after deposition of bBEBP molecules on Au(111) with post annealing at 320 K. (c, d) STM images showing the C–C coupled molecular chains after annealing at 425 K. (e) STM image showing the C–Au–C organometallic network after deposition of tBEP molecules on Au(111) with post annealing at 320 K. (f) STM image showing molecular networks after annealing at 450 K. (g–i) Close-up STM images of the C–Au–C organometallic network, the mixture of C–Au–C organometallic and C–C coupled motifs, the C–C coupled network, respectively [15].

7.3 Dehydrogenative Coupling Reactions Assisted by Metal Surfaces

Besides surface-assisted dehalogenative coupling, on-surface C–C coupling after dehydrogenation has also attracted wide attentions in recent years. In such surface-assisted reaction, the reactants should be hydrocarbons, and reactants covalently bind together by cleaving the C–H bonds and then generating a new C–C bond with the assistance of surfaces. Similar to dehalogenative coupling, dehydrogenative coupling of diverse kinds of hydrocarbon molecules has been studied on surfaces, in which zero-dimensional fullerenes and its derivatives, one-dimensional carbon wires and two-dimensional networks were successfully fabricated.

C–H bond activation is essential in many synthetic reactions, which has been widely applied for the synthesis of natural products, functional molecules and polymers. However, most of the reaction processes are poor at controlling selective reaction sites or are not energetically economical due to the high processing temperatures. In 2011, Zhong et al. reported the highly selective carbon-hydrogen (C–H) activation and subsequent dehydrogenative C–C coupling reaction of linear long-chain (>C20) alkanes on Au(110) [16]. Such reaction takes place exclusively at specific sites (terminal CH_3 or penultimate CH_2 groups) in the chains at intermediate temperatures (420 to 470 K). It indicates that Au(110) is an efficient catalyst to facilitate dehydrogenative coupling of alkanes and selectively fabricate specific longer alkyl chains without branches, as shown in Fig. 7.5a–d. Later, Cai et al. found that this reaction can also be achieved on Cu(110) with poor selection of reaction sites by [17]. Their results indicated that dehydrogenative coupling are preferentially occurred on alkyl groups compared to aryl groups when employing precursor DBP molecules which contain both alkyl and aryl groups as shown in Fig. 7.5e–n.

The aryl group is one important species for many conjugated molecules, which may provide potential application in nanodevices. In 2008, Otero et al. reported the first fabrication of triazafullerene $C_{57}N_3$ from aromatic precursors by employing a highly activated Pt(111) surface [18], which efficiently catalyzes the cyclodehydrogenation process of aryl groups, as shown in Fig. 7.6a–e. Later, Amsharov et al. further accomplished the fabrication of C_{60} fullerene on Pt(111) surface by deposition of precursor $C_{60}H_{30}$ and annealing at 480 K [19], as shown in Fig. 7.6f–h. Recently, Rogers et al. reported the preparation of peripentacene by deposition of 6,6′-bipentacene precursors on the Au(111) surface followed with post annealing to trigger the cyclodehydrogenation reactions [20] (Fig. 7.7). This result successfully showed a special synthesis process, which has not been realized by previous rational synthesis.

All these molecules shown above are fabricated by surface-assisted intramolecular cyclodehydrogenated C–C coupling of aryl group. In addition to these zero-dimensional organic materials, one-dimensional carbon chains and two-dimensional networks were also achieved by aryl-aryl coupling after dehydrogenation. By deposition of quaterphenyl (4Ph) molecules on Cu(110) with post annealing Sun et al. for the first time demonstrated

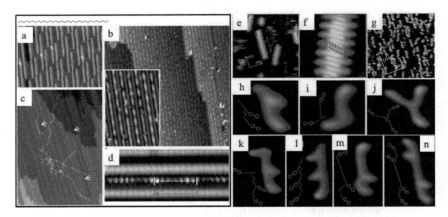

Figure 7.5 (a) STM image showing self-assembly of $C_{32}H_{66}$ on Au(110). (b) Parallel polyethylene chains formed in the Au(110) reconstruction grooves by heating at 440 K for 30 min. (c) Several chains are partially released from the grooves by the STM tip without breaking the chains, indicating the covalent properties of the polyethylene chains. (d) High-resolution STM image of gold atomic rows and polyethylene chains [16]. (e–f) STM images of self-assembly of DBP on Cu(110). (g–n) STM image showing the formation of various coupling products on Cu(110) after annealing the sample at 430 K and the corresponding models [17].

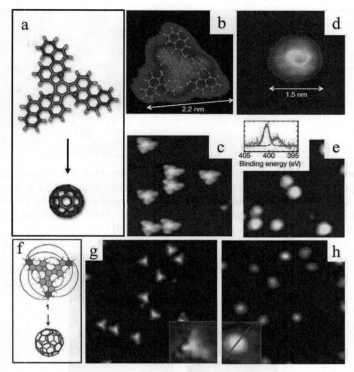

Figure 7.6 (a) The structure of precursor molecules $C_{57}H_{33}N_3$ (upper part) and its product molecule $C_{57}N_3$ triazafullerene (lower part). Blue balls represent nitrogen atoms. (b, c) STM images of the $C_{57}H_{33}N_3$ molecule deposited on Pt(111), the calculated structure is superimposed on the STM image in (b). (d, e) STM images after annealing the surface at 750 K. The appearance of different molecular orbitals and changes in the shape and size of the imaged molecule indicates that cyclodehydrogenation process occurred [18]. (f) Structures of the precursor molecule $C_{60}H_{30}$ and its products C_{60} fullerene. (g) STM image after deposition of $C_{60}H_{30}$ on Pt(111); (h) STM image of C_{60} fullerene after annealing the sample at 480 K [19].

unprecedented selective aryl–aryl coupling via direct C–H bond activation at the meta-C sites between the 4Ph molecules and further achieved the fabrications of one-dimensional carbon chains as shown in Fig. 7.8 [21]. Later on, aryl-aryl coupling is also achieved by selecting more complex precursor molecules, like CoPc and porphine molecules, as shown in Fig. 7.9 [22, 23].

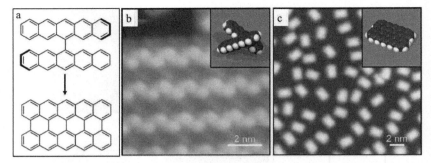

Figure 7.7 (a) The chemical structure of precursor 6,6′-bipentacene (upper part) and the corresponding product peripentacene (lower part). (b) STM image after deposition of 6,6′-bipentacene on Au(111) at room temperature. (c) STM image of peripentacene molecules on Au(111) after annealing the sample at 473 K for 30 min [20].

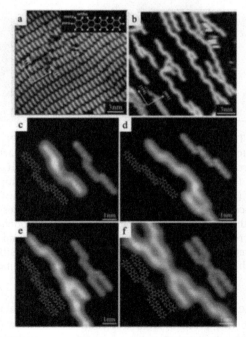

Figure 7.8 (a) STM image of self-assembly of 4Ph molecules on Cu(110) at a high molecular coverage. (b) STM image showing the chain-like structures after annealing the 4Ph-covered surface at 500 K. (c–f) The close-up STM images, the equally-scaled optimized structural models and the DFT-based STM simulations (the black and white ones) of the typical structural motifs within the chain structures [21].

Dehydrogenative Coupling Reactions Assisted by Metal Surfaces | 329

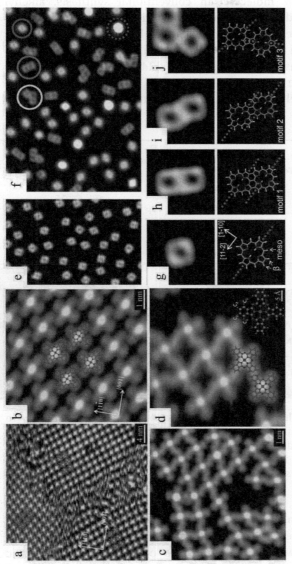

Figure 7.9 (a, b) Large-scale and close-up STM images of self-assembly of CoPc molecules on Ag(110). (c, d) STM images of the 2D polymer structure formed by annealing the CoPc covered surface at 680 K [22]. (e) STM image of individual 2H-P on Ag(111) after deposition at 345 K. (f) Overview image of distinct species (monomer (blue), dimer (green), and trimer (yellow)) after annealing the multilayer of 2H-P on Ag(111) at 573 K. The bright protrusions (dashed circle) are attributed to porphines interacting with Ag adatoms. (g) Monomer of 2H-P with structural model below; the first symmetry axis is indicated by red dotted line. (h–j) Three different binding motifs with corresponding structural models and symmetry axes below. C, cyan; N, blue; H, white [23].

Besides aryl-aryl coupling, the coupling of alkenes on metal surfaces can also be carried out by employing 4-vinyl-1,1′-biphenyl (VBP) as the precursor molecule. In conventional synthesis, alkenyl groups tend to dimerize into different kinds of butene moieties which is hardly separated from each other. However, VBP molecules tend to generate diene compounds on Cu(110) with a quite high yield (> 80%) as shown in Fig. 7.10 [24]. These novel findings exhibit the huge potential of on-surface chemistry on selective synthesis of novel compounds with high purities.

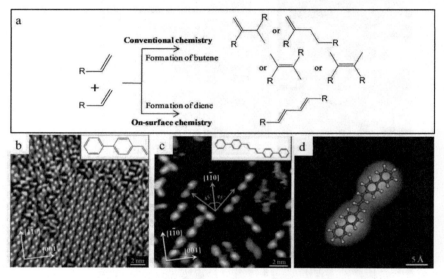

Figure 7.10 (a) The scheme of dimerization of alkene molecules. In conventional chemistry, four kinds of possible butene structures are formed (top). On Cu(110), a diene structure are predominantly synthesized (bottom). (b) STM image of self-assembly of VBP molecules on Cu(110). (c–d) STM images showing the dimerization of VBP molecules into diene molecules [24].

In recently years, C–H activation of terminal alkynyl group is extensively studied due to its potential application on fabrication of graphdiyne, which is predicted to possess excellent electric properties. C–H activation of terminal alkynyl and further C–C coupling has been studied on several surfaces, such as Ag(111), Au(111), and Cu(111). In 2012, Zhang et al. reported the first example of surface-confined dehydrogenative coupling of terminal alkynes on Ag(111) [25]. Through annealing 1,3,5 triethynylbenzene (TEB) covered on Ag(111) to 330 K, the formation of dimers was obtained by surface-assisted Glaser coupling (Fig. 7.11a–c). Similarly, larger molecules 1,3,5-tris(4-ethynylphenyl)benzene (Ext-TEB) were coupled to form 2D conjugated polymers by hierarchic reaction pathways as shown in Fig. 7.11d–f. Later on, by controlled selection of precursor molecules and substrate, the fabrication of one-dimensional graphdiyne wires was also succeeded on Ag(877) [26] as shown in Fig. 7.11g–i. In 2013, Gao et al. showed that the Glaser coupling of the alkyne molecule can be more efficiently achieved on Ag(111) than on Au(111) and Cu(111). The linear oligomer/polymer chains were shown to be generated in spatial confinement predominantly at Ag(111) surfaces [27], which is impossible in conventional solution synthesis. Interestingly, apart from the desired Glaser coupling, other coupling modes, for example, hydroalkynylation of the terminal alkyne, formation of dienyne products and alkyne trimerization, can also be occurred especially on the Au(111) surface, which indicates the diverse possibilities for coupling modes of terminal alkyne via C–H activation despite lack of selection. The different products on different surfaces clearly demonstrate that the C–C coupling via C–H activation of terminal alkyne is surface dependant. To further verify this speculation, Gao et al. further compared the surface-confined C–C coupling of terminal alkyne on Au(111), Ag(111), and Cu(111) by employing 1,4-diethynyl-benzene as a precursor molecule [28]. After detailed analysis based on both STM data and DFT calculations, they proved that silver surface is more efficient than gold or copper surface for Glaser-type coupling (Fig. 7.11j–l). Subsequently, Liu et al. achieved the cyclotrimerization of arylalkynes with high selectivity under the assistance of Au(111) [29], and the fabrication of two-dimensional covalently bonded polyphenylene nanostructures as shown in Fig. 7.11m–o.

Another possibility for C–H activation of terminal alkyne is forming organometallic covalent bonds as demonstrated by Liu et al. in 2015 [30]. Surface reactions of 2,5-diethynyl-1,4-bis(phenylethynyl)-benzene on Ag(111), Ag(110), and Ag(100) were systematically explored by using STM. On Ag(111), Glaser coupling reaction was dominant and 1D molecular wires are

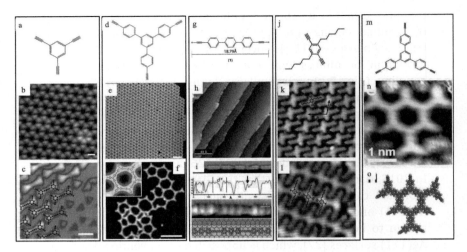

Figure 7.11 (a) Chemical structure of 1,3,5-triethynylbenzene (TEB); (b) STM image of self-assembly of TEB molecule on Ag(111); (c) STM image of an ordered dimer patch after annealing TEB-covered Ag(111) at 330 K by Glaser coupling. (d) The chemical structure of Ext-TEB. (e) STM image showing self-assembly of Ext-TEB on Ag(111); (f) STM image of covalent network patches through annealing the Ext-TEB-overed sample to 400 K which is also formed by Glaser coupling [25]. (g) Chemical structure of 4,4″-Diethynyl-1,1′,4′,1″terphenyl; (h) After annealing the 4,4″-Diethynyl-1,1′,4′,1″terphenyl-covered sample to 450 K a terphenylene-butadiynylene wire is present via Glaser coupling of terminal alkynyation at the (100) microfacets; (i) High-resolution STM image of a part of a erphenylene-butadiynylene wire placed and the corresponding models [26]. (j) Chemical structure of one alkyne molecule **1**. (k) STM image showing self-assembly of molecule **1** on Ag(111). (l) STM imaging showing 1D polymeric chains after annealing the sample at 400 K [27]. (m) Chemical structure of 1,3,5-tris-(4-ethynylphenyl) benzene (TEB); (n) STM images of TEB hexamer formed via cyclotrimerization of alkynynation; (o) Model of TEB hexamer [29].

generated. On Ag(110) and Ag(100), however, the terminal alkynes radicals covalently bind to surface metal atoms rather than other terminal alkyne radicals, which further results in the formation of one-dimensional organometallic chains, as shown in Fig. 7.12a–d. Detailed analyses revealed that the different reaction products should be attributed to the matching degree between the periodicities of the produced molecular wires and the substrate lattice structures. In addition, Sun et al. for the first time achieved the on-surface synthesis of metalated carbyne chains by zehydrogenative coupling of ethyne molecules and copper atoms on a Cu(110) surface under ultrahigh-vacuum conditions [31]. The length of the fabricated metalated carbyne chains was found capable of extending to the submicron scale (with the longest ones up to ~120 nm).

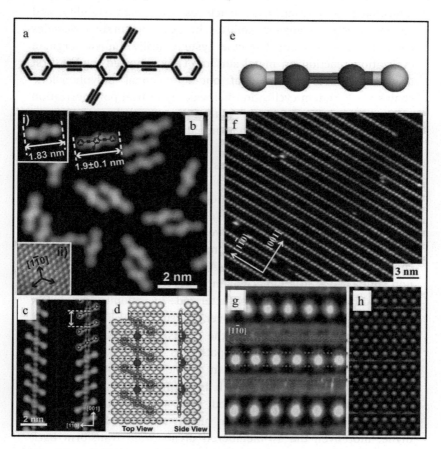

Figure 7.12 (a) Chemical structure of DEBPB molecule. (b) STM image of DEBPB on Ag(111). (c) STM image of an array of highly oriented organometallic chains on Ag(110) after annealing at about 350 K [30]. (d) Corresponding model of the organometallic chains [30]. (e) Chemical structure of ethyne molecule. (f) STM imaging showing metalated carbyne chain after annealing the ethyne-covered Cu(110) at 450 K. (h) The calculated model of metalated carbyne chain [31].

Bergman cyclization is another intramolecular reaction for organic molecules with multiply alkyne groups. During the process, actually a rearrangement reaction takes place when an enediyne is heated in presence of a suitable hydrogen donor [32] and benzene derivatives are produced. Therefore, Bergman cyclization is also one potential method for fabrication of covalently interlinked conjugated nanostructures with high stability and efficient electron transport abilities, by using precursor molecules containing alkyne group. Sun et al. systematically studied Bergman cyclization reaction on metal surface. They reported for the first time on-surface formation of one-dimensional polyphenylene chains via Bergman cyclization followed by radical polymerization on Cu(110) as shown in Fig. 7.13a–c [5]. Later on, by employing a precursor molecule with three terminal alkyne, Riss et al. achieved the fabrication of another kinds of one-dimensional chains on Au(111) with five-member rings rather than phenyl ring are cyclized from Bergman reactions, as shown in Fig. 7.13d–f [33].

On-surface Bergman cyclization could not only be used for construction of one-dimensional conjugated nanostructures, but also for in situ synthesis of novel organic molecules. In 2013, de Oteryza et al. investigated Bergman reaction of individual oligo-(phenylene-1,2-ethynylenes) on Ag(100) by heating. By using non-contact AFM, the molecular morphologies of several different complex products were characterized in atiomic resolution as shown in Fig. 7.14a. Moreover, based on DFT calculations, they further demonstrated the possible surface reaction mechanisms and the reaction pathways [34]. Besides heating, Schuler et al. found that Bergman reaction could be realized by atomic manipulation [35], which is verified by high-resolution, non-contact AFM. They revealed that an individual aromatic diradical could convert into a highly strained 10-membered diyne; moreover, the 10-membered diyne could transform back to the diradical form as shown in Fig. 7.14b–c. Because of the different physicochemical properties between the diradical and the diyne, the reversibility of the two reactive intermediates may open up the field of radical chemistry for on-surface reactions by atomic manipulation.

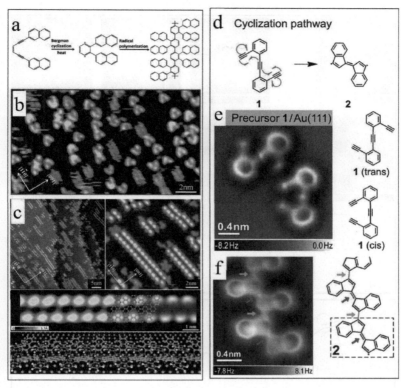

Figure 7.13 (a) Scheme of Bergman reaction of DNHD molecule. (b) STM image showing "heart"-shape molecular configurations of DNHD molecules on Cu(110). (c) STM images and corresponding models showing the one-dimensional molecular chains via Bergman reaction [5]. (d) Scheme of Bergman reaction of precursor 1; (e) AFM image of precursor 1 on Au(111); (f) AFM image of products 2 after annealing at 430 K via Bergman reaction [33].

7.4 Summary and Perspectives

In this chapter, we discussed recent achievements in surface-assisted chemical reactions of organic molecules. The studies reveal the following:

(1) Researchers could directly identify the products and even the intermediates, which facilitates to propose the possible reaction pathway.

336 | *Catalytic Reactions on Solid Surfaces*

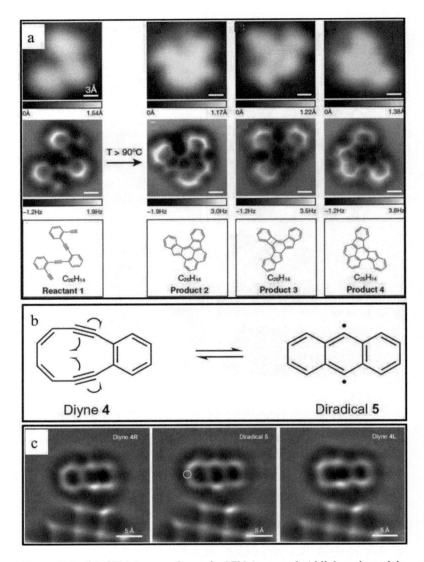

Figure 7.14 (a) STM images (upper), AFM images (middle) and models (lower) showing the evolution of molecular configuration via Bergman reaction on Ag(100) [34]. (b) Bergman cyclization of the cyclic diyne 3,4-benzocyclodeca-3,7,9-triene-1,5-diyne (4) to generate the 9,10-didehydroanthracene diradical 5. (c) Laplace-filtered AFM images of diyne 4R (left), diradical 5 (middle) and diyne 4L (right) on NaCl (2ML)/Cu(111) [35].

(2) Metal surfaces could greatly lower some reaction energy barriers and promote the fabrication of specific nanostructures.
(3) Metal surfaces could induce different reaction pathway of conventional synthesis, and achieve the fabrication of novel polymers or organic molecules.
(4) Different metal surfaces exhibit different catalytic activities, which further provide controllable selections on the species of products by simply changing the substrate.

Despite short-time development of on-surface chemical reactions, fantastic results have been achieved recently. In future, we hope that more reactions can be introduced to the 2D surfaces. The enriched database of the on-surface reactions may bring significant insights on control the chemical reactions with new pathways or in milder conditions, and the formation of novel molecular products or nanostructures.

References

1. Kawai, S., Sadeghi, A., Okamoto, T., Mitsui, C., Pawlak, R., Meier, T., Takeya, J., Goedecker, S., Meyer, E. (2016) Organometallic Bonding in an Ullmann-Type On-Surface Chemical Reaction Studied by High-Resolution Atomic Force Microscopy. *Small* **12**, pp. 5303–5311.
2. Held, P., Gao, H., Liu, L., Muck-Lichtenfeld, C., Timmer, A., Monig, H., Barton, D., Neugebauer, J., Fuchs, H., Studer, A. (2016) On-Surface Domino Reactions: Glaser Coupling and Dehydrogenative Coupling of a Biscarboxylic Acid to Form Polymeric Bisacylperoxides. *Angew. Chem. Int. Ed.* **55**, pp. 9777–9782.
3. Dienstmaier, J., Medina, D., Dogru, M., Knochel, P., Bein, T., Heckl, W., Lackinger, M. (2012) Isoreticular Two-Dimensional Covalent Organic Frameworks Synthesized by On-Surface Condensation of Diboronic Acids. *ACS Nano* **6**, pp. 7234–7242.
4. Zhang, J., Chang, C., Yang, B., Cao, N., Peng, C., Zhang, H., Tang, D., Glorius, F., Erker, G., Fuchs, H., Li, Q., Chi, L. (2017) Step-Edge Assisted Direct Linear Alkane Coupling. *Chem. Eur. J.* **23**, pp. 6185–6189.
5. Sun, Q., Zhang, C., Li, Z., Kong, H., Tan, Q., Hu, A., Xu, W. (2013) On-Surface Formation of One-Dimensional Polyphenylene through Bergman Cyclization. *J. Am. Chem. Soc.* **135**, pp. 8448–8451.
6. Bebensee, F., Bombis, C., Vadapoo, S., Cramer, J., Besenbacher, F., Gothelf, K., Linderoth, T. (2013) On-Surface Azide-Alkyne Cycloaddition

on Cu(111): Does It "Click" in Ultrahigh Vacuum? *J. Am. Chem. Soc.* **135**, pp. 2136–2139.

7. Ullmann, F., Bielecki, J. (1901). Ueber Synthesen in der Biphenylreihe. *Chem. Ber.* **34**, pp. 2174–2185.

8. Xi, M., Bent, E. (1992). Iodobenzene on Cu(111): Formation and Coupling of Adsorbed Phenyl Groups. *Surf. Sci.* **278**, pp. 19–32.

9. Hla, S., Bartels, L., Meyer, G., Rieder, K. (2000). Inducing All Steps of a Chemical Reaction with the Scanning Tunneling Microscope Tip: Towards Single Molecule Engineering. *Phys. Rev. Lett.* **85**, pp. 2777–2780.

10. Wang, W., Shi, X., Wang, S., Van Hove, M. A., Lin, N. (2011) Single-Molecule Resolution of an Organometallic Intermediate in a Surface-Supported Ullmann Coupling Reaction. *J. Am. Chem. Soc.* **133**, pp. 13264–13267.

11. Park, J., Kim, K. Y., Chung, K. H., Yoon, J. K., Kim, H., Han, S., Kahng, S. J. (2011). Interchain Interactions Mediated by Br Adsorbates in Arrays of Metal–Organic Hybrid Chains on Ag (111). *J. Phys. Chem. C* **115**, pp. 14834–14838.

12. Grill, L., Dyer, M., Lafferentz, L., Persson, M., Peters, M. V., Hecht, S. (2007) Nano-Architectures by Covalent Assembly of Molecular Building Blocks. *Nat. Nanotech.* **2**, pp. 687–691.

13. Cai, J., Ruffieux, P., Jaafar, R., Bieri, M., Braun, T., Blankenburg, S., Muoth, M., Seitsonen, A., Saleh, M., Feng, X., Müllen, K., Fasel, R. (2010) Atomically Precise Bottom-Up Fabrication of Graphene Nanoribbons. *Nature* **466**, pp. 470–473.

14. Sun, Q., Cai, L., Ding, Y., Ma, H., Yuan, C., Xu, W. (2016) Single-Molecule Insight into Wurtz Reactions on Metal Surfaces. *Phys. Chem. Chem. Phys.* **18**, pp. 2730–2735.

15. Sun, Q., Cai, L., Ma, H., Yuan, C., Xu, W. (2016) Dehalogenative Homocoupling of Terminal Alkynyl Bromides on Au(111): Incorporation of Acetylenic Scaffolding into Surface Nanostructures. *ACS Nano* **10**, pp. 7023–7030.

16. Zhong, D., Franke, J., Podiyanachari, S., Blömker, T., Zhang, H., Kehr, G., Erker, G., Fuchs, H., Chi, L. (2011) Linear Alkane Polymerization on a Gold Surface. *Science* **334**, pp. 213–216.

17. Cai, L., Sun, Q., Zhang, C., Ding, Y., Xu, W. (2016) Dehydrogenative Homocoupling of Alkyl Chains on Cu(110). *Chem. Eur. J.* **22**, pp. 1918–1921.

18. Otero, G., Biddau, G., Sánchez-Sánchez, C., Caillard, R., López, M., Rogero, C., Palomares, F., Cabello, N., Basanta, M., Ortega, J., Méndez, J., Echavarren, A., Pérez, R., Gómez-Lor, B., Martín-Gago, J. (2008) Fullerenes from Aromatic Precursors by Surface-Catalysed Cyclodehydrogenation. *Nature* **454**, pp. 865–868.

19. Amsharov, K., Abdurakhmanova, N., Stepanow, S., Rauschenbach, S., Jansen, M., Kern, K. (2010) Towards the Isomer-Specific Synthesis of Higher Fullerenes and Buckybowls by the Surface-Catalyzed Cyclodehydrogenation of Aromatic Precursors. *Angew. Chem. Int. Ed.* **49**, pp. 9392–9396.

20. Rogers, C., Chen, C., Pedramrazi, Z., Omrani, A., Tsai, H., Jung, H., Lin, S., Crommie, M., Fischer, F. (2015) Closing the Nanographene Gap: Surface-Assisted Synthesis of Peripentacene from 6,6'-Bipentacene Precursors. *Angew. Chem. Int. Ed.* **54**, pp. 15143–15146.

21. Sun, Q., Zhang, C., Kong, H., Tan, Q., Xu, W. (2014) On-Surface Aryl-Aryl Coupling via Selective C–H Activation. *Chem. Commun.* **50**, pp. 11825–11828.

22. Sun, Q., Zhang, C., Cai, L., Xie, L., Tan, Q., Xu, W. (2015) On-Surface Formation of Two-Dimensional Polymer via Direct C–H Activation of Metal Phthalocyanine. *Chem. Commun.* **51**, pp. 2836–2839.

23. Wiengarten, A., Seufert, K., Auwärter, W., Ecija, D., Diller, K., Allegretti, F., Bischoff, F., Fischer, S., Duncan, D., Papageorgiou, A., Klappenberger, F., Acres, R., Ngo, T., Barth, J. (2014) Surface-assisted Dehydrogenative Homocoupling of Porphine Molecules. *J. Am. Chem. Soc.* **136**, pp. 9346–9354.

24. Sun, Q., Cai, L., Ding, Y., Xie, L., Zhang, C., Tan, Q., Xu, W. (2015) Dehydrogenative Homocoupling of Terminal Alkenes on Copper Surfaces: A Route to Dienes. *Angew. Chem. Int. Ed.* **54**, pp. 4549–4552.

25. Zhang, Y., Kepčija, N., Kleinschrodt, M., Diller, K., Fischer, S., Papageorgiou, A., Allegretti, F., Björk, J., Klyatskaya, S., Klappenberger, F., Ruben, M., Barth, J. (2012) Homo-Coupling of Terminal Alkynes on a Noble Metal Surface. *Nat. Commun.* **3**, p. 1286.

26. Cirera, B., Zhang, Y., Björk, J., Klyatskaya, S., Chen, Z., Ruben, M., Barth, J., Klappenberger, F. (2014) Synthesis of Extended Graphdiyne Wires by Vicinal Surface Templating. *Nano Lett.* **14**, pp. 1891–1897.

27. Gao, H., Wagner, H., Zhong, D., Franke, J., Studer, A., Fuchs, H. (2013) Glaser Coupling at Metal Surfaces. *Angew. Chem. Int. Ed.* **52**, pp. 4024–4028.

28. Gao, H., Franke, J., Wagner, H., Zhong, D., Held, P., Studer, A., Fuchs, H. (2013) Effect of Metal Surfaces in On-Surface Glaser Coupling. *J. Phys. Chem. C* **117**, pp. 18595–18602.
29. Liu, J., Ruffieux, P., Feng, X., Müllen, K., Fasel, R. (2014) Cyclotrimerization of Arylalkynes on Au(111). *Chem. Commun.* **50**, pp. 11200–11203.
30. Liu, J., Chen, Q., Xiao, L., Shang, J., Zhou, X., Zhang, Y., Wang, Y., Shao, X., Li, J., Chen, W., Xu, G., Tang, H., Zhao, D., Wu, K. (2015) Lattice-Directed Formation of Covalent and Organometallic Molecular Wires by Terminal Alkynes on Ag Surfaces. *ACS Nano* **9**, pp. 6305–6314.
31. Sun, Q., Cai, L., Wang, S., Widmer, R., Ju, H., Zhu, J., Li, L., He, Y., Ruffieux, P., Fasel, R., Xu, W. (2016) Bottom-Up Synthesis of Metalated Carbyne. *J. Am. Chem. Soc.* **138**, pp. 1106–1109.
32. Jones, R., Bergman, R. (1972) p-Benzyne. Generation as an Intermediate in a Thermal Isomerization Reaction and Trapping Evidence for the 1,4-Benzenediyl Structure. *J. Am. Chem. Soc.* **94**, pp. 660–661.
33. Riss, A., Wickenburg, S., Gorman, P., Tan, L., Tsai, H., de Oteyza, D., Chen, Y., Bradley, A., Ugeda, M., Etkin, G., Louie, S., Fischer, F., Crommie, M. (2014) Local Electronic and Chemical Structure of Oligo-acetylene Derivatives Formed Through Radical Cyclizations at a Surface. *Nano Lett.* **14**, pp. 2251–2255.
34. de Oteyza, D., Gorman, P., Chen, Y., Wickenburg, S., Riss, A., Mowbray, D., Etkin, G., Pedramrazi, Z., Tsai, H., Rubio, A., Crommie, M., Fischer, F. (2013) Direct Imaging of Covalent Bond Structure in Single-Molecule Chemical Reactions. *Science* **340**, pp. 1434–1437.
35. Schuler, B., Fatayer, S., Mohn, F., Moll, N., Pavlicek, N., Meyer, G., Peña, D., Gross, L. (2016) Reversible Bergman Cyclization by Atomic Manipulation. *Nat. Chem.* **8**, pp. 220–224.

Chapter 8

Soft Matters for Future Catalysts: A Perspective

Qingmin Ji,[a] Harald Fuchs,[a,b] and Katsuhiko Ariga[c]

[a]*Herbert Gleiter Institute of Nanoscience,*
Nanjing University of Science and Technology,
200 Xiaolingwei, Nanjing 210094, China
[b]*Center for Nanotechnology,*
University of Münster, Wilhelm Klemm-Str. 10,
D-48149 Münster, Germany
[c]*WPI Center for Materials Nanoarchitectonics,*
National Institute for Materials Science, 1-1 Namiki,
Tsukuba, Ibaraki 305-0044, Japan

jiqingmin@njust.edu.cn

8.1 Multiforms of Catalysts

Catalysts come in different forms and can be gas, liquid, or solid. Although the form may not change the catalytic mechanism for forming or breaking of chemical bonds, it may affect the practical facility. A solid catalyst may allow easy separation and continuous process for the production, while a liquid or gaseous catalyst cannot be separated as easily for repeated use. Due to this feature, researchers have long made a distinction between heterogeneous

---

Soft Matters for Catalysts
Edited by Qingmin Ji and Harald Fuchs
Copyright © 2020 Jenny Stanford Publishing Pte. Ltd.
ISBN 978-981-4774-66-6 (Hardcover), 978-1-351-27284-1 (eBook)
www.jennystanford.com

catalysis (by solid catalyst) and homogeneous catalysis (involving other forms). However, it has been realized that heterogeneous and homogeneous catalysis can be merged for the design of new novel catalyst. Sophisticated analysis techniques and computational simulations also provide an in-depth understanding of the molecular behaviors in catalysis, which promote a more rapid progress in the design of catalysts.

Besides the activity of a catalyst, its stability has been much concerned for long-cycle usage. Thus noble metals, which are inert, have been widely studied for catalyst design. However, the resource deficiency and high price give rise to the studies on substituted catalysts with less amount of noble metals or from cheaper metals and organocatalysts. Zeolite, copper, and some organometallic liquids have performed well in certain reactions. The enormous possibilities from organic design and their combination with metals also allow endless variations to fine-tune for specific reactions.

A myriad of new substances have come out for potential catalysts, especially in past decade. For instance, MOFs show their flexibility on the constituents' geometry, size, and functionality. This variety has led to more than 20,000 different MOFs being reported and studied within the past decade, which offers ample opportunities to find a catalyst that suits specific reactions. Unfortunately, none of these MOFs has proved to be capable and useful in catalysts for large-scale production. More efforts are still required to develop new applicable catalysts.

The most promising inspiration for new catalysts might come from biology. The combination of chemistry and biology promotes the discovery of the synthesis of natural enzymes or their mimics with improved features. In the past decade, they have made great progress in unraveling nature's chemistry. This has resulted in catalysts that play an integral part in the production of pharmaceuticals and food.

8.2 Mimic from Nature

Nature shows the reaction processes to be more eco-friendly and also the precise formation of molecular structures. Nature presents fine-tuned processes, which inspire researchers to discover better systems for more precise control on production. The increasing capability for the precise design of molecules and functions allows the intricate regulation of the processes that mimic those occurring in living organisms. Enzymes can well distinguish enantiomers, which are exactly the same molecules with mirror conformations. Biological processes, which involve enormous variety of enzymes, always can make the reactions proceed at moderate temperatures and under very benign conditions. Through the collaboration of specific enzymes, the production in biological processes may proceed with a series of consecutive reaction steps under the same reaction environment. Natural systems may also adjust the reactions automatically according to the external environment. Whenever new bacteria or viruses enter our body, a completely new set of chemical reactions can be triggered by the immune system to prevent the new disease from spreading.

Although it seems impossible to synthesize the same smart and complex systems as living creatures, chemists and biologists have known how precise enzymes can perform their task and learned from their catalytic mechanisms for much more selective and active catalysts. An evolution for the design of catalytic system might be from self-assembly or self-organization, which enables the construction of complex structures by the incorporation of certain molecules or active sites in laboratory. From the insights from nature, the strategies of self-assembly offer new inspirations for chemistry and new perspectives on effective catalytic systems. Some self-assembled systems may also give rise to simple industrial processes. Although the integration of different processes in complex catalytic system is still a great challenge, the possibilities for molecular manipulations afford the adaption of the requirements for the needs. The mastering on some of the processes that underlie life may enable us to make the chemical production much more efficiently and sustainably.

8.3 New Challenges

It would be ideal if a catalyst could be effective for every chemical process. However, different catalysts may always show certain activity or selectivity to proper processes. There is almost no way to declare the effectiveness of a catalyst in a process without experimental verification. The process of trial and error now can be rationalized in a systematic manner. Computational technology is now applied for predicting catalytic behavior, thus enabling more directed experimentation for the trial-and-error process.

Computational modeling is based on different computational methods. However, it might be still difficult to predict the exact processes since the actual reactions have various unexpected variables and may react at different scales. Therefore, the validation by sophisticated spectroscopy is essential to study the complex chemical reaction process as it happens and in turn refine the modeling methods. Because catalysis takes place at an atomic scale, the catalytic materials must be controlled at an atomic scale in order to study their catalytic action. New spectroscopic instrumentation methods such as STM allow precise observation of ultrafast catalytic processes. This high accuracy makes it possible to better test computational models. The available large experimental data with ever-increasing details may also benefit the development of computational technology.

This combination and complementarity on real-time analysis and computational assumption may offer further fundamental insight into the mechanisms of catalysis by breakthroughs in the theoretical insight and computational methods. The greater precision and better understanding will make it possible to design catalysts compatible with demands and enable fine tuning of chemical bondings at ideal rates. This will bring us closer to the ultimate goal of predicting the performance of real catalysts under real reaction conditions. In turn, it will give us important tools for designing catalysts and controlling catalytic processes.

Developments in the fields of supramolecular and organometallic catalysts show great promise to simulate the functions from complex biochemical systems. The great developments in nanoscience and nanotechnology provide scientists an in-depth understanding of atoms, molecules and their arrangements under different conditions. Useful new chemical processes for a myriad of different molecules are developed by a high level of sophistication in arranging and rearranging atoms. However, despite the successful production of new molecules, it is still challenging to produce more efficiently, with less waste, under more benign conditions.

Mimic complex systems in nature, various catalysts are desired to integrate into a whole system, where the activity of each can be reinforced and multi-steps interactions under varying conditions can be adapted. The construction of such systems may demand the knowledge and skills for the assembled motifs, which possess different dimensional containers for the combination of active constitutes. Various such hollow morphologies have been successfully produced in recent years, such as microporous materials, mesoporous materials, and hollow capsules. The challenge of the integration of multi-functions into a whole system relies on how to well combine the isolated components or parameters with full realization on coordinated functions or interactions. The system might be capable of combining homogeneous, heterogeneous, and bio-catalysis in a truly profound manner. The new insights will make it possible to integrate processes in practical devices, with the aid of nano-engineering.

Besides the improved catalysis systems for large-scale production, more attention should be paid to the novel systems for micro-scale reactions, whose chemical changes may be involved in the devices for energy conversion, storage, sensing, parallel detection, etc. Progress in catalysis technology has made upscaling possible. Smaller-scale processes can proceed in more benign conditions, such as under moderate temperatures, with readily available feedstock, and less need to get rid of excess heat and to process wastes. Micro-devices can be constructed down to a precision of a few micrometers, which opens up new opportunities for small-scale chemical processes and producing molecules within nanometer precision. The tiny scale of devices will require substantial improvements in the organization of constituents and the use of catalysts in a very controlled way. This not only may present new ways to produce materials but also make it possible to study catalysis in a better way. It is a truly multidisciplinary task.

The combination of different components as a whole system or as a micro-device may also make the applications highly versatile. Different products may allow to be made using a single device. One day, one may be able to thread atoms at will to form a molecule, much like a printer that produces all kind of texts. Catalysis technology makes it possible to produce chemicals where they are needed and when they are needed.

Index

AC, *see* activated carbon
acid treatment 233, 257
activated carbon (AC) 124, 140, 230, 268
activated carbons 124, 140, 230, 268
AD, *see* Alzheimer's disease
alcohols 3, 36, 38–39, 42, 56, 66, 103, 149, 186, 207, 250, 256, 258, 306
aldehydes 3, 29, 32, 34, 44–45, 53, 110, 178, 180, 183, 301
aldimines 36–37, 44
aldol reactions 32–33, 180, 186, 188
alkenes 28, 30, 41, 112, 149, 330
alkoxide 32, 45–46
alkynes, terminal 22, 64, 331, 334
alkynyl halides 21, 322, 324
Alzheimer's disease (AD) 103–104
amino acids 32, 36, 95, 97, 179–180, 186, 188
amphiphiles 143–144, 146, 295
aniline, electrochemical polymerization of 237–238
arenes 29
aromatic aldehydes 54, 116
aryl chlorides 16, 18
aryl halides 24, 26, 322
azlactones 38–39, 194
azocarboxylate 23

BBR, *see* building block replacement

benzene 178, 182, 206, 331–332
benzyl alcohol molecule 257
benzylether 184, 188–190
Bergman cyclization 320, 334, 336
biocatalysis 5–6, 105, 292
biocatalysts 4–5, 8, 54, 57, 94, 100
biological processes 2, 103, 343
biomaterials 151
biotechnology, chemical 102
biphasic systems 50, 53–54, 204
boron 234, 250, 255
Brønsted acids 36, 44
  metal-free 36–37, 39
building block replacement (BBR) 118

camphorsulfonic acid (CSA) 39
carbenes 21, 41–42, 44
carbon 34, 43–44, 135, 229–230, 232, 239, 252, 261–262, 265, 268–270, 310
  hard 261–262
  porous 261–263, 266, 268, 270
  soft 261–262
  surface chemistry 270
carbon black (CB) 230, 243, 245, 249
carbon dioxide 4, 22, 40, 59, 61, 103, 106, 110–111, 147, 238, 242, 248–249, 306, 309
  supercritical 54
carbon electrode 270
carbon nanofibers, doped 126–127

348 | *Index*

carbon nanotubes (CNTs) 60, 123, 230–239, 270–271
carbon nanotubes, N-doped 234
carbonization 262–263, 265–266
catalysis efficiency 181, 192, 200, 202
catalysis reactions 15, 20, 96, 98, 178
catalysis strategies 184–185, 192
catalyst recycling 55, 178–179
catalysts
  active 68, 110, 194, 268, 305, 343
  carbon-based 271
  covalent organic framework–based 129, 131, 133, 135, 137, 139, 141
  DEN-based 202–203
  dendrimer 147, 149
  dendrimer-supported 186, 188
  graphene-based composite 239, 241
  heterogeneous 3–4, 60, 64, 67, 112, 121, 130, 134, 136, 140, 185, 193, 195, 197–199, 201–202, 302
  homogeneous 3, 13, 177–179, 192–193, 239
  membrane-embedded 143–144
  MOF-based 107, 109, 111, 113, 115, 117, 121, 123, 125, 127
  molecular 11–12, 35, 69, 130, 155
  multi-electron 147
  organometallic 3, 13, 15, 178, 229, 238
  palladium-phosphine 16
  PEG-bound 178, 180–181
  polymer-supported 59, 174–175, 203
  recoverable 59
  recyclable 114
  supermolecular 93–94, 96, 102, 104, 106, 108, 110, 112, 114, 116, 118, 120, 122, 124, 126, 128, 130, 132, 134, 136, 138, 140, 142, 144, 146, 148, 150, 152, 154, 156
  transition metal 14, 30
  vesicular 145
  ZIF-8 112
catalytic efficiency 17, 21, 59, 96, 99, 118, 122, 125, 128, 202, 207, 295–296, 298, 303, 305, 307
catalytic nanoarchitectures 306–307, 310
catalytic reduction, selective 268
CB, *see* carbon black
chemical vapor deposition (CVD) 240
chemistry
  biomimetic 95
  click 144, 301
  host–guest 94
  molecular 62, 93
  organometallic 2, 12–13, 64
  supermolecular 93–94
  surface-based 301
chemoselectivity 32, 36, 38
chiral ionic liquids 56
chromium 112, 115
CNPCs, *see* conjugated nanoporous polymer colloids
CNTs, *see* carbon nanotubes
COFs, *see* covalent organic frameworks
conjugated nanoporous polymer colloids (CNPCs) 139
covalent organic frameworks (COFs) 8, 129–134, 136–138, 140–142
covalent organic frameworks, active 130, 134, 136

covalent triazine frameworks (CTF) 140, 142
cross-coupling reactions 15–16, 18–20, 25–26, 110, 138
CSA, see camphorsulfonic acid
CTF, see covalent triazine frameworks
CVD, see chemical vapor deposition
cyclodextrins 97
cyclooctene 123, 126, 181

DEBPB molecule, chemical structure of 333
decarboxylation reactions 187
dehalogenative coupling 320–321, 323
dehydrogenative coupling 325, 327, 329, 331, 333
DEN, see dendrimer-encapsulated nanoparticles
dendrimer-encapsulated nanoparticles (DEN) 201–202
dendrimer-supported organocatalysts 185–186, 188
dendrimers 94, 143, 146–147, 149–150, 173–174, 184–192, 202
dendritic 146–148, 150
density functional theory (DFT) 45, 257
DFT, see density functional theory
dichloromethane 177, 179, 182, 190
diethyl ether 177, 179–180, 182

electrocatalysis 199, 209, 232, 239, 307, 309

electrocatalysts 235–236, 249–250, 252, 255, 258, 307, 309
  composite 251–252
  efficient 126
  N-graphene 253
electrocatalytic activity 64, 251–253
electrocatalytic performances 251, 254–255
electron mobility 242, 244–245
electron–hole pairs 241, 243, 245–246
electrophiles 15, 17, 34, 39, 44
enamines 31–32, 34, 38
enantioselective reactions 38, 41, 44
enantioselectivity 34, 36–37, 39, 43, 45, 54–55, 98, 115, 180, 182, 186, 188, 191
enzymatic activity 155
enzyme cascades 102
enzymes 2, 4–5, 8, 29, 31, 35, 94–97, 99–106, 151–155, 229, 259, 298–301, 343
  artificial 96–98, 100, 106, 155, 300
  immobilized 153, 298–299, 301
  inorganic porous nanoarchitectures 298–299, 301
  multifunctional 103
  natural 96–97, 342
epoxides 59, 66, 111, 141, 205
ethylene glycol 63, 177
ethylene oxide 177, 263

Fischer–Tropsch synthesis 233
formaldehyde 34, 264

fullerenes 231–232, 326–327

gas-phase catalysis 292, 302–305
GCE, *see* glassy carbon electrode
Glaser coupling 331–332
glassy carbon electrode (GCE) 260
glucose 103, 151, 155, 234, 256, 259
glucose detection 256–257
glucose oxidase 151, 154
graphene 106, 230, 239–247, 249–253, 255, 258–259, 270–271
  chemical properties of 255
  co-doped 255
  doped 239–240, 250, 252, 256
  N-doped 251, 259
  nitrogen-doped 252
  single-layer 240
graphene lattice 252, 254
graphene materials 243–245, 250–251, 258
  dual-doped 255
graphene nanoribbons 320–321
graphene oxide 239–240
  reduced 241, 243
graphene sheets 243, 245, 247, 249, 251–252, 255
graphene surface 245, 249
graphite 239–240
graphite oxide 240

heavy metals 30, 102
heteroatoms 234, 250, 252, 255, 268
heterogeneous catalysis 3–4, 13, 69, 110, 115, 125, 130–131, 134, 136, 138, 146, 174, 176, 192, 209, 229, 291, 302

homogeneous catalysis 2, 13, 62, 149, 176–177, 179, 181, 183, 185, 187, 189, 191, 204, 342
hybrid materials 66, 239
hydrocarbons 15, 63, 306, 325
hydrogen 4, 135, 241–243, 250, 306, 309
hydrogen bonding 33, 42–43, 49, 51, 95–96, 151, 301
hydrogen peroxide 99, 151–152, 186, 232, 234

ILs, *see* ionic liquids
imines 29, 31, 33, 37, 44, 132, 141, 148, 194, 257
inorganic materials 59, 107, 293, 295, 299
ionic liquid moieties 58–59
ionic liquid phases 56
ionic liquids (ILs) 7, 48–61, 205–206, 232, 237
  functionalized 56
  immobilization of 58
  polymer-supported 58–59
irradiation, visible-light 244–245

ketones 3, 20, 22, 111, 115, 180–181, 186, 189

lactams 36, 44
layer-by-layer (LbL) 151
LbL, *see* layer-by-layer
LbL technique 151–152
LCST, *see* lower critical solution temperature
ligands 13–14, 17–22, 24, 27, 29, 61–63, 65–66, 68, 95, 107,

119, 143, 176, 182, 184, 194, 196, 202, 238, 303
  bridging 107, 116, 303
  functional 62–63, 66
  stabilizing 68
LLC, *see* lyotropic liquid crystal
lower critical solution temperature (LCST) 203–204
lyotropic liquid crystal (LLC) 295

magnetic nanoparticles 58, 61, 266–267
magnetic porous carbon (MPC) 128
mesoporous materials 58–59, 62, 66–69, 292, 294, 299, 345
mesoporous nanoarchitectures 292–293, 295, 297
mesoporous organosilicas, periodic 58
mesoporous silica nanoparticles (MSNs) 66, 293–294, 299
mesoporous structures 295, 307–308, 310
mesoporous WO$_3$ 309
mesostructures 294–295, 298
metal-free organocatalysts 31, 33, 35, 37, 39–41, 43, 45, 47
metal nanoparticles 62–63, 66, 68, 124, 128, 130, 138, 208, 236
  catalysis of 65
  catalytic activity of 63–64
metal organic frameworks (MOFs) 8, 107–112, 114–115, 117–130, 155, 342
metal oxide nanoparticles 259
metal precursors 121, 123
methanol oxidation 309–310

methanol oxidation reaction (MOR) 251, 309
microorganisms 100, 103
MOFs, *see* metal organic frameworks
monomers 46, 134, 146, 174–175, 177, 199–200, 236, 329
MOR, *see* methanol oxidation reaction
MPC, *see* magnetic porous carbon
MSNs, *see* mesoporous silica nanoparticles

N-graphene 252–254, 258–259
N-heterocyclic carbenes (NHCs) 18–19, 27, 41–47, 117, 238
nanocatalysts 6–9, 62–63, 250, 292, 302–303, 306, 309
nanoclusters 62, 65–66, 206
nanoparticles, catalytic 60, 63, 197, 303–306
nanosheets 244, 296, 308
NHC catalysis 42, 45
NHCs, *see* N-heterocyclic carbenes
noble metals 4, 7, 20–21, 124, 250, 258–259, 302, 306–307, 342

organic polymers 174, 176, 178, 180, 182, 184, 186, 188, 190, 192, 194, 196, 198, 200, 202, 204, 206, 208, 269
organic solvents 23, 32, 47–48, 51, 53–54, 101, 112, 131, 148, 180, 182–183, 240
organic synthesis 8, 13–14, 18, 30, 33, 39, 43, 69, 136
organobimetallic nanoparticles 25

organocatalysis 31–32, 36, 41, 45, 150
organocatalysts 30–31, 35, 42–43, 45, 134, 173, 178, 185, 188, 208, 342
ORR, see oxygen reduction reactions
oxygen reduction reaction 60, 64, 126, 236
oxygen reduction reactions (ORR) 64, 126, 236, 239, 250–255

P-graphene 254–255
palladacycles 19, 28
palladium 16, 18–19, 182, 190, 196, 200
palladium-catalyzed process 18–19
PAMAM, see poly(amidoamines)
PANI, see polyaniline
PC, see propylene carbonate
PCR, see polymerase chain reaction
PEG, see poly(ethylene glycol)
periodic mesoporous organosilica (PMO) 58, 67
phosphines 17–18, 25, 27, 180–181, 189, 234
photocatalysis 125, 232, 239, 247, 268–269
photocatalyst-enzyme 106
  graphene-based 106
photocatalysts 142, 241–243, 245–249, 258, 262
  graphene-based 242–244
  semiconductor 242, 245–246, 249
photocatalytic efficiency 242
photocatalytic reaction 130, 138, 141, 242, 249

PMO, see periodic mesoporous organosilica
poly(amidoamines) (PAMAM) 184, 187, 191, 202, 238
poly(ethylene glycol) (PEG) 59, 177–182, 203
polyaniline (PANI) 199, 237–238
polyethylene chains 326
polymer-supported nanoparticles 194, 197, 200
polymerase chain reaction (PCR) 101
polymeric materials 174, 196–198
polymerization, chemical oxidative 236
polyoxometalates (POMs) 104, 120, 149
polystyrene 59, 123, 182, 194, 205
POMs, see polyoxometalates
POP, see porous organic polymers
porous catalysis systems 298, 303
porous nanostructures 292, 295, 299
porous organic polymers (POP) 196, 208
proline-catalyzed reactions 32
proline derivatives 186, 188
propylene carbonate (PC) 111
propyleneimine 184, 186–187
proteases 101, 104

reactive oxygen species (ROS) 104–105
ring-opening metathesis polymerization (ROMP) 146
ring-opening polymerization (ROP) 46
ROMP, see ring-opening metathesis polymerization

ROP, *see* ring-opening polymerization
ROS, *see* reactive oxygen species

SALE, *see* solvent-assisted linker exchange
SAMs, *see* self-assembled monolayer
SCR, *see* selective catalytic reduction
selective catalytic reduction (SCR) 268
self-assembled monolayer (SAMs) 63
semiconductors 241, 243, 245, 249, 262–263, 293, 310
silica 68, 192, 230, 296, 299–300, 303, 305
  mesoporous 58, 293, 295
  porous 196, 263, 303
SILP, *see* supported ionic liquid phase
silyl enolates 36–37
single-walled carbon nanotubes 236–237
solvent-assisted linker exchange (SALE) 118–119
stereoselectivity 13, 33, 45, 54, 63, 96, 186, 190

styrene 125, 182, 199
superoxide dismutase 99–100, 104
supported ionic liquid phase (SILP) 55, 58
surface-protected etching 305

task-specific ionic liquids (TSILs) 55, 57
thermoresponsive polymers 203–204
three-way catalysts (TWCs) 268
transition metals 12, 14–15, 18, 21, 24, 30, 42, 176, 180, 259
TSILs, *see* task-specific ionic liquids
TWCs, *see* three-way catalysts

UE, *see* urease enzyme
Ullmann reactions 24, 320–322
urease enzyme (UE) 153

zeolites 58, 107–108, 110, 112, 263–264, 268, 342
ZIF-8 112, 121–122, 124, 126
zinc carboxylate 67